A World on Fire

ALSO BY JOE JACKSON

NONFICTION

A Furnace Afloat

Leavenworth Train

Dead Run
(with William F. Burke)

FICTION

How I Left the Great State of Tennessee and Went on to Better Things

A World on Fire

A Heretic, an Aristocrat,
and the Race to Discover Oxygen

JOE JACKSON

Viking

VIKING
Published by the Penguin Group
Penguin Group (USA) Inc., 375 Hudson Street, New York, New York 10014, U.S.A.
Penguin Group (Canada), 90 Eglinton Avenue East, Suite 700, Toronto, Ontario, Canada M4P 2Y3 (a division of Pearson Penguin Canada Inc.) • Penguin Books Ltd, 80 Strand, London WC2R 0RL, England • Penguin Ireland, 25 St. Stephen's Green, Dublin 2, Ireland (a division of Penguin Books Ltd) • Penguin Books Australia Ltd, 250 Camberwell Road, Camberwell, Victoria 3124, Australia (a division of Pearson Australia Group Pty Ltd) • Penguin Books India Pvt Ltd, 11 Community Centre, Panchsheel Park, New Delhi – 110 017, India • Penguin Group (NZ), Cnr Airborne and Rosedale Roads, Albany, Auckland 1310, New Zealand (a division of Pearson New Zealand Ltd) • Penguin Books (South Africa) (Pty) Ltd, 24 Sturdee Avenue, Rosebank, Johannesburg 2196, South Africa

Penguin Books Ltd, Registered Offices: 80 Strand, London WC2R 0RL, England

First published in 2005 by Viking Penguin, a member of Penguin Group (USA) Inc.

10 9 8 7 6 5 4 3 2 1

PHOTOGRAPH CREDITS
Library of Congress: insert page 1 top; page 2 top and bottom; page 6 top; page 7 bottom. Edgar Fahs Smith Collection, University of Pennsylvania: page 1 bottom; page 2 middle; page 3; page 8 top. National Gallery, London: page 4 top. Joseph Priestley Collection, Dickinson College: page 4 bottom; page 5 bottom; page 6 bottom; page 7 top; page 8 bottom. The Metropolitan Museum of Art, Purchase, Mr. and Mrs. Charles Wrightsman Gift, in honor of Everett Fahy, 1977. (1977.10): page 5 top.

LIBRARY OF CONGRESS CATALOGING IN PUBLICATION DATA
Jackson, Joe, date.
A world on fire : a heretic, an aristocrat, and the race to discover oxygen / Joe Jackson.
p. cm.
Includes bibliographical references and index.
ISBN 0-670-03434-7
1. Oxygen. 2. Science—Europe—History—18th century. I. Title.
QD181.O1J33 2005
546'.721—dc22 2004065103

Printed in the United States of America
Set in Granjon with Snell
Designed by Carla Bolte

As always,

To Kathy and Nick

. . . there is a fire
And motion of the soul which will not dwell
In its own narrow being, but aspire
Beyond the fitting medium of desire.

—Byron, *Childe Harold's Pilgrimage*, Canto III

CONTENTS

A World on Fire

PROLOGUE

God in the Air

A GOOD MEAL CHANGES EVERYTHING. WE EAT AT MOMENTS OF drama, or when the group decides something must be etched in memory. As polymers break down, calories burn and oxygen use increases: the mind unfolds to new thoughts and opens to unseen relationships between people and ideas. The French, who turned the act of eating into the science of gastronomy, know that one must "rise above the table." Talk becomes convivial and the senses are engaged. We are at our most social and observant. Walls tumble. Secrets are revealed.

This evening's repast in the Paris of October 1774 is one such legendary meal. As European thinkers grew drunk on the wine of "enlightenment" and a tremor of revolt hung in the close distance like waves of heat, two similar yet remarkably dissimilar men of science faced one another across the table and plotted the end of ancient ways. They didn't consider themselves revolutionaries so much as seekers of a different truth, making free inquiry into the secrets and nature of reality. The rituals of dinner are like the rituals of science—precise, measured, and carefully weighed—and one rule of conversation during the French Enlightenment was that shop talk and "serious" discourse must not intrude on the table. "What?" sneered Michel de Montaigne of intellectual pretension during the sensual feast: "Would they try to square the circle while mounting their wives?" But talking shop in scientific circles was not pretentious—it was a puzzle, and thus a serious game. Tonight's subject took the form of a riddle. What was invisible, yet all around them? Nowhere, but everywhere? It was a vexing and ancient riddle, as old as science itself: What was the makeup of the air?

Ever since the ancient Greeks, the air was believed to be one of the four building blocks of existence, along with earth, water, and fire. That transparent ether breathed in by king and commoner was considered an elementary substance, mutable yet inviolate, as pure in composition as mercury or gold. This 2,300-year-old belief in the construction of the universe was older than Aristotle, but this night in pre-Revolutionary Paris, table talk gave birth to a science that envisioned matter's subjugation and control. Amid the clink of polished silver, the hush of liveried servants and chime of crystal goblets, what would one day be called the Chemical Revolution was given life because of a lab accident that happened two months earlier and a slip of the tongue occasioned by a pretty face and too much wine.

Science creates its own myths, as does any tribe or nation, for what is myth but a way of explaining existence and addressing change? In science, two particular myths hold sway. Change is the first—a radical change that forever destroys an old and comfortable means of perceiving reality, replaced instead by a dogma that is new, unfamiliar, and strange. A change that the modern historian Thomas Kuhn calls the death of a "paradigm," a brutal war of the intellect that forever reshapes the "new." Such change speaks of our view of mankind's place in the cosmos: Are we helpless before the forces of nature, or can we shape these ourselves like gods? This is the essential tension of myth: control of a mutable world. An apple probably did *not* fall on the head of young Isaac Newton, but the myth encompassed his theory of gravitation and his belief that God's mechanical laws could be understood and controlled by human beings. The German scientist F. A. Kekule *did* puzzle for months over the shape of the benzine molecule, but the night that he dreamed of the Worm Ouroboros—the alchemical symbol of the tail-biting snake—he realized on waking that benzine is arranged in a ring. The field of organic chemistry sprang from this vision, but the image was as old as the alchemists, their secret shorthand for unity and change.

The second myth is that of accident and anomaly in the lab, and its role in heroic discovery. Something is wrong, and only a clear-sighted hero can interpret the message and lead the new way. The mythical components of religion and science are similar, for what are both but faiths in a particular ordering of the world? The celestial bodies did not run smoothly in their courses until Copernicus's improved calculations ended the geocentric universe. Anomalies of heat and light in a controlled lab environment suggested unknown forces hidden deep in matter's core. When Marie and Pierre Curie took heed of the riddle, they unlocked the secret of radiation. Accident is portrayed as the first glimmer of truth, yet it takes a "colossus"—a Copernicus, Newton, or Einstein—to see above the herd.

From such myths, perhaps, we construct the deeper fables of our lives. Yet we fail to weigh all the consequences. "Scientists are maverick personalities," maintains Jacob Bronowski in *The Origins of Knowledge and Imagination*. "You do not invent a new world system by being satisfied with what other people have told you about how the world works. . . . The creative personality thinks of the world as a canvas for change, and of himself as a divine agent of change." Yet there's little mercy for the divine messenger. Consider the fate of Prometheus, who brought fire from Olympus in a dry fennel stalk to ease the plight of man. By controlling his environment, man approached the status of gods—and for that gift, Zeus directed his servants, Force and Violence, to chain Prometheus to a "high-piercing . . . rock" in the Caucasus, where each day an eagle would tear out his liver for giving a secret "to mortals not their due."

Tonight's elegant dinner is the stuff of one such mythical tableau. The guest was the English heretic Joseph Priestley, son of a "cloth-dresser" from Yorkshire, where that nation's Industrial Revolution had begun. At times shy, awkward, and brilliant, the forty-one-year-old Priestley was an enigma to the French academicians gathered around him; he stuttered when nervous, and he was stuttering now. He saw himself primarily as a radical theologian, a lonely Unitarian in a nation of Anglicans; his role as a "natural philosopher" was more of a hobby in his eyes. But to the rest of Europe, Priestley was known as the premier "pneumatic chemist" of the age, his fame springing from his discovery of several unknown gases—"airs," as he called them—from the atmosphere that all men breathed. The irony, of course, was that since air was one of the four basic elements, this was not supposed to occur. Yet two months earlier, he told his hosts, he'd teased something out of the atmosphere that seemed more vital than "common air" itself. If true, centuries of dogma were a lie.

The Paris *savants* fell silent as Priestley described in halting French what he had done. In June of that year, he'd purchased a new convex lens with a 12-inch diameter and 20-inch focal length that was rumored as once owned by the Grand Duke Cosimo III of Tuscany. A favorite trick of pneumatic chemists was to heat ash or metal in glass vessels and see what occurred: a gas often sprang forth, which was then captured in a flask inverted over a tub of water or mercury and its properties analyzed. On August 1, 1774, Priestley tried this with the brick-red ash left over when mercury was heated, a substance he called "mercury calx" but which today is known as mercuric oxide. He watched as globs of quicksilver formed in the focused rays of sunlight, as an odorless gas arose. On a whim he placed a lighted candle in the jar with the gas, expecting to see it extinguished. Instead, it flared brightly, and he was amazed. He inserted a red-hot wire; it glowed as white as the sun.

The excitement of the *savants* around him could barely be contained. As Priestley himself had written, few maxims "have laid firmer hold upon the mind" than that air "is a simple elementary substance, indestructible and unalterable," yet his experiments with air had led him to believe the opposite—that it was a mixture of some kind. The riddle of the air always hovered on the threshold of an answer: long before Priestley's birth, the secret of its composition was believed so vital that its discovery would reveal as much of Nature's inner mechanism as had Newton's laws of gravitation. The puzzle, which seems so simple today, was so confounding that it impeded the progress of chemistry for centuries, relegating the field to pseudoscience, a bastard child of alchemy. That air was necessary for both burning and breathing was already known; the link was obvious, if the process was not. Centuries earlier, the Chinese had written of *yin,* the active component in air that combined with sulfur and some metals. In the fifteenth century, Leonardo da Vinci wrote that air was composed of two substances, an observation married to the fact that "fire destroyed without intermission the air which supports it, and would produce a vacuum" if not for the in-rush of "other air." In 1535, the German alchemist and physician Paracelsus refined the metaphor by claiming that "man dies like a fire when deprived of air." One hundred years later, Robert Boyle said he suspected "that there may be dispersed through the rest of the atmosphere some odd substance on whose account the air is so necessary to the subsistence of flame." But what was this substance, and its role in life and ignition? Was Priestley on the cusp of solving one of man's greatest scientific mysteries?

Priestley noted as the room erupted that one *savant* watched with a greater measure of interest and calm. Antoine-Laurent Lavoisier was no amateur like Priestley, but a theoretician fearful that his discipline was in danger of becoming a handmaiden of medicine or, worse, a footnote of alchemy. Chemistry had reached a standstill, a state of confusion so great that even as new substances were discovered monthly and strange facts piled up, a slavish devotion to old theories prevented any comprehensive explanation. Nature's secrets seemed within man's grasp, yet the antediluvian insistence that all matter was made of four elements slammed shut like a gate across the threshold of understanding. All fresh discoveries must twist to conform to a worldview as dead as the Greeks: even the language of chemistry was tortured, packed with alchemic holdovers like "butter of antimony," "*materia perlata* of Kerkringius," and "oil of tartar *per deliqium*." If there was hope, it lay in this enigma of the air.

Lavoisier also had the wherewithal to do something about it. The handsome son of a wealthy Paris magistrate, Antoine, now twenty-nine, was by profession part of the despised *Ferme générale,* the King's tax collectors, making him one of the realm's wealthiest citizens. But he was also the youngest person ever elected to the *Académie des Sciences*. Beside him sat his beautiful wife, quietly translating the Englishman's lapses into French. Marie-Anne Paulze Lavoisier was only seventeen, yet in an age that savagely satirized marriage, the affection between the two was mutual and genuine. Lavoisier treated Marie as a partner, depending on her assistance in the lab and her translations of scientific papers to stay abreast of international findings. Priestley watched as Lavoisier hearkened to his experimental ruminations; what he did not know was that the younger man was as obsessed with the air as he.

In fact, two years before this elegant meal, the modern equivalent of a space race had begun. Priestley and Lavoisier were its leaders and unintended rivals, but where Priestley teased unknown gases from the atmosphere, Lavoisier focused on the process of burning. It was in the fire itself . . . the dancing flames and terrible destruction leaving fumes and ash . . . that the truth should lie. Of the several competing theories then in existence that tried to explain combustion, the one with the most adherents was "phlogiston," which held that all substances were made of water and three kinds of earth, one flammable. Yet phlogiston theory was just another variation of the ancient theory of elements and did not *predict* results, only explained them a priori. Without prediction, there was no science, and to Lavoisier, anything less was superstition.

With the characteristic arrogance that amused friends and infuriated enemies, Lavoisier wanted nothing more than to lay the foundations of a new science, with its basis in combustion. He personally would dump twenty-three centuries of accepted wisdom, he believed. In 1772, when he'd noticed that certain substances gained weight during heating rather than losing weight as predicted by phlogiston theory, he wrote in a sealed note to the French Academy that "this increase in weight comes from an immense quantity of air." Something in the air reacted with other substances during combustion—but what? And how? His note, dated November 1, 1772, was the equivalent of filing a claim on a gold mine: he was on the verge of some great truth that he suspected from the nuggets strewn about the surface. The dated note was an attempt to fight off anyone who might try jumping his claim.

But he had no idea where to start until this dinner with the tongue-tied Englishman. They sat on opposite sides of the table, separated by the glowing candles in their gilt holders. The secret lay between them in the flame. As Priestley struggled to explain himself, Lavoisier, thin and keen, smiled in recognition, for the key appeared as suddenly as Kekule's vision of the snake writhing in air. And Priestley, smiling back, wondered uneasily whether—as happened so often—he'd talked too freely again.

So it began. The English heretic who isolated oxygen would not recognize his discovery; the French aristocrat who saw the truth would start a revolt as far-reaching as any by Copernicus, Newton, Darwin, or Einstein. Both became famous for their part, and twenty years after this friendly dinner, both would pay the tab. As Priestley fled England aboard the ship *Samson*, Lavoisier fell victim to the Reign of Terror and was brought before its chief executioner, of the same name. "The Republic has no need of scientists," the court pronounced. The catalyst of their fame and ruin was oxygen, the mystery they sought in the air.

▪ ▪ ▪

O.

Symbol of nothing, symbol of everything, a circle encompassing all. In logic, a negative stance; in alchemy, the Worm Ouroboros, mystic symbol of the world. In serology, the most common blood type, defined by what it lacks: the agglutinogen A or B. The shape of the lips and sound of the voice in moments of awe or sadness. "Praise the Lord, O Jerusalem," declaimed the psalmist. "O, for the touch of a vanished hand," lamented Tennyson.

O, symbol of oxygen, without which the flame won't ignite, or will snuff out, exhausted. The vital principle over which Greek philosophers bickered for centuries, the secret for which the alchemists searched and yearned. The fuel for our own vital spark, without which we die.

We breathe in, we breathe out, kept alive in that gasp by an odorless, two-atom pairing that, true to its symbol, is nowhere and everywhere. "Man is an obligate aerobe," Hippocrates wrote 2,400 years ago, and in all the philosophical pronouncements of all the Greek thinkers, that short statement is perhaps the most true. The minute-by-minute exchange and absorption of oxygen in our bodies is the true secret to existence, the focal point for all research into the mechanics of life and the starting line from which art, science, philosophy, and theology are conceived by the conscious mind. Each single breath sparks a slow combustion, the essential burning that provides the heat needed to keep us alive.

But we always need more. Oxygen pours down the lungs, through the bronchioles, into the tiniest air sacs, the alveoli; it passes through these walls to the attached capillaries, where oxygen molecules link to hemoglobin, the protein pigment of the red cells. It is this chemical bonding of oxygen to the iron atom in its center that turns the blood red. Oxyhemoglobin, the new compound, is so reactive that 1 liter of blood can dissolve 200 cubic centimeters of oxygen, much more than will dissolve in the same volume of water. The charged blood cells rush from the lungs to the left heart, then out through the aorta and into a labyrinth of steadily shrinking pathways until reaching the tissues and their cells. There hemoglobin passes oxygen to another enzyme with an iron core: monooxygenaise acts as a catalyst for the body's various oxidation processes, including building new molecules and detoxifying others. In the cell's mitochondria, the oxygen is processed and carbon dioxide released, the latter carried away by the blood for destruction or release through the liver, kidneys, and lungs. If this exchange is life's basic economy, oxygen is its currency.

As in any economy, a dearth of currency bankrupts the system. A reduction of oxygen levels below the normal 21 percent found in the air will have a drastic effect on the body. At less than 15 percent, judgment deteriorates; at 10 percent or less, there is reversible damage. Less than 6 percent is lethal. It was one such deficit that prematurely ended the Biosphere Project set up in Tucson, Arizona, in 1993. Eight human guinea pigs had been sealed into the glass-walled ecosystem to see if it was possible to sustain life in such an artificial environment as on Mars or the Moon, but within a month all eight were gasping for breath as the oxygen content fell to 17 percent. Somehow, 30 tons of the gas had vanished; only later was it discovered that the oxygen probably reacted with the iron in the soil—much as it does in the blood.

We swim through oxygen like fish, taking for granted the fluid that washes over and through us in constant baptism, an elixir lingering, without taste, on the tongue. Even in those forms of life that do not use hemoglobin as an oxygen carrier—lobsters and spiders, anaerobic bacteria and some sea worms—the gas is still essential for life since it is a constituent of DNA and almost all other biologically important compounds. It comprises nearly two-thirds of the human body, and as the most common element on Earth, accounts for nearly half of most rocks and minerals, one-fifth of the atmosphere, and nine-tenths of the mass of the world's lakes, rivers, and oceans. It is the third most abundant element in the universe, after hydrogen and helium; in stars like the Sun, its formation is part of the carbon-nitrogen fuel cycle, a process intrinsic to stellar energy production and the Big Bang. It is the most promiscuous substance on the atomic chart, mating with nearly half the known elements to spawn new compounds. The world is an orgy of oxidation, and like love and lust, the combustion can be as rapid and violent as fire, or as slow and gentle as the energy exchange in cells. Either way, the result is identical: O_2 oxidizes carbon compounds to form CO_2. The result can be fatal, as when oxygen kills certain bacteria on contact, or destructive, as when it binds with iron or steel to form rust and what once seemed impregnable simply flakes away.

Anywhere claims are made for energy, freedom, or health, oxygen is invoked as a totem and symbol. It is *Oxygenium* in Latin, *Kyslik* in Czech, *oxyène* in French, and Кислород in Russian, but the metaphors are similar everywhere. There are 4 million–plus hits for "oxygen" on the Internet, ranging from Oprah Winfrey's cable TV network to sites memorializing oxygen bars, defunct radio stations, and experimental rock bands. Pocket-sized oxygen canisters are pitched as a cure-all for hangover; a private company plans to harvest oxygen from the soil of the Moon. The "House of Oxygen" promises to "treat, heal, eradicate and cure" most illnesses with an oral oxygen regimen; forcing pressurized oxygen into injured tissues in hyperbaric oxygen therapy is said to "wake up" dormant or damaged cells. The Massachusetts Institute of Technology's "Project Oxygen" envisions a Utopian techno-future where "pervasive human-centered computing" will be "freely available everywhere, like . . . oxygen in the air we breathe."

Although oxygen occurs in the lab as a violet solid and as a strangely magnetic, pale blue liquid (the latter essential for space exploration and mass destruction as a prime component in rocket fuel), it is in the air that we think of oxygen first and in the air that we personally feel its benefit as O_2. It is in the air that the balancing act so important for life occurs. Oxygen comprises 20.96 percent by volume (or 23.15 percent by weight) of Earth's atmosphere—halfway between 17 percent, below which breathing becomes difficult, and 25 percent, above which organic compounds spontaneously burst into flame. An

Earth with an atmosphere of one-quarter oxygen would be a hellish place, a constantly burning world, yet it took billions of years before this balance was achieved. By end of the Permian era, 500 million years ago, plants and the aquatic cyanobacteria called blue-green algae had pumped so much oxygen into the atmosphere that its level swelled as high as 35 percent, making possible a giantism producing roaches as big as house cats and dragonflies the size of magpies. Present levels settled out about 150 million years ago, and only after animal life developed the ability to generate energy by breathing oxygen.

Just as oxygen molded life, it molded thought as well. Beginning with the Greeks, Western philosophy returned to the idea that an unseen essence fed the fire while sustaining the soul—and this essence lurked in the atmosphere. God was in the air. But how could one describe the properties of the unseen and unknown? How could one debate its existence when no one agreed on a name? The naming of this mystery became as much an issue as its discovery. The "first principle" of the ancients became the *prima materia* of the alchemists; the natural philosophers sought life's *pabulum,* which would somehow unite the world and man with God. Thales of Miletus, the first Ionian philosopher of historical record, believed that all natural phenomena were different forms of one fundamental substance, which he believed to be water. Anaximander, his disciple, maintained that this "first principle" was like the air itself—an invisible, infinite, and indestructible essence he called *apeiron,* the "boundless," which found equilibrium in both matter and energy. Anaximenes, the third great Ionian philosopher, came next: he returned to Thales' assumption that the first principle was familiar and material, but borrowed from Anaximander to place it in the air. To Anaximenes, all change was explained by atmospheric flux, an endless cycle of condensation and rarefaction, in which nothing, not even the gods, was permanent.

The search for a "vital spirit" connecting all life to the world around it was as old as the mind itself, and made its way to modern times through various paths and means. In a compromise with his forefathers, Aristotle suggested that air was only one of the four basic elements, all co-equals. Keeping these four in balance was necessary for health, a medical view that prevailed for two thousand years.

Newton believed that gravitation was the vital spirit, but called it his "unifying force"; Edison chose electromagnetism; the Curies, an unseen but pervasive radiation. The quantum physicists aimed for a "world equation" that described all forces in a "unified field." Today, cosmologists posit a universe filled with "dark" matter and energy to account for the fact that the galaxies fade from each other with increasing acceleration, a process beyond their abilities to explain; the percentage of dark matter to dark energy is roughly the same as that of oxygen and nitrogen in the air. The language becomes more technical and arcane, but the heart of the problem remains in the unseen.

The history of science seesaws back and forth like this, between the quest for finite explanations and a spiritual return to the ineffable "dark" linkage always suspected but never perceived. With the slightest push, scientists invoke God. Newton was a mystic; the Curies signed affidavits expressing their trust in certain spiritualists; Einstein trusted in God never to gamble with creation. Edison imagined building a ghost machine, a cylinder "so delicate that if there are personalities in another existence or sphere who wish to get in touch with us, the apparatus will at least give them a better opportunity," he told *Scientific American*. The essence of life left the body after death, but lingered in space until ready "to go through another cycle."

Life sputters . . . the soul breathes out . . . but somehow returns from the air. Life is like the air around us. The only permanence is change.

▪ ▪ ▪

Why shouldn't a man dream of decoding the essence of matter in such an enlightened age? It was, after all, an age of revolution on many fronts and in many fields. Isaac Newton's scientific revolution, announced in his 1687 *Philosophiae naturalis principia mathematica* (commonly called the *Principia*), was followed politically in 1688–89 by Britain's Glorious Revolution with its Bill of Rights for the common man. Newton's revolt has been called the first to be recognized as such by contemporaries, while the Glorious Revolution has been labeled the first major political revolution of modern times. The

American Revolution broke loose in 1776, followed by the French Revolution of 1789. It was a time when discoveries followed one another in dizzying succession, and our understanding of reality could change in a few short years.

This period is also known as the age of Enlightenment, generally situated between the discoveries of Isaac Newton and ascension of William and Mary to the British throne in the late 1680s and the *coup d'état* of November 9, 1799, when Napoleon Bonaparte took power. Some end it earlier, with the July 14, 1789, storming of the Bastille. The French mathematician Jean Le Rond d'Alembert called his era *l'age des lumières,* the age of illumination, and the name stuck like flypaper. Thinkers and writers, primarily in London and Paris, became convinced they were more enlightened than their contemporaries and set out to change Europe and the world. "To enlighten" had religious connotation: one brought spiritual awakening to the masses as Prometheus brought flame and light, removing the blindness of centuries of misrule by the Church and Crown. Democracy and science were being established as secular faiths to replace the misplaced beliefs in failed gods and systems: if there was any agreement among so many voices, it was in the power of reason and liberty to conquer ignorance and build a better world.

But in this new world, who were the heroes and villains? Centuries of aristocratic privilege and fanatical clericism had led to repression, persecution, superstition, and poverty, but could science, the child of reason, improve the human condition and bring happiness to the world? Science could "overturn in a moment the old building of error and superstition," Priestley announced, "extirpating all error and prejudice," overthrowing all forms of misused authority. The sciences progressed so quickly, wrote Benjamin Franklin, that he often regretted "that I was born so soon. It is impossible to imagine the heights to which may be carried . . . the power of mind over matter." Franklin even imagined the conquest of death—a day when "all diseases may by sure means be prevented or cured, not excepting that of old age, and our lives lengthened . . . even beyond the antediluvian standard."

Such hope lay everywhere. One short year or so before this dinner, in June 1773, the ancient and deadly problem of determining longitude was solved by John Harrison, a Yorkshireman like Priestley, with his accurate and elegant chronometer. All of Nature was being ordered and catalogued, as seen by Sir Joseph Banks's return from the South Seas with Captain Cook and the one thousand new species of plants, hundreds of drawings, uncounted insects, and five hundred bird skins he brought home with him. New substances were emerging from the labs of "chymists" so quickly it seemed a fire had been lit underneath the researcher, too. In 1774 alone, not just oxygen but chlorine and manganese were discovered; in the twenty years before that, researchers isolated cobalt, nickel, magnesium, hydrogen, and nitrogen. There were new fluids, systems, and cosmologies—"phlogiston," "caloric," "animal heat," "inner mold," and the "fire soul."

Yet a glance at the era's scientific journals illustrates a central irony: there were so many systems clamoring for attention that it was acutely difficult to tell the valid from imaginary. One man claimed the secret of life lay in a vitalistic "vegetative force"; another unveiled a "motionless astronomy" that was "the key to all the sciences." Which system was correct? Whose reality was real? The war of competing realities that developed between Lavoisier and Priestley exists at the spear tip of the culture wars still being waged today. Without a central authority, who has the right to reshape the world? A profound and radical individualism lay at the heart of Enlightenment thought, establishing the individual—not God, the King, or the Church—as the arbiter of all meaning and truth: René Descartes might doubt a lot, but he didn't doubt himself. Since the individual mind refined and defined all experience, that meant it lay at the center of all systems. "I observed that while I thus desired everything to be false, I, who thought, must of necessity be something," he wrote in his *Discourse on Method* (1677). "I think, therefore I am" was the most famous and ardent battle cry of Enlightenment thinkers: the mind filtered order from the chaos around it, constructing the world anew.

But if this age of intellect produced marvels and wonders beyond any prior imagining, it also gave the public real cause for fear. In 1773, one year before this dinner party, Joseph-Jérôme Lalande, France's foremost astronomer, calculated the end of the world. He predicted that a comet might swing close enough to Earth in 1789 to push the seas from their beds and devastate civilization, a catastrophe a million times greater than the 1755 earthquake and tidal wave that devastated Lisbon, snuffing out in a matter of minutes 30,000 to 50,000 souls. The Archbishop of Paris recommended forty hours of prayer per Frenchman to stave off the apocalypse; the King's chief of police begged the Academy of Sciences to repudiate Lalande, but was told it could neither repudiate nor reverse the laws of astronomy. Ladies of the court confessed their sins; stillbirths abounded; in the country, public displays of repentance flooded market squares. When the astronomer finally published his findings, it turned out that his estimate of the actual occurrence of such a cataclysm was only 1 in 64,000, but by then the apocalyptic fervor was so entrenched that the public assumed the truth was being suppressed by the government. Later, a commentator wrote: "The time of prophets has passed; that of dupes will never end."

Was it truly an age of reason, or of insanity? At times it seemed as if man himself were the experimental subject, just another mouse in an air pump, and all these new systems and truths were merely tests designed to see how well he would respond. Forget the great thinkers; to the average Frenchman or Englishman, this age of great hope seemed more like an age of insecurity. The threat of gigantic international conflict always seemed to loom on the horizon, a fear that finally bore fruit with the terrible Napoleonic Wars. This "equality of man" that Rousseau's disciples preached—where was this fleeting thing? In real life, men and women were born into vast fortunes or fantastic poverty, and few bridges spanned the gulf between the two. The chasm was most evident in Paris and London, seat of Enlightenment dreams. While the Prince Regent spent nearly £1 million on his Turkish Royal Pavilion in Brighton, the gin taverns of Whitehall advertised: "Drunk for a Penny! Dead drunk for two pence! Clean straw to lie on provided free!" In the first half of the century, fashionable rakes wore red high heels, patted blue powder on their faces, and dressed their hair in thirty-six curls, while in the sewer-running streets, the poor took a "rat's-eye view of life," one writer said. Armed press gangs charged with filling the navy's ships rounded up men like trappers beating the brush. When the Duke of Wellington looked over a new batch of recruits, he remarked, "I don't know what effect they will have on the enemy, but by God they frighten me!" Undergirding all talk of democracy and rights was a very real fear of "King Mob," a beast that even royalty dared not rouse. Riot was the voice of the crowd. Ben Franklin reported that he saw in England "riots about corn, riots about elections, riots of colliers, riots of coal-heavers, riots of sawyers, riots of Wilkesites, riots of chairmen, riots of smugglers." Ultimately, all riots were the same, just like fire, and it was during this era that the crowd's enduring symbol, fire and combustion, gained legitimate currency.

Twenty years after this dinner, during his flight to America to escape the mob's vengeance, Priestley would think of such things. Chaos sprang from dreams of order, yet new dreams sprang from the ruins. He remembered the "Church and King" mob that broke into his lab near Birmingham and burned down his home. He learned soon enough that Lavoisier, who had sat across from him at dinner and became his rival in the race for oxygen, was arrested and handed to the crowd. Lavoisier would face the Revolutionary Tribunal and hear the verdict: "The Republic has no need of scientists." He urged Marie to go into hiding, then stepped into the tumbril and rolled to the guillotine.

Priestley was born ten years before Lavoisier, and died in exile ten years after his rival's execution. His life brackets Lavoisier's, yet both were bound by fate and cruel intellect, and both were claimed as martyrs by the new faiths of science and liberty. In the end, neither could have advanced science so far without the conflict with the other, yet if either could have seen into the future that night as he studied his rival, would he have turned away? Across the table in 1774, their eyes strayed to the burning candle as others talked: the fire that evokes excitement, arousal, or panic, that has symbolized through the ages desire, passion, knowledge, comfort, anger, and revenge.

At the heart of all their decisions lay a drive to understand the mechanics of the flame. Put a match to the wick, and the carbon and hydrogen atoms that are bonded in the molecules of wax move vigorously and break loose from their ordered positions. The wax at the base of the wick melts as the temperature approaches 400°C, soaking into and up the open-weave cotton as the liquid wax turns to incandescent vapor. The molecules bend and stretch; some vibrate so violently that electron pairs binding the atoms together are severed.

This results in *pyrolysis,* the process that shakes the molecule into individual fragments, a heat-inspired apocalypse that if acting on larger bodies would register off the Richter scale.

Consider a world like this, so different from the ordered one of a fraction of a second earlier, where the solid infrastructure of carbon and hydrogen atoms was arranged in an elegant lattice, then blown to pieces. In a cataclysmic instant, what was stable is reactive—electrons flash in and out of existence, molecules in the vapor are chopped apart or broken in half. Hydrogen atoms convert to water vapor, further heating the gases instead of quenching the heat; stable molecules "crack" as temperatures reach 1200°C–1400°C. More oxygen-charged air rushes into the vacuum, and we watch the flame build.

The colorless and odorless gas that Priestley and Lavoisier discovered transformed the world. Its isolation and identification led to a new science, the discovery of ninety-two naturally occurring elements, and entry into the realm of the atom. Oxygen's central role in combustion unlocked the process of respiration and energy exchange in the human machine. Their insights, often at odds, dethroned an idea that had dominated science for twenty-three centuries and changed human status on this earth in ways never before imagined; it gave mankind a control over its world that it had always lacked, and unlocked the potential for creation and destruction that had belonged to the gods alone.

Thomas Kuhn observes that, like artists, scientists must "be able to live in a world out of joint." No one illustrated this more precisely than Priestley and Lavoisier. Like artists, they turned first to metaphor to explain what was not understood. The intellectual adventure is couched in metaphor—it is the first organizing principle, the fuel that feeds further inquiry. Newton saw the Moon as a ball thrown around the Earth, but long before that, *gravitas* represented the binding nature of God's love. Einstein saw time's fluidity in the tale of two men who would never meet, one standing by the railroad tracks, the other on a passing train. Priestley and Lavoisier searched for the secret of life as the world burned around them. All the time, the answer lay in the flame.

PART I

Problem

Then said Christian to the man, "What art thou?"
The man answered, "I am what I was not once."

—John Bunyan, *Pilgrim's Progress*

CHAPTER 1

The Cloth-Dresser's Son

THE HERETIC EVENTUALLY CALLED "GUNPOWDER JOE" COUNTED death, fever, and loneliness among his first memories. He was seven, a cloak of cold enveloped the earth, and inside the small homestead in Yorkshire, Joseph Priestley's mother "lay like a lamb" on her deathbed.

It was the Great Frost of 1739–40, when snow and ice rolled from coast to coast, killing indiscriminately. It was hard to imagine that something so white and soft, soft as the white-cloth "dressed" by his father, could be so bitter and deadly. The snow fell on Christmas Day and lay until March: in London, people froze to death in the streets, while here to the north, in the West Riding of Yorkshire, postboys not much older than he was died of exposure while waiting on their coaches. The Thames froze over above London Bridge, and the grinding ice crushed and sank several ships. Joseph would wake in the morning with his younger brother Timmy shivering beside him, his breath frozen on the sheets. The fish were frozen stiff as knife blades in the Birstal beck, the clear stream outside their six-house hamlet of Birstal Fieldhead.

This was an austere landscape, the kind that bred austere beliefs, beautiful in its severity. The village was seven miles from Leeds on the old coach road to Manchester, perched at the head of fields that pitched into the southern valley below. Behind the hamlet, to the north and northeast, lay a high, dark moor. Joseph was surrounded by nature and open space, perceptions that shaped his critical thinking. Nature's harshness was evident as entire flocks of birds dropped frozen from the trees, a slaughter so great that some species would not be seen in the fields for another three years. The harvest had been poor that spring, accompanied by rising grain prices and famine. Death was familiar that winter, and here in the three-room stone house of his birth, with the thick rafters and whitewashed walls, Priestley watched and listened as his mother died.

"It was my misfortune to have the idea of darkness, and the idea of malignant spirits and apparitions, very closely connected in my infancy," he later said in memoirs written for his own children. Because his mother Mary "was having children so fast," he did not see much of her when she was alive. Joseph, the eldest, was born on March 13, 1733; his brother Timothy was next, on June 19, 1734; a little sister Mary would die by age three. The birth of Joshua, the last, killed their mother: as if in penance, he stayed in the village his entire life.

Soon after his birth, Joseph was taken in a wicker basket to the farm of his maternal grandfather, Joseph Swift; he stayed in that hamlet until he turned four. His return to his mother and father was occasioned by the death of his younger sister. His scant memories of his early childhood are all dark. Of his father, Jonas, he would say little; of his mother, his memories were poignant, but stern. Though raised as Calvinists, his parents were Presbyterians for whom life was but a brief stop. Mary Priestley saw it as her duty to prepare her children for eternity. She taught Joseph the Presbyterian Catechism, and by age four and a half, when he finally came home, he gained her approval by repeating it perfectly at her knee. At age six and a half, he asked Timothy to "kneel down with him while he prayed; this was not at bed time (which was never neglected) but in the course of the day," the younger brother said. Joseph had one other memory of this time: while playing with his cousins, he picked up a pin and brought it home, but his mother made him take it back. Joseph learned from this "the distinction of property." As the nursery rhyme warned, "It is a sin, to steal a pin."

We do not know what killed his mother: death was too common for the recording of such details. During the eighteenth century, one in five babies died in their first year, and one in three of fever before age five. Ten of thousands of mothers like Joseph's died in childbirth. Although plague left England the century before and smallpox was battled with inoculations, there was still typhus, dysentery, measles, influenza, and tuberculosis. Joseph looked around him and saw that few people seemed old: less than half lived to fifteen, while the average life expectancy at birth was thirty-five. For Joseph to die before adulthood would be mourned. But it would not be unexpected.

His mother died instead. Mary Priestley lay like a sacrifice as she waited for sickness to carry her off; she hallucinated and cried out, her moans inescapable in the small house with whitewashed walls and low, dark rafters. Shortly before her death, she dreamed she was in a "delightful place," which she insisted was Heaven. "Let me go to that fine place," she said when she awoke, then drifted back to sleep, and died.

Such suffering was a fact of life, and death came out of the blue. As late as 1771, there was a terrified acceptance of death that seems alien today. "Last Thursday night, a most Melancholy Accident happened," reported the *Northampton Mercury*:

> Two Farmers went to a Neighbor's House to view a sick Calf; the Weather was then most intensely cold; while they were handling the Calf it went mad, and slaver'd on each of them; alarmed at this, they by Advice of Friends, set off to bathe in the Sea; but on the Tuesday following they were brought back dead in a cart, to the Terror of the whole neighborhood.

Joseph was lost "without a mother" in a world that made no sense: motherless children learn too early that the world is unforgiving, and hang on like grim death to what little security they have. He stayed at home, helped his father dress wool, and preferred not to remember the next two years.

▪ ▪ ▪

The death of Mary Priestley occurred at the cusp of what is now judged the end of old, premodern England and the painful birth of the new. In 1700, the England outside London was considered a minor rustic nation of barely 5 million inhabitants, with nearly 80 percent of that number living in a countryside abandoned to heath, bog, and fen. Society and the economy had reached an unpleasant equilibrium: jobs were limited, couples married late, and birth rates were held in check by epidemic and famine.

Yet by 1740 this was changing, nowhere more than in Yorkshire. For hundreds of years the North Country had been the center of the nation's woolen industry; wool was brought from throughout England to be woven into cloth, then sold in market towns like Leeds and Halifax for shipment to the four corners of the world. In these narrow valleys that mounted quickly from sea level to bleak hills, small hamlets like Birstal Fieldhead were hives of cottage industry and generations of Priestleys were a fixture. Joseph's grandfather had employed ten men in the workshops adjacent to his house: a "cloth-dresser" also, he treated the woven material until it had an even texture, then singed or cut off the loose ends of fiber and ironed it to a finish. When Daniel Defoe passed through Yorkshire in 1725, he said that Birstal was known for the quality of its white-cloth; part of that reputation was due to the Priestleys. The grandfather passed the trade on to Priestley's father, though on a smaller scale. Now Jonas Priestley hoped to pass it to his sons.

As important as work was religion, the second mainstay of the Yorkshireman. The independence of the area was famous, and the sympathies of West Riding had been with Cromwell and the Roundheads during the Civil War. Independence and Dissent went hand-in-hand, "Dissenters" the catch phrase for all non-conforming Protestants who refused to submit to the Church of England's Thirty-nine Articles of Faith. In addition to Presbyterians, Puritans, Calvinists, and Unitarians, it came to mean Quakers, Catholics, and Jews. Yorkshire was a center of dissent, and although William and Mary in 1689 signed the Toleration Act that gave Dissenters the right to exist beside the Church of England, there were restrictions. The word "church," for example, was reserved for the official faith, while all others referred to their place of worship as "meeting house" or "chapel." Most important to one's livelihood and status was the fact that Dissenters could not hold public office, serve in the army or navy, or attend public universities. The nation's highest sinecures could never be theirs: if they wanted to shape the course of empire, they either bowed to authority or obtained power through some other means.

It was in this stern mix of God and work that the "spirit of capitalism" was born. As early as 1746, there was a general belief that one of the best ways to make money in this life was to be a Dissenter. The Priestleys were prime examples. Although they were what today would be called lower middle class, they hoped to rise by hard work, and one spur to that ethic was their grim creed. Calvinists held that because of Original Sin, all mankind had "lost communion with God, are under His wrath and curse, and so [were] made liable to all miseries in this life, to death itself, and to the pains of hell forever." One either belonged to the "elect," those predestined for Heaven, or the "preterite," those passed over and damned, no matter how many good works and acts of faith were performed. Yet there was a loophole. Wealth was a sign of election. If you worked hard and grew rich, you must be blessed, a popular philosophy still heard in such outlets as the 700 Club. The "Protestant ethic," as the sociologist Max Weber later dubbed it, allowed rewards in this life *and* the hereafter—but only as a product of rigorous discipline, thrift, and pursuit of property.

Nonconformists like the Priestleys contributed something else to the ethic, and this was scruple—an earnest desire to exercise control over themselves and others, for righteousness' sake as well as earthly gain. They discovered "deserving causes," and as the century progressed led the pack in educators, doctors, and reformers. The fight against slavery was a favorite cause: many gave up sugar, since it came from the slave plantations of the West Indies, or, like the Quakers, mobilized opinion against the slave markets in Liverpool. Others campaigned for better public health, hospitals, prisons, and morals, while a third front championed religious and civic liberty.

Everywhere Priestley turned as a child, he was inundated with the creed. His grandfather at his death called his workers to his bed and declared, "See how a man of God dies." His grandmother, Sarah Healey, was a "strong-minded" Calvinist who drove her faith into her children and grandchildren as if driving a nail into wood. Although now a Presbyterian, Jonas Priestley never completely abandoned his childhood faith and returned to it in old age. Joseph never mentioned early ancestors, but tales of Yorkshiremen who died for their faith during the Civil War were common in Birstal Fieldhead.

But most important for Joseph was Sarah Keighley, his father's eldest sister, who adopted the boy in 1742, when he was nine. Aunt Sarah's entrance marks the point where Joseph is no longer passed from hand to hand. She was rich, married for seventeen years to a man of property, and childless; she "took me entirely to herself, and considered me as her own child, having none of her own." Although Aunt Sarah was a strong Calvinist like her mother, she was also remarkably open-minded for the time and place, opening up her farmhouse for Dissenting ministers of all stripes as a kind of religious salon.

Once a week, or more, she held a "club" for religious debates. Clubs were everywhere—for singers, politicians, sexual adventurers, and pudding makers, so why not for Dissenters, too? Sarah's husband, John Keighley, had prosecuted Dissenters during his youth, hunting down traveling ministers who held services in the dead of night and sending them to jail. One night, when he'd hidden in a chapel to gather evidence against heretics, he'd listened closely to the sermon and converted to the faith instead. His large brick farmhouse at Heckmondwycke, three miles from Birstal, was called Old Hall and had a chapel and cemetery attached. John Keighley died three years after Joseph's arrival and left his property to Sarah, "who knew no other use for wealth, than to do good," Joseph wrote. She opened Old Hall as a "resort for all the dissenting ministers in the neighborhood, and those who were most obnoxious on account of their heresy were almost as welcome to her, if she thought them honest and good men."

It was as if a door suddenly opened in Joseph's mind. He listened and participated, and soon showed an inquiring turn of mind. She must have suspected that the boy was smart, for all her efforts as his guardian were directed at education. Although religion was the meat of conversation and first order of existence, politics and religion were inextricably linked, and Joseph heard bandied the issues dividing Tories and Whigs. Georgian England at all levels jealously guarded its freedoms: Montesquieu wrote as early as 1729 that "this nation is passionately fond of liberty," while exiled or imprisoned French writers envied English free speech and often first published in London before their works appeared in Paris.

One frequent visitor was John Wesley, who came to Birstal forty times during his career, the first time in 1745, when Priestley turned nine. By then Wesley had been holding his open-air meetings for six years, sometimes traveling 5,000 miles annually and delivering five sermons a day, and it is hard to imagine that Sarah Keighley would not have opened her doors to this famous, indefatigable man. It might have been Joseph's first contact with the idea of a more personal faith and the assurance that all people—not just the elect—were children of God.

But there were tensions and contradictions in the household that flew in the face of such open-mindedness and new ideas. Sarah Keighley might love her adopted son, but there were conditions to that love. John Calvin ruled within her four walls. Joseph's new world opened up to him a vast flowering of books and fresh ideas, but by revealing an intellect, Aunt Sarah expected him to prepare for the Calvinist ministry. The doctrines of experiences and new births, elect and preterite took their toll. His aunt sent him to schools to learn Latin, Greek, and Hebrew; any time for recreation was spent in study, and the only amusement Priestley recalled from this period was reading *Robinson Crusoe*. He snatched away from Timmy a book on "knight errantry," declaring the subject and his brother too frivolous; he developed such a horror of hearing an oath that he'd quiver with revulsion. He was hypercritical, self-absorbed, and filled to the brim with his gloomy creed, and must have seemed like an old man living in the body of a child.

▪ ▪ ▪

This is, with certain cavils, the opposite of Joseph Priestley the man. A 1765 likeness called the "Leeds" portrait, now in the possession of the Royal Society of London, shows a thirty-two-year-old adult with little connection to the somber child. Though he stands in ministerial garb, Bible in hand, he looks at the viewer "with sparkling yet kindly eyes." He was married when this sitting occurred, and loved by a host of students; he was within a year of election to the Royal Society and the protégé of Benjamin Franklin, a man who didn't suffer pompous fools kindly. He was an educator, historian, and soon to become one of England's premier men of science. What saved him from becoming a prig?

In 1745, two things occurred, though Priestley never acknowledged them. The Jacobites under Charles Edward Stuart, better known as Bonnie Prince Charlie, almost conquered the nation. And Priestley almost died.

He was twelve going on thirteen. In 1744, England declared war against France—yet another chapter in the ancient enmity between nations—and although its four-year course would seem little more than the distant rumbling of kings to most Englishmen, the Jacobite Rebellion began in Scotland, England's own backyard. Two years of panic and bloodshed ensued. The Young Pretender landed in the Hebrides in July 1745 with French support and soon took Edinburgh. This was a Catholic invasion, a bitter religious and civil war left over from the last century, all over a young man whose grandfather, James III, had been King of England sixty years earlier. Panic ran before his Highland allies, clansmen with their long plaid kilts, tartan waistcoats, and basket-hilted broadswords; the boy's army swept across the border and south through Newcastle, Leeds, and York, taking Derby in December, 130 miles northwest of London. Residents fled the city, and Parliament dithered, ill-prepared. But Charles lingered, strangely reluctant to finish the business, and the invaders faded back to the Highlands. What was left of his army was butchered in less than thirty minutes in April 1746 at Culloden, near Inverness; in late September, Bonnie Prince Charlie and a handful of followers were rescued at dawn by French frigates after five months of hiding, and the Prince was said to have wept as the hills of Moidart fell astern.

It was a bloody, melancholy affair, but victory forged a Protestant nation with stubborn Hanoverian kings, patriotic rhetoric, and a virulent jingoism that lasted the rest of the century. It also reignited a hatred for Catholics that spread to include any Dissenting religion not considered loyal to "Church and Crown." A few days after Culloden, mobs streamed through London crying, "Death to all Papists!" They burned down a Catholic chapel and four nearby houses. All that autumn of 1746, rebels were hanged, disemboweled, and beheaded in London, York, Manchester, and Newcastle; the reports were carried in the newspapers beside advertisements for "exceedingly good LIMONS and BITTER ORANGES, the Bitter Oranges are fit for marmalade." The heads of the Jacobins were spiked before the public gates of each city; the skulls grinned down on the streets until the end of the century.

During those bloody years of the rebellion, Joseph "was nearly carried off by a complaint of his lungs." The cause was then blamed on "bad air," but the true culprit was pulmonary tuberculosis, the "consumption" that lingered in the Birstal household and almost certainly "carried off" his mother and younger sister. The rod-shaped *Mycobacterium tuberculis* bacillus was mankind's first reported plague, its ravages found in bones dating from 5000 BC. Its advent was slow and fastidious, and the northern European environment was perfect for its transmission, especially during the winter months when people shut their windows and gathered around the fire. The microscopic droplets exhaled by a carrier could float in the air for days, and the only relief was to open the windows and let in the cold. But researchers have since determined that one good flushing of air like that would purge 63 percent of the airborne particles, and ultraviolet light from the sun would kill others. Timothy noted that his brother's first interest in science occurred as he struggled for life, and Joseph's first experiments concerned the mystery of air. "The first he made was on spiders," Tim remembered. "By putting them into bottles, he found how long they could live without fresh air."

More than anything else, Priestley was certain he'd die young, and soon. He was under the influence of Calvinism: Man's estrangement from God meant that Joseph would die and be damned. "I felt occasionally such distress of mind as is not in my power to describe, and which I still look back upon with horror," he wrote.

> Notwithstanding I had nothing very material to reproach myself with, I often concluded that God had forsaken me, and that mine was like the case of Francis Spira, to whom, as he imagined,

> repentance and salvation were denied. In that state of mind I remember reading the account of the man in the iron cage in *The Pilgrim's Progress* with the greatest perturbation.

References to Francis Spira and John Bunyan speak volumes of Joseph's terrified state of mind. Both were well-known stories of the fate of one abandoned by God. Spira was an Italian lawyer who converted to Lutheranism in 1548 but returned to Catholicism under threat of torture. His lapsed faith led to despair, a great Christian sin: he refused to eat, and despite entreaties, died of hunger or suicide. The man in the iron cage is thought to be drawn from Spira, and is considered the grimmest passage in Bunyan's metaphorical tale: Christian sees a man who once had been a "fair and flourishing professor," now sitting alone in a cage of despair. He asks the caged man whether it is not possible to repent and return to God. "I have crucified [Christ] to myself afresh. I have despised his person; I have despised his righteousness," the professor replies. "Therefore I have shut myself out of all the promises and there now remains to me nothing but threatenings, dreadful threatenings, fearful threatenings." Learning is a deadly thing.

If Priestley was terrified, Aunt Sarah was more so. She doctored him through the crisis, no doubt by hearty doses of the standard "physicks": morphine, opium, and belladonna; rum, tar, quinine, and buglewood tea. Like Edward Gibbon, who lost six siblings to fever, Priestley could honestly say, "I swallowed more physick than food." Timothy recalled that "never was human creature more tenderly nursed; his aunt was fond of him in the extreme." It was also during this time that the two brothers, living apart, grew closer: "About this time," recalled Timothy, "the school we went to was one mile from his habitation, and two miles from mine; whenever he was kept at home by illness, or we did not see each other for a whole week, it seemed a long period to either of us; few brothers ever felt such a particular attachment at so early an age."

But by releasing Joseph from fever, Aunt Sarah released him from Calvinism, too. She thought at first to get him out of sodden, unhealthy England, and arranged for him to live with an uncle in Holland and enter the commercial world. As preparation, he foreswore classical languages and studied modern ones; he trained in composition by writing out sermons; his skill increased when he learned shorthand. Yet just before departing, his health took a turn for the better and he decided to attend college in hopes of entering medicine or the ministry.

But he'd survived the physical crisis only to open himself up to a spiritual one. His wider, more contemporary study of the world gave his opinions a liberal slant, out of sync with his aunt or her Heckmondwycke congregation. Yet he tried to please one more time. "Before I went from home," he recalled, "I was very desirous of being admitted a communicant in the congregation which I had attended, and the old minister, as well as my aunt, was as desirous of it as myself." It was a last gift before leaving, a kindness to those who'd shown such kindness to him. But if Priestley was to have trouble later in professing the Anglican Thirty-Nine Articles of Faith, he had just as many problems now with the ten articles of Calvinism. He suffered guilt because he had not had an *experience,* the conversion that came from a direct visitation by God. Worse, he had trouble mustering sufficient repugnance for Original Sin. The elders of the Church interrogated him on both points and found him "not to be quite orthodox." They refused him admittance to his childhood Church; by doing so, in their eyes, they banished him to Hell.

Priestley was enraged, and on the spot rejected Calvinism—as it had rejected him. Almost immediately, he called himself an Arian. Arianism, one of the beliefs debated in Aunt Sarah's parlor, was a doctrine that denied Christ's divinity; its namesake, the Mediterranean ecclesiastic and poet Arius, was born in AD 250 at a time when there seemed as many forms of Christianity as there were believers. His poem *Thalia* claimed the Son of God had not been endowed with eternal life: when Arianism appeared in England in the mid-seventeenth century, some believers went farther, also rejecting the Holy Trinity. Mere mention of the sect was a sore point for many God-besotted people in Britain.

It was the first of Joseph's many heresies, but at this point he tried to keep it quiet, apparently out of love for Aunt Sarah. Young men try on beliefs as easily as new hats, and one donned in anger can just as quickly be abandoned. But Priestley's rejection by the elders meant he could not attend the Calvinist colleges. Aunt Sarah had planned to enroll him in nearby Mile-End, or possibly Plasterer's Hall in London, but to be accepted he would have to describe an experience, assent to the ten Articles of Faith, and swear to the articles every six months—things that Priestley now refused to do. At the advice of her minister, a man more open-minded than the elders, Sarah sent her brilliant but hardheaded nephew to the more liberal academy at Daventry in Northhamptonshire, where at least he might fit in.

▪ ▪ ▪

Priestley's baggage was formidable when he entered Daventry, but so was his preparation. He'd been rejected by his childhood faith, and so rejected it. Since he was too proud to convert to the Church of England, most promising avenues for a young man were closed to him. He'd nearly died of tuberculosis, and in the process convinced himself he was damned; he'd watched his mother and sister die and believed he lived on borrowed time. Yet he could speak or read Latin, Greek, Hebrew, Chaldean, Syrian, and some Arabic; had studied algebra, geometry, and other branches of mathematics; had read most of John Bunyan's works, Willem Gravesande's *Mathematical Elements of Natural Philosophy*, Isaac Watts's *Logic: The Right Use of Reason*, and John Locke's *Essay Concerning Human Understanding*, all advanced texts for the time. He was the most prodigious scholar ever to enter Daventry, a young man haunted by the fears of old age and bursting to make his mark in the world. At the least, he was an enigma.

He also seemed likable, if somewhat awkward. At nineteen, he was thin and wiry, standing nearly five feet eight, with quick, birdlike movements and darting, friendly eyes. He had a high forehead, pointed nose, and pronounced, dimpled chin; he stooped slightly when standing, but could explode into bursts of activity that took everyone by surprise. The reserved Calvinists at home called him "giddy," and he probably would have been louder and more demonstrative if not for the stutter, a trait that first appeared after his battle with tuberculosis and grew more pronounced each year. It greatly embarrassed him, and though he often smiled, he rarely laughed out loud.

One thing that both Aunt Sarah and Timothy noticed was that after his brush with death, he seemed obsessed with efficiency. He wanted to pack the most experience and learning into the shortest possible time, and was fascinated by ways to "economize." Complaining that he never had time to write things out, he practiced the shorthand system first developed by Peter Annet, a poor schoolteacher and member of the Robin Hood Society, a London debating society that met every Monday evening and where, for sixpence, any man or woman could speak for five minutes on religion, politics, commerce, or the law. After that, debate ensued. Joseph corresponded with Annet about shorthand and theology, suggesting improvements on each; the only poetry he is known to have written was a single verse entitled "To Mr. Annet on his new Short Hand."

Another example of Priestley's "rigid economy" was his method of walking. Since he did not like to linger, he devised a system whereby he spoke a word at every step, thus creating a kind of cadence. By constant repetition, this pace became habitual; his whole life sped up, hitting the pillows late, rising early, rarely sleeping more than six hours at a time.

Daventry had never seen anything like Joseph, and his depth of knowledge was such that he was allowed to skip several introductory courses required of incoming students. But then, he had never seen anything like Daventry either, the kind of institution Edmund Burke would later call "the new arsenal in which subversive doctrines . . . are formed." Yet as Max Weber and other sociologists have pointed out, such heresy is the engine of great powers: it lifts societies from their comfortable ruts, challenging their basic orientation to the world.

Priestley loved Daventry, and felt he had come home. "Three years, from Sept. 1752 to 1755, I spent at Daventry," he later wrote, "with that peculiar satisfaction with which young persons of generous minds usually go through a course of liberal study, in the society of others engaged in the same pursuits, and free from the cares and anxieties which seldom fail to lay hold on them when they come out into the world."

Dissenting academies filled an important niche, and Daventry was one of the most important of its day. By midcentury, English public education was a sorry affair. The basic curriculum was Latin and classical Greek, thought to promote politeness, the graces, and a habit of superiority. Freethinking and the questioning of authority were frowned upon. The two English public universities—all-male, celibate, and Anglican—were lacking in matriculants: Christs College in Cambridge had just three new students in 1733, the year of Priestley's birth, while by 1750, Oxford's entire intake had fallen to fewer than two hundred a year. Enrollment was primarily made up of young gentlemen spinning their wheels or curates' sons seeking admission to the Church, a perfect career for what Lord Chesterfield called "good, dull and decent" young men. The dons themselves were poor scholars, devoted to "port and privilege"; Horace Walpole, England's first prime minister, called Oxford "that nursery of nonsense and bigotry." The great thinkers of Georgian England were not dons or students; men like Edward Gibbon, Jeremy Bentham, Priestley, and Henry Cavendish all came from Dissenting academics like Daventry.

Dissenting academies also employed forms of teaching that encouraged doubt and skepticism, but that made enemies. Daventry, founded in 1729 by Philip Doddridge, was one of the oldest and most successful schools, yet its survival and success had not been easy. It outlasted attempts to close it down for not having a bishop's license, religious persecution, an attack by an angry mob on Doddridge's house in 1733, and the ill-health and eventual death of its founder in 1752. Joseph arrived just as Doddridge's successor, Caleb Ashworth, took over: the building lay on the site of an old Roman camp at the northern edge of town. One of England's oldest highways, running from London to Holyhead, passed right outside its gates. There was a sense of history about the place; an expectation that, after graduating, a young man would go somewhere.

Joseph's curriculum was modeled on a medieval scholar's: since one could not see the entire world in a lifetime, the compromise was to read all the books. His first years at Daventry included courses on political theory, anatomy, astronomy, elementary mechanics, trigonometry, "celestial mechanics," psychology, and natural and experimental philosophy. These were taught in English rather than Latin, another non-traditional detail. Priestley rose before six every morning, when most students rose, to study Greek with a friend. After that, his day followed the regular schedule: prayers and private reading till 8 a.m., breakfast and lectures from 10 until 2. The afternoon was free, followed by evening prayer at 7 p.m., tutorials until 9 p.m. when supper was served, and lights-out at 10:30. Anyone caught outside his room after that was fined.

Priestley loved it all. Students were expected by their third year to specialize for the ministry, law, medicine, or commerce, and Joseph decided that the life of a Christian minister was the "noblest of all professions." This is not surprising, considering his childhood, yet he would be one of the most unusual clergymen to emerge from this freethinking academy. In the frequent debates, he took extreme positions that incorporated science, religion, and psychology into a unified field of science and God. He became a disciple of David Hartley, whose psychological *Observations on Man, his Frame, his Duty, and his Expectations*, published in 1749, tried to apply to the inner world the same mechanical laws that Newton applied to the outer. Hartley argued that man was born a clean slate, and if educated correctly could be led to a state of "perfectability." All things, good and bad, were necessitated by the will of God and were links in the chain of cause and effect that would culminate in a glorious end, a doctrine called "philosophical necessity." The role of the scientist, then called a "natural philosopher" in England, was to decipher God's causal laws as Newton had done, in one stroke singing God's praise while making sense of an otherwise random world.

Joseph was inspired. Where Calvinism portrayed man's efforts to improve his life as pride, and thus an insult to God, philosophical necessity urged man to change his circumstances so that he might fulfill and approach God's perfection. Nothing was random; everything was for the best. Here in a nutshell was the "new birth" that had eluded him as a Calvinist, provided unexpectedly by a disciple of Newton. It seemed a great weight lifted from his shoulders, and the gloom and guilt of Heckmondwycke finally disappeared. Instead, he enjoyed himself immensely, as shown by his shorthand diaries for 1754. On April 25, "at the club, the chimney took fire, which raised the mob of the town about the house, cursing us," he wrote. But "no harm ensued, as it was a wet night." On April 26, he attended a Friday night party in which "several enjoyed too much liquor." On

May 2, "at our club, talked first about ladies, particular ones," before buckling down to studies. During that summer, he fell for a "Miss Carrott": on June 10, he spent "all the forenoon with Miss Carrott," but when he called on her later in the day, "Miss Carrott not at home." On Tuesday, October 8, "when we got up after prayers, had a full view of the cuddliest creature I ever beheld. I did nothing but stare at her, but could not find where she went after the service." His hormones were working overtime, with no Calvinist guilt to intercede. He studied the Koran, practiced his sermons, watched a magic lantern show. Over ten months, he dissected a rat, a dog, and a cat. He wrote long letters to Peter Annet and David Hartley. And slowly, almost imperceptibly, his love of science grew.

Afterward he would say that experimentation was the purest form of intellectual adventure; one might guess instead that he saw it as a release from the more serious and endless polemics over God and man. Hartley himself claimed that science could be the route to God's truth, and brother Timothy noticed a change in Joseph whenever he came home. Sometimes they would walk the fields, arguing long and hard about doctrine: after Joseph's apostasy, Timothy had slipped into Aunt Sarah's good graces by studying for the Calvinist ministry. These days, Timmy worried more and more about the state of his brother's immortal soul; but when Joseph talked about stars or the mysteries of the blood, the argumentative side of him disappeared.

"My brother, when he began to learn astronomy, would be in the fields with his pen and papers," Timothy later recalled. This spread his notoriety in their little hamlet, "as it was at the time a science very little known." He told Timmy how the planets rolled around the Sun in elliptical tracks; at other times, he seemed in a positive lather about the mechanics of Archimedes. He'd lived too long in nightmarish doubt, but the certainties of science could serve as a shield. "I well remember how much these experiments [of Archimedes] pleased him," Timmy recalled. "He soon told me how many men he could lift from the ground in a moment."

Priestley, like Archimedes, knew that with science in his arsenal, he could move worlds.

CHAPTER 2

The Sums and Receipts of Parallel Worlds

AS THE TWENTY-YEAR-OLD JOSEPH PRIESTLEY WOOED MISS CARrott and prepared for life after Daventry, his future rival entered the most exclusive boarding school in Paris and inherited, at the age of eleven, a legacy of 45,600 *livres,* or about $2.7 million in current dollars. Such a fortune so early in life, a biographer would later suggest, "was bound to instill a taste for sums and exact accounts" in the man whose "new science" would depend so much on weights and measures.

Issues of class and wealth would always divide Priestley and Antoine-Laurent Lavoisier. Yet in matters of mortality and upbringing, the two were remarkably similar. Antoine was born in Paris on August 26, 1743, the only son of Jean-Antoine and Emilie Lavoisier. His birth occurred in the thirty-fifth year of the reign of Louis XV; that same year, Voltaire turned fifty and Rousseau, grand prophet of the Revolution, thirty. The War of the Austrian Succession was at its midpoint, and England would declare war within a year. Priestley had moved in with Aunt Sarah: after the deaths in his family, he was showing signs of the tuberculosis that nearly killed him, too.

Antoine's parents were staunchly Catholic, and he was baptized in the Parisian parish of Saint-Merri. Both were middle class, his mother's background wealthier and more solidly urban than his father's, and an appreciation of money and culture flowed in Antoine's veins. His father hailed from Villers-Cotterêts, a rural town fifty miles northeast of Paris, where an ancestor had been a groom in the King's stables. The groom's son snagged a position in the postal service, and over generations the family climbed the social ladder, advancing from hostelier to sheriff's officer to landowner, then finally from small-town lawyer to Parisian official. In 1741, Antoine's father was appointed *avocat au parlement*—a law officer of the crown, with an annual income of 10,000 *livres,* more than enough to provide the household with servants and carriage. Fourteen months before Antoine's birth, Jean-Antoine married the beautiful if frail Emilie Punctis, daughter of a fellow lawyer; it was an arranged marriage, as were many in the middle class, a form of networking and way to secure position. Yet the partnership, as was often hoped, developed into love. The young couple settled in a cul-de-sac surrounded by mansions and a monastery, and led the bourgeois existence that would produce the judges, poets, scientists, and intellectuals who defined the French Enlightenment and catalyzed the Revolution.

Two years after Antoine's birth, Emilie Lavoisier gave birth to a daughter. She never regained her health, and in 1746, she died. Antoine was five, two years younger than Priestley when his own mother died, and like Priestley, it would not be the only death of his childhood. Like Priestley, he also lost a sister, but Marie-Margaret-Emilie Lavoisier was fifteen when she died of smallpox, compared to the three-year-old Mary Priestley. Antoine and his sister were close; her death struck him during that same period in his life that Priestley was rejected by Calvinism, and his reaction would be as dramatic. A seriousness that began after the death of his mother suddenly deepened and took hold of his mind.

In 1748, Jean-Antoine left the house of his marriage and moved with his children into the Paris mansion of his mother-in-law. Madame Punctis lived there with her second daughter, Constance, who although only twenty renounced marriage to raise and educate Antoine and his sister. Aunt Constance would be to Antoine what Aunt Sarah was to Priestley: a surrogate mother, fussy and kindhearted, who made the advancement of this motherless child the center of her world. It was a highly protected environment, but still the adults worried, especially after the death of Marie-Margaret-Emilie. The boy seemed overly solemn for one so young, preferring study to games.

Antoine's second home was the family's rural estate at Villers-Cotterêts, and it was here that he saw firsthand another side of France—the destitution and misery that would topple the *ancien régime.* Although French agriculture had prospered in midcentury, aided by a series of good harvests, rural poverty was the realm's most visible evil. Peasants dressed in rags, went barefoot, and lived from hand to mouth; most existed almost exclusively on a cereal diet, without milk, eggs, or meat, lacking in the protein necessary for manual labor. Their houses were leaky huts, their lives devoid of beauty or dignity. Much of the misery was caused by the government: the peasant paid half of what little he earned in taxes to the Church and Crown, while the *seigneur,* or hereditary landowner, possessed feudal rights over his farm. Many landlords—noblemen, or landed gentry like the Lavoisiers—rarely visited their estates, since the sight of such misery was depressing. There was always talk of reform, but never action: inertia was woven into the social fabric as tightly as the poverty.

When Lavoisier turned eleven, two things occurred. He inherited the fortune from his maternal great-grandfather, Christophe Frère, a meat merchant and burgess of Paris. And that same year, his father enrolled the serious, studious boy as a day student at the Collège Mazarin, "the most beautiful and prestigious as well as the wealthiest of the Paris secondary schools," a contemporary crowed. Though the elder Lavoisier hoped his son would continue the family tradition by one day entering the law, the new school directed Antoine's interests elsewhere.

The Collège Mazarin was a special place; in addition to providing traditional courses in classics and rhetoric, it had one of the best science faculties in Paris. In France, this meant the *exact* sciences, physics and mathematics, with a smattering of chemistry. In the year of Lavoisier's admission, the college's most famous *savant* returned to teaching after an extended leave. The abbé Nicolas-Louis de Lacaille, assistant astronomer at the French Academy of Sciences, returned from a three-year expedition to the Cape of Good Hope; while there, equipped with a quarter-circle with a 3-foot radius and a 32-inch telescope, he measured the arc of the meridian, located the position of 9,766 new stars, and named 14 new constellations. He did two things upon coming home: describe his findings in his *Coelum Australe Stelliferum*, and calculate his expenses at 9144 *livres* 5 *sous*.

The Parisians found the 5 *sous* hilarious, but Lavoisier understood. Whether counting stars or pennies, accuracy made sense of the world.

▪ ▪ ▪

Just as Priestley and Lavoisier were born into different classes, they were heir to competing scientific traditions—quality versus quantity; a deep search for "essence" versus a faith in things that could be measured. One referred to God, the other to man and his machines.

By the mid-eighteenth century, two divergent approaches to the study of the physical world had developed; as in so many dichotomies of that time, these faced each other across the English Channel. The British "pneumatic chemistry" focused on the collection of unknown gases with the help of the recently improved air pump; it emphasized sensory properties like color and smell, a holdover from the alchemists, who delved into the "essence" of all things. Guided by the British tradition of utility and natural theology, chemistry had "ends" as well as "aims," a view that was probably best expressed by Priestley's direct successor, Humphry Davy, who said that chemistry should be known for "increasing the comforts and enjoyments of man, and the demonstration of the order, harmony, and intelligent design of the system of the earth."

The Continental approach, best personified by the French, began from the opposite direction. French chemists emphasized the balance and weight to describe a substance, a less subjective indicator than the researcher's eye or nose. Research was focused on more palpable substances like minerals and their "salts," rather than invisible and mysterious gases and airs. Labwork began self-consciously with observations, not metaphysics or "first principles." Some of Lavoisier's later success has been attributed to his merging the two traditions, yet his method remained fixed in the empirical traditions of Continental researchers who emphasized analysis over "truth," and saw their work as part of a well-defined craft stretching back for centuries.

Part of the split could be blamed on the paths taken by France and England over religion. In the more liberal atmosphere of Georgian

England, religious tolerance and the freedom to publish dissenting views flourished. The opposite could be said of France. More new truths about the external world were discovered in France during the last half of the eighteenth century than in all preceding periods put together, yet the role of an almost rabid anti-clericism and anti-mysticism cannot be ignored. When Louis XIV in 1685 revoked the Edict of Nantes, the limited protections granted to French Protestants, he kicked off decades of autocratic rule. This harsher atmosphere of repression gave the French Enlightenment a more bitter, uncompromising stance toward God and religion than that seen in Britain, a rejection that lodged within the very structure of science.

It also led to a rigor in the French system that was not present in the English—a rigor, or lack of it, reflected by the scientific academies of the competing nations. Society epitomizes in its institutions what it values; they become signposts of where the culture is headed. The first truly scientific society was the Royal Society of London for Improving Natural Knowledge, incorporated in 1662; its stated purpose was to share information and ideas between members, communicate these findings to the world through publications, and encourage experimental investigation by the posing of problems and granting of prizes and awards. Each year, the Royal Society awarded the Copley Medal in honor of research or an invention that solved a particularly vexing problem: in 1749, for example, the society awarded the medal to John Harrison for his solution to the deadly problem of determining longitude; in 1753, the medal was awarded to Benjamin Franklin for his discoveries in electricity, thus making him a kind of honorary Englishman.

The early society claimed to be a democratic body that did not discriminate on the basis of religion or class, but there were inherent problems. Since there was no state funding, annual dues were imposed that deterred all but the wealthy. By midcentury these strictures had eased, but the society still had a "clubbish" feel. There were no paid positions, most members saw themselves as learned amateurs, and it was so easy for those with wealth and status to join that fellowships lacked prestige. Although the society fulfilled its charter of disseminating knowledge, its greatest use to many was as a means of networking. "I should pity the man who expected, without other advantages of a very different nature, to be well received in a brilliant circle at London, because he was a fellow of the Royal Society," observed the agronomist Arthur Young as late as 1789. "But this would not be the case with a member of the Academy of Science in Paris; he is sure of a good reception anywhere."

The French Academy was a different world. Established four years after its English counterpart, its purpose was to couple the King's rule with science and sponsor practical research that would benefit the nation: one of its first charges was to assemble an accurate map of Louis XIV's kingdom. Since the Academy was state-funded, it held a more official position than the Royal Society. Senior fellows were salaried: science was their life, and for the first time scientists were "professionals." As there were a limited number of seats, admittance was more prestigious, but also more political and competitive. In addition to eighteen senior fellows, or salaried *pensionnaires,* there were a dozen honorary members drawn from the clergy and nobility, a dozen unpaid associates, or *associés,* and at the bottom, another dozen unpaid assistants, or *adjoints.* Members moved up the rank as vacancies appeared in their discipline. In return for state support, the Academy was expected to produce fairly tangible results, serving in essence as a state-funded research institution rather than a learned club of friends.

The exacting nature of science in France was shaped and reflected by research from the Academy. Funds were directed to such projects as the development of the metric system, the measurement of the arc of the meridian, and the attempt to standardize the endless variety of weights and measures used throughout the realm. These were huge projects involving commitment of both capital and manpower, and academicians were assigned to boards that kept the research on track. In addition to defining the problem, a board chose teams of *savants* to find a solution, an administrative innovation that had never before been established in any way. This French innovation of teams of scientists working toward one goal never caught fire in England during the eighteenth century; instead, the norm was the solitary researcher working alone in his privately funded lab. The British tradition of private initiative that yielded empire-building results in the industrial sector shackled any collaborative spirit in the sciences.

The French Academy also funded research by qualified experts and instrument makers, an early form of outsourcing to private contractors that stumbled upon a way to accelerate innovation. As France's foreign policy changed to a war footing, the Academy funded and directed research on improved gunpowder production: in time, Lavoisier would sit on a board for gunpowder research, direct gunpowder production at the state arsenal, and own a private research lab. The Academy funded development of the Borda repeating circle, the most precise surveying instrument then available for measuring the angular distance between two points, and refinements in the thermometer and barometer, made possible by the increasing ability of state-funded instrument makers to manipulate and purify mercury and eliminate gaseous occlusions.

Yet nothing symbolized the precise nature of French science like the balance, and in turn the balance determined the philosophy and methodology of French research. The art of balance making was well developed by the end of the seventeenth century for use by financiers, money changers, and goldsmiths, and the work of physicists during that period pointed out improvements such as positioning the center of gravity, making parallel knife edges, and aligning the supporting planes. But only the development of a finer-quality steel and advances in the art of fine adjustment could lead to the production of the great precision balances of the eighteenth century. Initially there was a race between the English and French: the first to be called a "balance" was John Harrison's, the lone genius of chronometer fame, and that was made for the private laboratory of the wealthy chemist Henry Cavendish; another made for the Royal Society also won renown. Yet these were solitary efforts, unsupported by funding, and as the century progressed, French instrument makers—especially Alexis Mégnié and Nicolas Fortin, who made balances for the Commission on Weights and Measures and in 1788, one for Lavoisier with a beam 3 feet long—led the way. Their secret was a series of adjusting screws, constructed with such precision that a spirit level was placed on the balance for adjustment. The oscillations of the balance could be read off a graduated sector, and readings were aided by a magnifier. For the first time, chemists were not limited by the scale of their measurements: weights as heavy as a kilogram (with an accuracy within

1 milligram) or as light as some tens of grams (with an accuracy within 1/100th of a milligram) were now achieved. Such levels of accuracy had never been possible in human technology, and excitement over the seemingly endless possibilities defined the French school.

If there was an analogous technical advance that defined the English pneumatic chemistry, it was the air pump, refined by Robert Boyle. The frontispiece for Thomas Sprat's *History of the Royal Society of London*, published five years after its founding, showed Sir Francis Bacon, the society's ideological figurehead, pointing at the instruments that would be a source of knowledge; most prominent were a giant telescope and the vacuum pump of Boyle. Bacon's experimental creed was simple, if overwhelming in its scope: only through the collection and organization of massive amounts of data could the laws of Nature be uncovered, approaching truth from the outside and working in. Significantly, the telescope and vacuum pump were etherous instruments, designed for discovering new celestial bodies or the properties of unknown airs.

Boyle's air pump was largely an improvement on the original, a fire pump used for sucking water from a pond to fight fires, modified in 1650 by the German physicist Otto von Guericke. In turn, Guericke's ideas came from a number of sources: he'd heard of the experiments in air pressure by the French scientist Blaise Pascal and the Italian scientists Galileo and Evangelista Torricelli (inventor of the barometer); he'd read the speculations of William Gilbert, physician to Queen Elizabeth I, who said that space was a vacuum that kept the celestial spheres apart. Although critics countered that a vacuum would never be possible on earth, Guericke decided to try. By modifying a fire pump, he sucked air from a variety of containers, but these always collapsed from the weight of surrounding air. He then hit on the idea of a metal sphere as the best possible shape to contain a vacuum, since curves like domes and arches were the strongest load-bearing structures known. It worked, but the success led to other questions. Why did bells fail to ring in a vacuum? Why did mice die and candles go out? In 1654, he demonstrated before the Imperial Diet at Regensburg the famous experiment of the Magdeburg sphere: two hollow bronze hemispheres were fitted together, then one of his pumps sucked the air from the resulting sphere. Two eight-horse teams straining in opposite directions could not pull the sphere apart, yet the halves separated with the touch of a finger when he readmitted air.

Such were the twin systems as Priestley and Lavoisier schooled and stumbled separately into research. In France, there could be no truth without analysis. In England, one collected and observed phenomena until God—or an emergent Newton—revealed all mystery.

▪ ▪ ▪

As Lavoisier entered the most prestigious school in Paris, Priestley entered a hard, unforgiving world. The England of 1755 was about as peaceful and ennobling as the American West of another hundred years. Although the British saw themselves as the envy of starving, onion-eating Europe, home to lecherous Papists and mincing French courtiers, the Continent saw a much different picture. England "is different in every respect from the rest of Europe," noted Casanova on his travels, not the least for the violence he observed. Highways were strung with gibbeted corpses, and a man could be hung for stealing a sheep. "Anything that looks like a fight is delicious to an Englishman," said Henri Misson, and reasons to riot ran the gamut from the rising price of bread to the wrong play printed on a playbill. Force, rather than consensus, was the way to achieve results. "Violence was as English as plum pudding,"states one historian, while Christopher Hibbert, in *The Roots of Evil*, observes that pity was a rarity. Cat-dropping, bull-baiting, bear-baiting, and cockfighting were popular sports, while "unwanted babies were left out in the street to die or were thrown into dung heaps or open drains."

One cause of such cruelty could be blamed on chemistry. Colin Wilson makes a good case in his *History of Murder* that although life was always cheap, a strange callousness to suffering rose among the British when alcohol was affordable for the rich *and* poor. In Elizabethan England, people drank beer, wine, sherry, mead, and cider because the water was unfit, but since such wellsprings were costly, truly rampant alcoholism was rare. That changed in 1650–60, when a Dutch chemist named Sylvius discovered that a distillation of juniper berries produced a potent spirit called *geneva,* French for "juniper." The drink became popular in Holland, and when William of

Orange became England's king in 1689, *geneva* began to flow in great quantities and its name was shortened to "gin."

English distillers soon realized that an even cheaper and stronger spirit could be distilled from low-grade corn. When an Act of Parliament in 1690 allowed anyone to make and sell spirits without a license, gin shops filled the towns and cities and one in six houses sold gin. By 1699, the crime rate had risen so alarmingly that an act was passed making the theft of goods worth more than 5 shillings punishable by death; by 1734, 8 million gallons of gin were consumed in England annually, while in London alone, the consumption rate was 14 gallons per head. Horror stories popped up everywhere. In 1734, Judith Dfour was hanged for murdering her baby: she collected it from the workhouse, where it had been freshly clothed, then stripped the infant of its garments, strangled it, threw its body in a ditch, and sold the garments for the price of gin. The Mohocks, a club of young gentlemen dedicated to "doing all possible to hurt their fellow creatures," drank themselves beyond pity before boring out the eyes of old women and prostitutes with their swords. A report in a 1748 issue of *The Gentleman's Magazine* described how a nurse became so drunk that, instead of laying her charge in its cradle, she placed it in the fire. When examined before a magistrate, she testified that "she was quite stupid and senseless [with drink], so that she took the child for a log of wood; on which she was discharged."

It is all the more remarkable that not a hint of such violence is mentioned in Priestley's memories of his childhood, youth, or early manhood. It was as if he "dwelt in some far serene planet, whose inhabitants were wholly given up to study and prayer." Bessie Raynor Belloc, whose mother had known Priestley, commented on his "inwardness," as if his interests lay "otherwhere." "He impressed those about him as a being from another sphere. . . . Yet his own life was really one of the first to be swept into the vortex," she noted.

Still, his first position in 1755 as assistant pastor in Needham Market, Suffolk, was far from otherworldly or ennobling. He was about 160 miles away from home, the farthest he'd strayed in his life, and 90 miles from the companionship of Daventry. He felt isolated, physically and spiritually, and problems soon arose, though it is hard to tell whether they began with him or his parishioners. A minister either pleased his congregation or he did not get paid, but Priestley's flock seemed a potpourri of varying and sometimes conflicting beliefs—a true clutch of "autem cacklers," as Dissenters were called by outsiders—all thrown together in an isolated "autem cackle tub." There is evidence that mutual dislike existed from the start. Joseph certainly didn't fit the picture they'd had of a new pastor—he was thin and nervous, with a long nose, bulging eyes, and a profile that seemed to change from different angles—an "arch duke" for sure, as eccentric fellows were called. He walked about the rural parish with "a kind of disjointed bird-like trot," and when he got excited, he rattled on disjointedly, the word flow chopped and mutilated by his stammer. For his part, he refused aid from those with whose beliefs he differed, and this struck the Needhamites as a form of pride. If nothing else, it put a crimp in his purse, driving his annual salary below £30 and adding to his eccentricity the pinch of hunger.

Poverty, or its threat, would be a lifelong goad for Priestley, and this was its beginning. The average yearly salary for lower clergy was £40–£50, and he didn't even make that. There were some pockets of aid: a friend's uncle was generous, and the Queen granted small sums to help indigent ministers through their first year. But Joseph counted pennies, and two blows fell. It quickly got out among his congregation that he was Arian, and one Sunday a parishioner asked whether his views on the Trinity were "sound." They weren't, at least not by their standards. Not only was he a Deist, but he doubted Christ's divinity, believing him an inspired and great man but not a god. That year he'd begun his *Institutes of Natural and Revealed Religion*, a book that would bring him much sorrow and notoriety when published years later. In it he argued that all the trappings of contemporary Christianity—the deification of Jesus and the Virgin Mary, the doctrine of the Trinity, and belief in saints, angels, an afterlife, and the soul—were corruptions of Jesus' teaching by the early Church in its struggle for ascendancy. His mind was filled with it these days: when confronted like this, his anger flared. He gave a series of sermons on the "Institutes." In response, his parishioners dubbed him as a "furious freethinker" and stayed away.

This reduced his salary even more. In desperation, he made it known he was starting a school for half a guinea per student per quarter and would board the students in his house for 12 guineas a year. Clergymen often taught the catechism, but "because I was not orthodox," he stuck to more neutral ground, proposing a course of classics, mathematics, and geography. It didn't matter, since no one came.

Then he was hit from an unexpected quarter. Aunt Sarah still sent an allowance, but now that ended, too. This time the knife in the back came from his brother Timothy. By now, Tim had entered a Calvinist academy to prepare for the ministry, and the brothers argued furiously over doctrine every time they visited. Timothy was convinced (as he would be throughout his life) that Joseph was damned. During one visit, he watched as Joseph shook the hand of one of the freethinkers who'd argued doctrine in Aunt Sarah's "salon" and said, "You are the man who brought me out of the dark hole of Calvinism." Afterward, the man approached Timothy and confessed, "Have I spoilt the finest genius in the world?"

This worried Tim, and he wanted to save his brother. So he acted. Aunt Sarah, "though one of the finest characters in the world, was not aware how much her nephew differed from her, till I informed her," he confessed later, with just a hint of sibling rivalry. So they set a trap. During a visit, Aunt Sarah sat opposite her adored nephew so she could see his face as she handed him a copy of *Hervey's Dialogues*, a mainstay of the Calvinist catechism. "Joseph," she asked, "will you read to me these passages I've turned down?"

"I cannot, Aunt Sarah," he begged, feeling his horror of the old, abandoned creed. "Please don't ask this of me."

"Joseph, I insist you read those pages!"

Since he could never knowingly hurt his aunt, he stumbled through the words. But Aunt Sarah could see how his face paled as he read; she noticed the strain about his lips and his refusal to meet her eyes. That night she cried, and afterward, cut off the allowance. Joseph would later say that she diverted the funds to support a deformed niece, but Timothy said otherwise. And sadly, the incident turned into a permanent estrangement. Aunt Sarah had always told Joseph she would leave him enough in her will for financial independence, but now "she was no longer able to perform her promise." In the end, Aunt Sarah willed him a silver tankard, and that was all. "She has spared no expense in my education," he rationalized, "and that was doing more for me that giving me an estate."

These three years in Needham seem almost as dreadful as the decade earlier when he was certain he would die and be damned. A neighboring minister, distressed by Joseph's plight, advised that all his troubles would end if only he would conform to the Church of England. But Priestley would not hear of it. His family had rejected him because of Calvinism. He was scorned because he was a bachelor, but what local girls would look twice at a stuttering heretic who walked like a turkey? Some days it was all he could do to get out of bed.

He rarely wrote about his isolation of those years, but a picture of that "low despised situation" sometimes leaks through. "Even my next neighbour, whose sentiments were as free as my own . . . declined making exchanges with me," since to do so would make him a pariah, too. Joseph's stammer grew worse, so debilitating that every time he stood before what remained of his congregation, he could not even preach. The Dissenters fixed their unsympathetic gaze upon him; he broke into a sweat and the words clutched at his throat. "Without some such check as this, I might have been disputatious in company, or might have been seduced by the love of popular applause as a preacher," yet the stutter so ruined his sermons that it "took from me all chance of recommending myself to a better place." In a newspaper he saw a "treatment" advertised by one "Mr. Angier" of London; he begged the 20 guineas from Aunt Sarah and went to the metropolis for the first time in his life. The treatment consisted of a series of breathing exercises and practice in reading slowly and distinctly, but within a month it had relapsed and he "spoke worse than ever." He dove deep into himself, deeper than ever before, and during this desperate period he decided that his only hope of making ends meet was by becoming a writer.

▪ ▪ ▪

Everyone thinks they can be a writer, and no doubt the phrase "my life should be a novel" was as common during the Enlightenment as it is today. There were precedents for Joseph's new literary direction, if just barely. By midcentury, a self-conscious intelligentsia was stumbling into being, though they wouldn't see full flowering until the next decades. They were mostly young, disaffected, and at this point, alone; they made a living through a hodge-podge of teaching, writing, journalism, lectures, editing, publishing, translations, and reviews; their articles were often astringent in tone and belief, not surprising since they hailed from the Dissenting crowd. They seemed unsure whether others like them blindly struggled forward, and would have thought it fantastic that they were forming the British avant-garde that would be such a cultural force during the Victorian era. By the end of the century, writers like Priestley, Richard Price, William Godwin, Thomas Beddoes, Mary Wollstonecraft, and William Hazlitt would form this cadre of professional thinkers and communicators whose subjects included the primacy of ideals, the rapacious privilege of the mighty, and the corruption of the state. But right now, most, like Priestley, usually thought as far ahead as the next meal.

Despite the difficulties, however, it was growing evident by the 1750s that money could be made from the product of one's intellect. William Herschel, before becoming the famous astronomer, made his living as a professional musician in Bath, earning £500 annually from conducting, composing, and giving lessons. Henry Fielding took home £800 from his novel *Amelia*, while Samuel Johnson pocketed the unimaginable sum of £1,575 for his dictionary. In his time, Priestley would write over 150 volumes on subjects ranging from government and religion to science; he held forth on history, language, and pedagogy, making him the premier "book-maker" of his day. He never took to heart Martin Luther's complaint two hundred years earlier that "the multitude of books is a great evil. . . . Everyone must be an author, some for some kind of vanity to acquire and raise a name, others for the sake of lucre or gain." Priestley did concede once, when asked how many books he'd written, that there were "Many more, Sir, than I should like to read." Contemporaries were stunned by his output; but Joseph had a habit of working on a draft as the idea hit him, then returning to it years later. He'd already written the first draft of his *Institutes* in 1755, but would not publish it, in three volumes, until 1772–74. Hints in his letters and memoirs show that at his lowest point in Needham, he was already toying with his rejection of the idea of the soul, the subject of his later *Disquisitions Relating to Matter and Spirit*. He was starting it all in Needham, when he was angry, depressed, and alone.

During his incarceration in Needham, Joseph developed what could best be called a compulsive *need* to be heard. He wrote at great length about almost anything that entered his head. Maybe the need arose from that moment when he felt rejected by God and existence itself; certainly he felt rejected by society and family here in Needham. Yet writing is a two-edged sword: the more time we spend alone with the muse, the more eccentric and odd we become. As the neurologist Alice Weaver Flaherty has noted, "This is most obvious in situations where society no longer keeps us in line: the eccentricity of the very rich, or of castaways."

After Needham, friends would often advise Joseph to curb his more explosive literary enthusiasms, but he rarely showed any interest in keeping himself in line. He wrote to save himself from Needham, and those three years changed him more than he would ever admit. He would always hereafter be grateful for contact and acceptance, but it was only in isolation that he was most productive and creative—that he was most truly himself.

▪ ▪ ▪

Finally the exile in Needham ended, and he emerged as if from a wasteland. Though few personal letters remain from that time, his friends from Daventry were apparently worried: through friends, he preached a trial sermon in Sheffield, and although he did not get the post, he was recommended for another at Nantwich in Cheshire. He didn't think twice when offered the church, gathering up his books, globes, and manuscripts and traveling by sea to save money—and, perhaps, avoid the ever present highwaymen and corpses in irons. He took up his new post in 1758, at age twenty-five.

Again he seemed reborn, another sea change like that from Calvinism to heresy. Part of his rise in spirits came from the fact that Nantwich was a preindustrial manufacturing region like his home in Birstal. His congregation was composed of traveling packmen, often

Scots, who were not opposed to his doctrines. Their trade was in Cheshire salt, the "whie" salt made only in the "Wiches" (Nantwich, Norwich, and Middlewich) from rock salt and brine. The technique depended on tides: sea water passed into large lead cisterns, to which the rock salt was added; when this dissolved, the strong brine left was poured into 800-gallon pans and evaporated even further by fire. These refined salts were sold throughout Britain, and in Ireland and the Americas: the salt, like the white-cloth of Priestley's childhood, was essential for England's overseas trade.

It has been suggested that Priestley's first interest in chemistry began with this move. He would have read *The Art of Making Common Salt*, published by physician William Brownrigg in 1748 and a common text in the area. In his preface, Brownrigg quoted Francis Bacon, who stressed the importance of learning the "mechanic arts" for studying natural science, a goal "which is not taken up in vain and fruitless speculations, but effectively labours to relieve the necessities of human life." Early in his own career, Priestley distinguished between "pure" science, a theoretical endeavor he dismissed as sophistry, and the practical use that "applied" science could provide. During the summer, he could watch as packmen and heavy wagons carried the practical results of such a pretechnical art from Cheshire to the Midlands; he witnessed his parishioners' incomes rise slowly during his tenure in Nantwich. Life was still hard, but at least they put some distance from the grinding poverty they'd known when young.

Priestley was overjoyed to be out of Needham, and the only fault his congregation found with him was his "gay and airy disposition." He had a habit of jumping over the counter of the greengrocer's shop where he boarded, or forgetting himself and whistling as he strode the lanes. Though both were thought unseemly for a minister, his eccentricities amused rather than repelled. His landlord taught him to play the flute, and though Joseph had a tin ear, he didn't seem to care. In time he would recommend the flute to all "studious persons," especially if they had no "fine ear, or exquisite taste," since like him they would "be more easily pleased, and be less apt to be offended when the performances they hear are but indifferent."

One reason for his acceptance in Nantwich may have been his genuine affection for the young. The feeling seemed mutual. Soon after arriving, he started a school and shortly had thirty-six students—thirty boys and six girls, schooled in two different rooms. Educating girls in a public setting was innovative for that era. He began to earn a decent living, and with this bought books and scientific instruments—a microscope, air pump, telescope, and an electrical machine. Priestley was not yet a "scientific person"; these instruments were solely for teaching purposes, not his own experiments. Yet their use was still unusual, and his students seemed excited and enormously grateful for this glimpse into the scientific brave new world. He taught his oldest students how to use and maintain the equipment, and parents were entertained by the experiments and lectures given by their children. Word spread, and his reputation grew.

He also began teaching English, math, and history. This too was innovative: in traditional grammar schools, the curriculum was almost entirely Greek and Latin, since mastery of the classics was required for admission to the public universities as well as positions with the Church or State. Since Dissenters were excluded from that future, Joseph created a more practical preparation for a life of commerce, medicine, or industry. His instruction in rhetoric, which he published in 1761 as *Rudiments of English Grammar*, abandoned the finer points of grammar and oratory for what Dr. Johnson called "use," and eliminated the technical terms of other languages that had no bearing in English. He used humor to illustrate usage, a habit his students loved. As an illustration of a double contrast, for example, he included the couplet:

Beneath this stone my wife doth lie:
She's now at rest, and so am I.

Joseph's attractiveness to his students cannot be undervalued, for it played a great part in his future. In the gray world of a stern God and often sterner families, he seemed a breath of freedom. He was open to questions, and he questioned the status quo. His vision of a better future was deeply held. He began to teach history—not Greek and Roman history, like everyone else, but a history of mankind, and especially of England. He developed a wall chart of biography as a teaching aid, and hoped to include everything that made "the great societies of mankind happy, numerous, and secure, with which young men of fortune cannot be too well acquainted."

Yet before he could finish these projects, he was offered a position as tutor of languages and belles lettres at nearby Warrington Academy. He was not yet a scientist, and except for the purchase of instruments "for my own taste" had no idea that he was headed there. Letters of recommendation mentioned his "unexceptionable character," "steady attachment to the Principles of Civil and Religious Liberty," range of "Critical and Classical Learning not common in one so young," and notably, his "singular genius for the management of youth." The last detail seemed to clinch the appointment, and he arrived at Warrington, at age twenty-eight, in September 1761.

▪ ▪ ▪

Now the world began to catch him up, or alternately, catch up to him. In Needham and Nantwich, he'd sunk into himself and was only beginning to climb out, yet in the England around him, momentous change was underway. The Seven Years' War had raged since 1756, one more endless conflict between the European powers, and especially between England and France, yet this could truly be called a world war. In 1759, "the Year of Miracles," the English fleet had swept the seas; an obscure clerk named Robert Clive and 350 men conquered Bengal, expelling the French from India and claiming a colony that would bring in revenues of over £2 million a year. The French were defeated at Minden and the British captured Guadeloupe and Quebec. By 1760—the year George II died and George III ascended the throne—England controlled nearly one-third of the earth's surface: from India to parts of Africa and the West Indies; from Canada to America, from which a flood of tobacco, fur, cotton, sugar, and timber promised to flow.

The year of George III's ascension was also a watershed in England's transformation from an agrarian to an industrial society. That year, the factory system and use of water-driven machinery were firmly established in the silk industry; blast furnaces were perfected for producing cast iron; the Duke of Bridgewater completed his cross-country canal, an improvement that halved the price of coal in Manchester. By 1764, the spinning jenny was introduced, transforming English textiles into the international standard for the emerging Industrial Age. The marriage of science and practicality that Francis Bacon envisioned a century earlier was becoming a reality. The Machine Age was here to stay.

Yet with this new power came new relationships between the state and its citizens. That part of the "body politic" increasingly labeled the "labouring class" was already referred to by its most useful part—its hands. Bernard de Mandeville's *Fable of the Bees* calculated that "the hard and dirty Labour throughout the Nation requires three Millions of Hands"; a letter published in a 1751 issue of *The Gentleman's Magazine* said the writer's village had "upwards of 100 hands." When not engaged in work, the growing middle class saw the "bodies" of the "idle" poor as "grotesque objects of fear and revulsion"; in return, the laborers rejected Priestley's class, which had in essence grown out of them. In the days before trade unions and strikes, the most common response to such anger was a good fight or the downing of greater quantities of gin.

Simultaneously, in Dissenting academies like Warrington, an important development in English political thought was taking hold. The philosophy of political "candor" originated in the 1760s–1780s as a result of the plight of the Dissenters, and in places like Warrington the idea of religious toleration and political liberalism was being born. It was at Warrington and similar academies that the free thought of Presbyterians crystallized into Unitarian theology; it was during his stay at Warrington that Joseph laid the foundations for his political philosophy, turned to science, and removed any shreds of lingering belief in the divinity of Christ or the Trinity.

At first, though, Joseph was simply happy. The town of Warrington in Lincolnshire was a manufacturing center, where sailcloth made of hemp and flax was imported from Russia and converted into nearly half of the heavy-grade material used by the Royal Navy. Other local products included pins, locks, and hinges. In addition, copper smelting, glassmaking, sugar refining, foundry work, and brewing and malting were all growing industries. Warrington was the most bustling place Joseph had lived in yet, a good example of the kind of capitalism Englishmen saw growing around them, a primitive globalism that wed self-interest with an expanding worldview.

The Academy itself was separated from the town, perched on the south bank of the Mersey, a stream then filled with salmon. The austere brick buildings were gated from the lane outside. There were small houses on the grounds for tutors, with whom students were expected to lodge. Yet this stern appearance belied the student life inside. Since Warrington was patronized by local glassmakers, linen weavers, and merchants, there was very little separation between town and gown. There were dances, card games, and parties, and the tutors took part in arranging plays. Although no female students were officially enrolled, plenty of tutors had daughters and students from town had sisters. The separation of the sexes was so inconsequential that Anna Laetitia Aiken, daughter of the theological tutor John Aiken, could honestly report that "we have a fine knot of lassies as merry, blithe, and gay as you could wish, and very smart and clever."

Joseph fit right in. In 1761, he published his *Rudiments of English Grammar*, followed next year by his *Theory of Language, and Universal Grammar*, his *First Chart of Biography*, and *New Chart of History.* All were very popular, remaining in circulation for half a century, and were the beginning of his wider reputation. Priestley's teaching was clear and direct; he asked his students to come round his house for tea, let them borrow material to read at home, and was overjoyed whenever a student challenged him. "I do not recollect," wrote one student,

> that he ever shewed the least displeasure at the strongest objections that were made to what he delivered; but distinctly remember the smile of approbation with which he usually received them nor did he fail to point out in a very encouraging manner the ingenuity or force of any remarks that were made, when they merited these characters. His object . . . was to engage the students to examine and decide for themselves, uninfluenced by the sentiments of any other person.

He seemed to like his students, too, relating in a letter to the rector, John Seddon, that one pupil was being moved by his father to a rival academy at Exeter. "We both shed tears, I could not help it," Priestley wrote. "He is leaving us presently."

One student who followed Priestley from Nantwich was William Wilkinson, son of the Wrexham ironmaster Isaac Wilkinson. In time, William and his brother John would become the wealthiest ironmasters of the age, but right now of greater interest to Priestley was William's younger sister. Mary Wilkinson was seventeen and one of Anna Aiken's "knot of lassies." She was also quite interested in her brother's freethinking tutor. So interested, in fact, that by Priestley's second year at Warrington, he was a married man.

It is easy to see how he fell in love, though with his awkward shyness, Mary Wilkinson probably had to make the first delicate moves. She was described as "extremely intelligent and original," and the latter speaks volumes. She was tall, prone to speaking her mind like her fierce-tempered brothers, endowed with a penetrating glance and a sudden smile that was "like the sun in January." Friends called her well-read, generous, and "a fiendish organizer." It had been seven years since the attentions of Miss Carrott at Daventry, an eternity since any woman had looked twice at him. Joseph may have thought he was doing the wooing, but Mary hooked and played the tutor as expertly as any salmon swimming in the Mersey nearby.

Mary also brought a rapprochement between Joseph and his estranged brother Timothy. "When he was near being married, he told me he thought he had found out a method of bringing us to a nearer union in opinion," Timothy recalled. And how, asked Tim, did he plan to do what both considered impossible?

"I am going to marry a very orthodox lady." Joseph grinned. "You marry a heterodox one."

Timothy had to see such a woman for himself. What woman who called herself orthodox could possibly marry such a wild heretic as his brother? He rode the fifty miles to Warrington: on learning Joseph's schedule, "I took the advantage of speaking to her alone." But before Timothy could speak, Mary gazed at the visitor and said, "I know who you are."

That took him aback, and being a Priestley, he stood there looking awkward. He told her what Joseph had said, and that he'd come to test her orthodoxy for himself.

Mary's eyes flashed, and she looked at this contentious brother whom Joseph loved but by whom he'd been betrayed. "I give no preference to any sentiment if a man's conduct is moral and agreeable," she declared. Timothy blinked in surprise and, after a moment, probably laughed, for these were the words Aunt Sarah had used about visitors to her freethinking salon.

Joseph soon arrived. "Brother, you have deceived me," Timothy announced, trying to look stern.

"How is that?" Joseph asked, wondering if his brother had judged Mary as harshly as he always did him.

"Mary is neither orthodox, nor heterodox," he answered, unable to suppress a grin. "She is no dox." Joseph and Mary started laughing. "It diverted them not a little," Timothy recalled. Henceforth, although the brothers still argued religion, the ice between them thawed and they were close again.

And so, on June 23, 1762, Joseph and Mary were married in her childhood chapel at Wrexham. Joseph was thirty, Mary seventeen. A Presbyterian divine called in to officiate became so immersed in a Welsh Bible that he had to be found and reminded of his duties. Given his recent poverty, Joseph worried about supporting a wife and prepared for ordination as a way to increase his income. "The hazard of bringing a person into difficulties which she cannot possibly have any idea or prospect of, affects me . . . very sensibly," he noted. But this day, he only saw happiness and release from the loneliness that had plagued him for so long.

CHAPTER 3

The Gas in the Beer

ON DECEMBER 18, 1765, PRIESTLEY CLIMBED INTO A WAITING coach for the 175-mile trip to London. Mary held up their infant daughter Sally, born the previous year, for a kiss, then stood waiting by the coach until Joseph pecked her cheek good-bye. Displays of affection were more common in this dawning age of sentiment, but Joseph didn't know if its public nature was quite proper for an ordained minister, even a Dissenting one. He was easily shocked, and a woman's bare foot or an overheard curse could turn his face pink or see a return of his stammer. But Mary would have her way: she was a modern girl. The "knot of lassies" watching his departure giggled, and Mary shooed them away before Joseph became unglued. The coachman snapped his whip and the black coach disappeared into the hush of the falling snow.

He settled against the hard cushions and contemplated his plans. Life had been good to him at Warrington; it was as if he'd been tested, and this was his reward. Sally's birth had been free of complications. Every man in those days faced the possible death of his wife and child during birth, and until the midwife opened the bedroom door and presented him with his baby girl, he'd lived in fear that Mary would not survive. He had to wonder how he would carry on without her. There was an immensely practical side to Mary: although their house in Warrington lacked locks, grates, and wallpaper when they moved in, she had everything in order so quickly that he barely had to lift a hand. The Yorkshire middle class believed in a practical education for both sons and daughters, and Mary was so well-read that she could hold her own in the academic atmosphere. He'd found an intellectual companion, and every night he would sit with her for a game of chess or backgammon. She arranged tea parties where students could talk or seek advice from Joseph; she drew a line, however, on those wishing to copy out his lectures, since in the days before copyright, these could be printed and pirated. She encouraged him in his ambitions while she managed the details: the trip to London was at her urging, even if it meant he'd be away for the holidays.

Priestley's interests were expanding, too. In 1764–65, he revised and published his educational charts, and on the basis of these was awarded an LL.D. from the University of Edinburgh. Thereafter he was known as "Dr. Priestley," a title that pleased him. Being part of Warrington's faculty gave one a certain cachet: he met Josiah Wedgwood, the potter whose Ivy House Works mass-produced Creamware for the fashion-minded, and Matthew Boulton, inventor and owner of the Soho Manufactory. He also met the surgeon Matthew Turner, whom he persuaded to serve as lecturer on chemistry at Warrington. From 1763 to 1765, Turner's lectures on "practical and Commercial chemistry" were Joseph's first real introduction to the discipline. True, he'd learned celestial mechanics and math at Daventry, and had dissected cats to observe the body's machinery, but now he was introduced to practical labwork and put in charge of preparing quantities of "Spirit of nitre," or nitric acid, for Turner's demonstrations. Also known as *aqua fortis,* or "strong water," its preparation was time-consuming, but in this guise as a lab assistant, or "Laborant," the infant science got under his skin.

It's lucky the *aqua fortis* didn't get on his skin, for nitric acid is an extremely corrosive compound, a colorless, toxic acid that can cause disfiguring injuries similar to second- and third-degree burns. It is a powerful oxidizing agent, so that, like oxygen, it reacts with virtually every other substance but gold. Such reactions can be explosive, especially when mixed with carbides, cyanides, or metallic powders, often so quick as to be "hypergolic" or self-igniting. The safest way to store and handle nitric acid is with glass equipment, expensive in those days, and how Priestley avoided injuring himself during his first excursion into chemistry is a mystery.

Perhaps because of this reactivity, nitric acid is an important industrial compound, used in the production of explosives like nitroglycerin and TNT, in fertilizers like ammonium nitrate, and in metallurgy and refining. The standard method of preparation is by mixing nitrogen dioxide (NO_2) with water, but a very pure form can be made from distillation with sulfuric acid. The method Priestley used for preparing Spirit of nitre was never recorded, but in all likelihood it followed the process described by Robert Boyle in his "experimental notes and accounts" of 1686–90, a chemical cookbook that was one of the most widely used manuals then available in England. If so, he mixed 1 pound of "purify'd salt peter" with 3 pounds of "good White Clay." He ground these together in a large mortar, then placed the mixture in a stoppered retort and heated it over a flame, gradually increasing the intensity. Joseph loved the way the flame curled around the bottom until the glass glowed white-hot; as he watched, red fumes curled tentatively from the mixture, then gushed out in a scarlet flood. Following Boyle's instructions, he distilled the heated mixture with scalding water, obtaining a precipitate of yellow crystals whose form "was that of a feather . . . a more beautiful appearance can hardly be imagined." For every pound of saltpeter, he obtained 5 or 6 ounces of nitric acid.

Joseph was dazzled by the transformations he observed. Clay was the common dirt he rolled in as a child and nitre was scraped from damp cellar walls, yet mix, grind, and heat the two together and red fumes curled up like living vines. He never tired of preparing Spirit of nitre: there was a delicious suspense in waiting for fumes to erupt, and he always wondered whether it would happen again. That it did spoke of secrets hidden deep in all matter, and his mind drifted to other mysteries and associations. He wondered if there was a connection between the ancient belief in the soul, which he considered a heathen corruption, and these etherous mysteries trapped in solid bodies. There were many such examples: the "spirit of the liquor" was the unknown gas that bubbled from a fermenting liquid—small animals died if trapped within it, and strong wooden casks sometimes burst during fermentation. There was the *spiritus,* or essence, of a compound prepared by distillation, just as he did now with saltpeter. Humans associated life with breath, and in Hebrew, Greek, and Latin the words for "soul" and "breath" were often the same.

Most biographers place Priestley's first steps in science at a later date, but it seems evident with this trip that his enthusiasm began in Warrington. Of all the physical sciences, however, chemistry was considered the least exact and most tainted by ancient association. Chemistry has always had an image problem, and still does today. Its very familiarity in the kitchen, workshop, and factory breeds, if not contempt, a certain ho-hum attitude. Its goal is always practical: while scores of Nobel Laureates have won honors in chemistry and biochemistry, chemists (unlike physicists, who co-opted the idea of the atom from the ancient Greeks) do not generally think in epistemological terms. Jean Jacques, in *Confessions d'un chimiste ordinaire*, lamented that chemists rarely write about their discipline in the same awed tone as physicists or astronomers: "Chemistry," he wrote, "is like the maid occupied with daily civilization: she is busy with fertilizers, medicines, glass, insecticides." In 1923, the Toulouse physicist and Nobel Laureate Henri Bouasse enraged the chemical community when he sniffed that chemists merely *"faire le cuisine."* Chemistry was kitchenwork, or child's play.

The discipline also could not shake its medieval associations with magic and alchemy. Francis Bacon was an alchemist, as was Newton: while laying the foundations of modern astronomy and physics in his *Principia*, Newton searched throughout Europe for the philosopher's stone. It was a futile search, but one that was centuries old. Alchemists believed that the specific character of a metal was determined by the amount of sulfur and mercury in its basic makeup. If one could subtly alter these proportions, "baser" metals like silver, iron, tin, and lead could be turned into the "noblest" metal of all: gold. Alchemists believed that all matter strove toward a higher state, and the philosopher's stone was the quintessence of this state. Remove all impurities, and the "stone" would be revealed.

The importance of the idea that there was a physical key or actual shortcut to this higher state cannot be minimized. Much of early physical science was devoted to discovering and unlocking this key. Aristotle's belief that all Nature strove toward a perfect state expressed the meaning of existence: find the stone, or "essence," or "vital flame," and you'd hold the secret of perfection. You could cure disease, renew youth, prolong life, and confer immortality—you would be a god. Just as the alchemist sought the stone in an attempt to learn this secret, Priestley's contemporary, the natural philosopher, sought God's natural laws to edge man closer to His divine perfection. Both were ennobling occupations, and both were firmly seated in the search for an "ideal."

By the time Priestley boarded the London coach, he'd imagined a new project that wedded his old enthusiasm, history, with science, his new one. What could be more useful than showing how the new discoveries of Nature were linked to social progress? It was a huge undertaking, but doable if approached in discrete packages. He was proposing to write a history of the new science of electricity since "the principal actors in the scene . . . were still living"; such a history would show that knowledge could not be stopped, a subtly subversive comment on freedom and liberty. Armed with his optimism and a letter of introduction from his rector, Priestley drove into the snow to meet the "electricians."

▪ ▪ ▪

In the first half of the eighteenth century, electricity, more than any other discipline, was the marvel of the age. As early as the sixth century BCE, Thales of Miletus noted that amber rubbed against animal fur picked up small bits of material, and for 2,300 years all studies of electricity involved simple friction. That changed with technology, and suddenly "electricians" were earning a living all over Europe. Some gave simple demonstrations—shocking audience members, or killing birds and small animals—but others were more imaginative. In 1746, the abbé Jean-Antoine Nollet sent a current through a line of 180 royal guards standing before Louis XV; soon afterward, he connected a kilometer-long row of white-robed Carthusian monks with an iron wire and reportedly made them leap into the air at once when he discharged a spark. In 1747, the English physicist William Watson transmitted a spark through a wire strung across the Thames at Westminster Bridge.

Was electricity the long-sought "soul of the world"? Priestley's friend Matthew Boulton said electricity was the fabled "animal spirit" itself, a secretion of the brain thought to be the source of all motion and sensation. Others called it an imponderable fluid, one of the "effluvia" that filled the world and connected all things. Again, as with the air pump, the inventive genius behind the excitement was Otto von Guericke. In 1663, he developed a crude device for generating a spark by static electricity, yet his generator could not store a charge. It wasn't until 1745 that the Dutch physicist Pieter van Musschenbroek invented the "Leyden jar," an early storage battery named for the University of Leyden in the Netherlands, where most of electrical research was then performed and published. The Leyden jar was nothing more than a glass jar filled with water, with a heavy wire of high conductivity running through a cork that sealed the opening. A spark-generating machine—such as a gun barrel hung by a silk thread and charged by a spinning metal globe—was held close to the protruding tip of the wire. The transferred charge was retained by the wire so effectively that with a simple improvement (removing the water and wrapping the sides of the jar with metal foil), a shock could be administered that was powerful enough to knock people off their feet, paralyze their arms, and give them nosebleeds. Van Musschenbroek warned his colleagues: "I would like to tell you about a new but terrible experiment, which I advise you never to try yourself, nor would I, who experienced it and survived by the grace of God, do it again for all the kingdom of France."

The most famous "electrician" was Benjamin Franklin. In his day, Franklin was the international apotheosis of "enlightened" man. In 1748, he'd sold his printing press, newspaper, and almanac to become one of the wealthiest men in the Commonwealth of Pennsylvania, then funneled that wealth into the new science of electricity. Within a short time, he'd invented (and marketed) the "Franklin rod" for protecting buildings, made improvements on the Leyden jar, and developed the one-fluid theory of electricity which dominated the field for one hundred years. Before this, researchers thought electricity consisted of two fluids—one positive, one negative—that flowed between poles; but Franklin countered that it was two states of the same fluid, which was present in everything. A substance containing an unusually large amount of the fluid would be "plus," or positively charged, while one with less than the normal amount would be "minus." In 1753, Franklin was awarded the Copley Medal, the Nobel Prize of its day, and in 1756 he became a member of the Royal Society in London.

But it was Franklin's work with lightning that made him an international star. Noticing that lightning was similar to the charge he could create in his lab, Franklin hypothesized that the two phenomena were the same. In June 1752, he reportedly performed his famous experiment with the silk kite and brass key to capture electricity from a thundercloud; the current was carried to the ground via the kite's wet twine, and this was attached to a Leyden jar. "Reportedly," because the actual performance of the experiment by Franklin is today in some dispute: another researcher, Professor G. W. Richmann, was killed shortly afterward in St. Petersburg while performing the same procedure. And then there was the manner in which Franklin described it. In a letter dated October 17, 1752, he claimed that once struck by lightning, "the kite, with all the twine, will be electrified, and the loose filaments of the twine, will stand out every way, and be attracted by an approaching finger. And when the rain has wetted the kite and twine, so that it can conduct the electric fire freely, you will find it stream out plentifully on the approach of your knuckle." Since Franklin wrote in the second person, that and some discrepancies have led historians to doubt the authenticity of his feat, believing instead that what he wrote was actually a set of instructions describing what "should" happen, based on theory.

Priestley believed in Franklin, as events would show. When the two men met in London that December, the older scientist was sixty and had a lurid reputation as a genius, entrepreneur, diplomat, rake, and raconteur. He was a member of the notorious Hellfire Club, famous for its Gothic trappings and sybaritic excess; he was said to have sired two illegitimate children and authored such tracts such as *Advice to a Young Man on Selecting a Mistress* and *The Technique of Farting*. In time there would be a cult of Franklin, especially after the American Revolution—English baby boys were named for him, and French artists painted him as a mythic philosopher-scientist who "seized fire from the heavens and the scepter from the tyrant's hand." At the moment, however, the picture was more down-to-earth: seated in the place he called "the Club of Honest Whigs," a club of philosophers and politicians meeting in a coffeehouse near St. Paul's Cathedral, Franklin was surrounded by fellow electrician William Watson; John Canton, council member of the Royal Society; and Richard Price, the theologian and mathematician whose career would be closely tied to Priestley's. Joseph described to the four his idea of a comprehensive history of the sciences beginning with electricity, and they responded by taking him to the January 9, 1766, meeting of the Royal Society.

The meeting was a turning point in Joseph's life, marking his change from teacher and preacher to scientist and political firebrand. Much of the transformation can be attributed to Ben Franklin. Joseph was little more than half his age, and he looked like a stick beside the taller, more imposing Franklin, but a friendship sprang up quickly between this mismatched pair. Years later he called Priestley an "honest heretic," and seemed intrigued by the rage of thoughts that bubbled in the younger man's brain. "Do not mistake me," he wrote. "It is not to my good friend's heresy that I impute his honesty. On the contrary, 'tis his honesty that has brought upon him the character of heretic." Franklin sent to Priestley his lab and field notes and collected electrical works to aid in the treatise. On June 12, 1766, Priestley would be elected a fellow of the Royal Society to make reception of the history that much easier.

Franklin would be one of Priestley's two most important patrons; perhaps the most important, since it was he who first encouraged Joseph's talent in the lab. Patronage was a two-way street in the eighteenth century: one benefited and suffered from its effects. A system that relied upon informal personal power rather than formal institutions, patronage also served as a shield and a leg-up where necessary. One was set in the social strata of one's connections, and since there was no welfare state or government safety net, people had to fend for themselves. As a Dissenter on the outside of the social structure, Priestley needed all the connections he could get.

He returned to Warrington on February 14, and set to work on his history. One room was especially equipped for experiments to liven up the text. He spent hours charging Leyden jars, growing "fatigued with the incessant charging of the electrical battery, and stunned with the frequent report of its explosion." He deduced the theorem that electrical attraction is proportional to the inverse square of its distance, and discovered that coke and charcoal made excellent conductors, upsetting the given that only water and metal conducted an electrical charge. On November 12, he melted iron wire with electricity as Timothy looked on: "When the metal melted, he called, 'Oh, had Sir Isaac Newton seen such an experiment!' and his pleasure on those occasions cannot well be described."

During this period he made a discovery that had nothing to do with electricity, but is appreciated by anyone who ever made a mistake on a test or while paying the bills. While working on his manuscript, Joseph found that the product of a certain South American tree sap just being introduced into Europe was perfect for rubbing out stray pencil marks—of which he made many. He called the sap "rubber," and so it has been known ever since.

At one point, he put his life in Franklin's hands. Franklin's episode of the kite, fixed so firmly in scientific lore, was actually a mystery: Did he or didn't he fly his kite on an uncertain date in June 1752? Fifteen years after Franklin's first description of the experiment in the *Pennsylvania Gazette*, a second account appeared in Priesley's *History of Electricity* with more details of the old electrician's thoughts and specific actions. In fact, Priestley believed in his mentor enough to try the experiment himself. Timothy, now a carpenter and Calvinist minister in Manchester, was called upon to make the kite. Joseph wanted it to be like Franklin's—large, and of colored silk—and Timothy was finished in March 1766. "It was 6 feet, 4 inches wide," he said. "Joseph could put the whole thing in his pocket, for the frame would take to pieces, and he could walk with it as if he had no more in his hand than a fishing rod. The string was composed of thirty-six threads," surrounding a metal wire, "this to bring the electrical fire from the clouds."

Timothy cautioned his brother against the wisdom of such an experiment, as did Mary in no uncertain terms. But Joseph went ahead anyway. On an unspecified stormy day in March 1766, he stood in a field as had his hero. Nevertheless, he did take precautions: a chain trailed behind the kite to ground it, and, according to Timothy, Mary "would not suffer him to raise" the kite any higher than his head.

One wonders whether he remembered her pleas. Priestley could be fearless and oddly naive when it came to mortality. According to an apocryphal tale, a neighbor's curious goose plucked at the trailing chain as the lightning struck, and the goose was martyred to science. It isn't recorded whether Joseph compensated the owner.

▪ ▪ ▪

Priestley's *History and Present State of Electricity, with Original Experiments*, was published in 1767 to brisk sales. It launched his career as a researcher, spread his name outside the circle of pedagogy, and established Franklin as a scientific hero. Priestley followed this up with *A Familiar Introduction to the Study of Electricity*, meant as a more popular work, and "An account of Rings consisting of all the Prismatic Colours, made by Electrical Explosions in the Surface of Pieces of Metal," read before the Royal Society in March 1768. The rings formed by an electrical spark when it was discharged on a metal surface had been observed before, but never explained: his paper was a sensation for its time, and henceforth the phenomenon was called "Priestley's Rings." "New worlds may open to our view," he wrote, "and the glory of the great Sir Isaac Newton himself, and all his contemporaries be eclipsed, by a new set of philosophers, in a quite new field of speculation."

One wonders if he included himself among this "new set of philosophers" who would change the world. His *History of Electricity* suddenly made him more known than he'd ever dreamed, and the effect must have been heady. When the sixteen-year-old Jeremy Bentham, the future social reformer, toured northern England with his father, he dreamed of meeting his idol. "We went to Stockport, Liverpool, Chester, Macclesfield, and the Wiches, where the salt is made," Bentham recalled. "Warrington was the classic ground. Priestley lived there. What I would not have given to have found courage to visit him!"

But the hard accounting of daily life intervened. By now Warrington was feeling financial strain, due largely to disagreements between conservative donors and the academy's liberal attitude. Priestley wondered how much longer the college would survive; of more immediate worry was the ill health that had wormed into his family.

Sally was a sickly child, while Mary complained that the banks of the Mersey "did not fit her [own] constitution." He could not help but remember the "bad air" that had nearly killed him as a child.

In fact, he'd thought often of air quality during his electrical studies. The discovery that burning charcoal made an excellent conductor called up old thoughts about the "goodness" of air. "Having read, and finding by my own experiments, that a candle would not burn in the air that had passed through a charcoal flame, or through the lungs of animals, or in any of that air which the chymists call mephitic, I was considering what kind of change it underwent, by passing through a fire, or through the lungs, &c., and whether it was not possible to restore it to its original state, by some operation or mixture." What was the nature of the change air experienced as it passed through fire or the lungs? Could such "vitiated" air be returned to its former wholesome state by some natural or man-made means? He placed small animals in jars of "mephitic" air through which he'd sent a spark, but they always died—he concluded that mephitic air was unique, a "fluid *sui generis,*" and Franklin reminded him of those medicinal springs in Bad Pyrmont in Lower Saxony (now Germany), where ducks and other birds died mysteriously when landing on what seemed a safe haven.

These were sinister phenomena. The world was filled with unexplained forces that could kill you in a heartbeat and you'd never know why. Such was the case of the Pyrmont Springs. The source of the "mephitic" air, of course, was carbon dioxide bubbling up from a depth of 300 meters and leaking through fissures in the rock; this caused the water's famous effervescence, turning its volcanic springs into one of Europe's most fashionable spas. But the flip side was death, for at Bad Pyrmont there existed a former quarry where workers would go into a faint: it was now called the *Dunsthöhle*, or "fume cave." In 1720, the spa's doctor, Johann Philip Seip, proved the existence of an invisible "lake" of fumes that floated over the water itself, flowing up from the surface to spill onto and cover the quarry floor. This "lake" could not be seen: the only way Seip could prove its existence was by lighting a candle at the end of an iron rod. The candle burned normally until reaching the invisible lake, and then it went out. That was the closest anyone could go in safety, for to step further into the *Dunsthöhle* meant suffocation. The locals had known this for decades. Bad Pyrmont's dirty secret was that in addition to being a draw for the rich and famous, it was also a magnet for suicides. Death was as easy as walking into the quarry and sitting down.

Franklin also sickened for a related reason. In 1764, when passing through New Jersey on his way to New York, he'd "heard it several times mentioned, that, by applying a lighted candle near the surface of some of their rivers, a sudden flame would catch and spread on the water, continuing to burn for near half a minute." This had the opposite effect, of course, of water saturated with carbon dioxide; in this case, the unknown gas was methane released by the decomposition of organic material in the turgid water. Franklin stirred the water with his walking stick and watched the rising bubbles; he applied a candle, and the flame "was so sudden and so strong" that it lit the ruffle of the man who'd led him to this place and "spoiled it, as I saw."

The mystery didn't end there. From New York, Franklin caught a ship to England, the beginning of the journey during which he met Priestley in December 1765. On arrival, he mentioned the strange incident to some of his disbelieving philosophical friends. "I suppose I was thought a little too credulous," too much the colonial bumpkin, he said. During his stay, he tried the experiment twice more. The first time, nothing happened. The second time, he stirred the stagnant water at the bottom of a deep ditch, probably in London. "Being some time employed in stirring the water," he later told Priestley, "I ascribed an intermitting fever, which seized me a few days after, to my breathing too much of that foul air, which I stirred up from the bottom, and which I could not avoid while I stooped, endeavoring to kindle it."

What were these "airs" that rose from the earth itself and changed the nature of water or poisoned the atmosphere? Were they connected to the bad air that nearly killed Priestley as a child and now, according to Mary, rose from the Mersey to sicken their child? The thought was merely passing as in September 1767 they left Warrington in sadness for Joseph's new post as minister at Mill-Hill Chapel in Leeds, abandoning the academic sanctuary that had given him such joy to reenter the contentious world of religion. Their new house was on a quiet road called Meadow Lane; behind the house sat the public brewery of Jakes & Nell. Another minister might rail at the wellspring of public drunkenness out his back window, but Priestley studied the vats of fermenting liquid—and from them rose slow bubbles of mephitic gas that lingered on the surface, then grew bigger and bigger, like his questions.

Joseph was never one to question Providence. In his powdered wig, black overcoat, and bright shoe buckles, the new and angular minister asked his neighbor if he might be allowed to make some tests on that strange gas in the beer.

▪ ▪ ▪

Leeds was a busy and earnest place, a growing industrial town of over twelve thousand inhabitants and three thousand families. Like other cities of the North, it was making that transition from a market town to a center of industry. The fields crept up to and into town; cattle and sheep wandered in the common green spaces. The early manufactories and mills that depended on waterpower for their spinning machines had settled along the river; the waters of the Aire were darker from dyes and chemicals than during Priestley's childhood. Though St. John's Cathedral still loomed on its hill over town, it overlooked more gin shops and Dissenting meetinghouses than it had just a few years earlier. Leeds seemed an intensely public place to Joseph: the streets bustled with flower girls, piemen, street vendors, street urchins, and bands of apprentices hooting at passing milkmaids. These were the decades when the British supplanted the Dutch as the commercial movers and shakers of Europe. The new discipline of political economy blessed the pursuit of profit, and in places like Leeds, Manchester, York, and Liverpool, that pursuit was loud.

For Joseph, Leeds was also home. He was now only seven miles from the place of his birth, and a ride out the rutted post road to Manchester took him back to Birstal Fieldhead. Aunt Sarah was dead: she'd never come to terms with his rejection of her faith, and Old Hall at Heckmondwycke seemed a mausoleum to her love. Sources are unclear on the year of his father's death, but his youngest brother Joshua had taken over the family cloth-dressing business on the Birstal farm. Timothy lived farther down the road—thirty-five miles away in Manchester. Though staunchly Calvinist with his own congregation, Joseph and he had learned that sibling loyalty was as important as rigid devotion to a creed.

Joseph was thirty-four upon his return, and for all his quirks, he landed in "a liberal, friendly, and harmonious congregation." As a way to overcome his stammer, he'd developed an oratorical style at the pulpit that was quiet and personal, as if he talked in private with each parishioner. He had an ingratiating personality, milder now than in Needham Market; he was lively in conversation and fond of telling stories, often at his own expense. He also had the rare facility of making the best of adversity, and it was said he could charm "away the bitterest prejudices in personal intercourse." His earliest portraits were painted during this period and show an alert and intelligent face, with sharp nose, dimpled chin, and eyes "irradiating hope and enthusiasm." He paced around town in his ministerial black coat with lappets and large cuffs, ruffles, knee breeches, large shining shoe buckles, and powdered wig, a uniform he would keep to, with few variations, until a few years before he died.

Leeds changed Priestley in a number of ways. On the domestic front, two sons—Joseph and William—would be born; intellectually, it was in Leeds that scientific threads in his life and thought came together, and in Leeds that his religious transformation became complete so that he would be one of the modern founders of Unitarianism. It was here that a two-decade string of chemical research began that would see him discover nine unknown gases and make him the premier pneumatic chemist in the world. It was in Leeds that he wrote and published *Experiments and Observations on different Kinds of Airs*, which earned him scientific honors throughout Europe, as well as the *Institutes of Natural and Revealed Religion* and the *Essay on the First Principles of Government*, which earned him the hatred of the Church of England and the enmity of George III. America, which figured prominently in his religious and political thought, was becoming a constant issue in the British press: in 1767, the year of Priestley's move, the Townshend Acts were passed, imposing new taxes on the colonies for lead, paint, paper, glass, and tea. That year, Daniel Boone crossed the Cumberland Gap despite King George's edict, setting the stage for westward expansion. As Priestley matured as a thinker and researcher, America adopted the rhetoric of war.

Finally, it was in Leeds that science and religion became wedded in Priestley's mind. The seed was planted early, almost as soon as he arrived. One day a woman in his church told him she was possessed by Satan and begged him to cast the demon out. Priestley tried to reason with her, but her mind was set, and the next day she returned demanding an exorcism.

During the night, however, he'd had an idea. His electrical apparatus—his spark generator and Leyden jars—had been transferred from Warrington and were assembled. He looked long and hard at her and said, "My dear woman, I think I know a way to relieve you of your incubus." This pronouncement was so grave, so different from his earlier attitude, that at first she was taken aback. "Stand on this," he directed, indicating a small stool with glass legs, then placed in her hand a brass chain connected to the Leyden jar. He spun the friction globe and charged her with electricity. "Pay very close attention, this is important," he intoned, then touched her arm with the discharger. There was a crack and blue flash. With a shout, the woman was knocked off the stool.

For a moment, Joseph was certain that he'd killed the poor woman. He rushed to her side. "Are you all right?" he asked, lifting her to her feet. "That was a powerful shock."

The woman's hair stood up; she swiped it back, but it sprang up again. She patted her sides and hips. "There," she said, "the devil is gone. I saw him in that blue flame, and he gave me such a jerk as he went off. I have at last got rid of him and I am now quite comfortable."

During those first two years in Leeds he was drawn in several directions and unable to get a fix on any one. There were religious duties, a growing family, the month-long trips to London, final revisions on *Institutes* and *First Principles*. He continued teaching, as well as writing about education, and won financial backing for a public subscription library. He continued his electrical research and presented papers to the Royal Society on the force of electrical explosions.

During this time, he proposed a partnership with his brothers to make and sell electrical machines. Tim was good with a lathe and tools, so he turned the legs and did brass work; Joshua was deft with his fingers and in charge of assembly. Joseph designed improvements and was the salesman, "for though he had such natural abilities, he could scarcely handle any tool," Timothy recalled. The partnership sold machines to Joseph's friends and scientific associates in London, and at least one to the Leeds Infirmary, where electric shocks were administered for some nervous disorders—John Wesley mentioned such uses, and the press wrote stories of remarkable cures.

The partnership helped heal old rifts in the family. With an annual income of £100, Joseph was one of the best-paid Dissenting ministers in England, yet he was feeling the financial pinch that governed his course so often. Timothy was in tougher straits: there was sickness in his growing family; he added a child "to my number almost every year"; and his income was a meager £60 per annum. Yet the partnership was dangerous work. At one point, Tim was standing near a large battery when, "had [Joseph] not called out, I should have been struck through the head in a moment with electrical fire." Nevertheless, the extra money was appreciated. "Though we then differed so wide in sentiments on religion, we agreed in this business, and this occasioned us to be frequently together," Tim later recalled.

With these many duties, Joseph was drawn closer and closer to the enigmatic air.

He could not escape it physically. The "mephitic" air, which he and Franklin had spoken of at length, bubbled right behind his house in the fermentation vats of Jakes & Nell. A shift in the breeze brought its yeasty aroma into their kitchen or bedroom, and Mary complained that the air here was as bad as on the banks of the Mersey. During the autumn of 1767 and winter of 1768, Joseph read all he could about earlier research in air, but most was inconclusive. Oliver Goldsmith, in a posthumous book, summed up the existing confusion when he said that air was still thought of as a simple elementary substance, while the 1765 edition of the *General Dictionary of Arts and Sciences*, the latest encyclopedia then available in England, lamented: "There is no way of examining air pure and defecated from the several things that are intermixed with it: consequently there is no saying what is its particular nature."

There were some hints, however. Francis Bacon showed that air had "elasticity," which meant it could expand or be compressed. Robert Boyle obtained an "elastic fluid" from ripe fruit and rotting animals, as well as from fermenting liquids like that outside Priestley's yard: he even wondered whether such airs could be produced artificially. Newton, in his 1704 *Opticks*, noted that heat and fermentation gave rise to airs that seemed to have different properties. More modern chemists were beginning to refer to these varied airs as "gases," a term coined in the 1600s by the Flemish physician and chemist Jan Baptista van Helmont. Such airlike substances, without fixed volume or shape, reminded van Helmont of the Greek term *chaos:* unshaped and unordered, the original material from which the universe was formed. But to an Englishman, "chaos" sounded like "gas" in van Helmont's heavy Flemish, so "gas" it remained.

The most practical reading, from Priestley's point of view, was the research of the Scottish physician and chemist Joseph Black, fifteen years earlier. Black's discoveries were possibly the most important chemical advances so far in the subject: he'd found that "air" could form part of the composition of dense bodies. This was not his original intention; instead, he'd been trying to find a remedy for gout by heating limestone, taken from a secret recipe of calcined snails that had reportedly cured England's first prime minister. Though he didn't find the cure, he did discover that when he heated the limestone, a gas he called "fixed air" (known today as carbon dioxide) was given off, leaving behind lime. By recombining the fixed air with the lime, he once again formed limestone, proving for the first time that gas was an important component of a solid substance.

Black's findings were important for a number of reasons. First, he showed that fixed air could be formed by heating a mineral; second, he showed that gases were not merely given off by liquids and solids, but actually combined with them to form new substances. When the lime was allowed to stand in the air, it slowly turned back to limestone, which led him to conclude that small quantities of fixed air must be present in the air we breathe. Here was the first clear demonstration that air was not a simple substance, but was made of at least two distinct substances—"fixed air" and ordinary air.

What a fascinating idea! A mouse died in Black's fixed air; candles would not burn. It was like the choke-damp lurking at the bottom of mines to kill unwary miners; like the volatile ingredient that collected so lethally in the *Dunsthöhle* and in the Pyrmont Springs.

In the summer of 1768, Joseph told Mary that he'd decided to look into the fixed air in the neighboring vats "for my own amusement." He asked the workmen at the vats if he could make some tests; they were used to the ways of their amusing if amiable neighbor, and agreed. Though Priestley would often say he knew little chemical theory, he was a careful observer of things around him. He watched as he floated burning chips of wood on the surface of the fermenting liquor: the flames were extinguished. Smoke that rose from candles and the wood chips lingered in the gas; if he swirled it with his hand, the now-visible gas spilled over the side of the vat and onto the floor, which must mean it was heavier than common air.

Despite the warnings of the maltsters, Joseph tried to breath the gas: the fixed air didn't have a scent, but his lungs were stifled and he felt lightheaded. He placed butterflies and other insects in the gas: they grew torpid and fell over, but revived if he took them out immediately. A poor mouse had convulsions. A large frog swelled up: after six minutes in the gas it seemed quite dead, but revived when taken out. The only casualty was a snail, which when "treated in the same manner, died presently."

The next day, Joseph tried mixing water with the fixed air. If he placed water in a shallow dish and held this over the vat, the water acquired a pleasant acidy taste that reminded him of the Pyrmont and Seltzer waters he and Ben had discussed. He gave some to the maltsters and they smacked their lips; he gave some to Mary, and she agreed. Maybe he was on to something here. Next he found that by holding beakers beneath the level of the gas and pouring water back and forth for a few minutes, he could create the taste more rapidly. The water absorbed the gas more quickly with simple agitation. He'd invented soda water, that weak solution of water and carbon dioxide, by doing little more than playing around.

Now he grew ambitious and tested what might happen if he mixed other fluids from beaker to beaker. The unfortunate thing about Joseph was that he tended to grow clumsy when following a line of thought; if he was not careful, accidents occurred. Such was the case of the goose and his kite; such was the case now. The first fluid he apparently tested was a solution of ether, and there are two versions of what happened: either the ether bubbled up and its fumes leaked into the fermenting liquor, or he spilled a beaker into the beer. In either case, the entire batch of beer was ruined, and Joseph was forever banned from experimenting at Jakes & Nell.

Not that it mattered. He'd learned from a close reading of Black's experiments that fixed air could be produced artificially if he poured some "oil of vitriol," or sulfuric acid, on chalk and water. In his workshop, he shook a glass vial containing water, chalk and acid and watched it begin to froth; with a bladder, he pumped this fixed air into a beaker of water. In about thirty minutes, he was able to "impregnate" the common water with fixed air.

Then his attention wandered elsewhere. He was often like that, abandoning one line of inquiry as others arose. "My manner has always been to give my whole attention to a subject till I have satisfied myself with respect to it, and then think no more on the matter," he once noted. Thus it was that for four years he set aside the brief amusement that would be his doorway to fame, yet by then his direction was set. "My experiments on air I find will run out to a great length, several new circumstances have occurred, which I cannot yet ascertain," he wrote to Richard Price in the winter of 1768. "Some of the operations require several weeks before they are completed."

For once, he was wrong. His investigations of air would absorb the rest of his life.

CHAPTER 4

The Prodigy

WHILE PRIESTLEY PUZZLED OVER BUBBLES OF GAS, LAVOISIER PREpared for a journey. He was bound for the east, to the shrouded Vosges Mountains, as assistant to Jean-Etienne Guettard, France's premier mineralogist and one of it most contentious *savants.* They were adding information to Guettard's *Atlas minéralogique de la France*, an ambitious inventory of "the quarries, excavating mines, mineral springs, and all main materials contained in the earth" of the realm. Although it was never finished, the *Atlas* would bring his young assistant to the notice of the King, the nation, the "tax farmers" who made him rich, and the woman he loved.

Antoine was journeying into a land still crippled by the Seven Years' War. On February 10, 1763, when the great powers signed the Treaty of Paris, Britain walked away with the spoils and France lost its resource-rich colonies in India and North America. Before 1763, the nation's stocks of saltpeter, essential for gunpowder, came from India; its coal and coke production was scattered throughout the country in small mines; its native steel was inferior, so France relied upon Germany, England, and Sweden, all frequent enemies, for its needs. France had come into the century suffering from the prolonged conflicts of Louis XIV and the financial debacles of the regency; Louis XIV's persecution of 2 million Huguenots, his worst mistake, saw the flight of the core group that drove the Industrial Revolution elsewhere. Papermakers, textile workers, and other craftsmen took their secrets to Germany and England; merchants and bankers took their capital; printers, shipbuilders, and the educated classes—doctors, lawyers, pastors—all fled. Now, in midcentury, a dearth of strategic resources was added to that burden.

By 1767, France was stuck at a crossroads and had to choose its way. The nation entered a period of industrial growth but was slow to mechanize, and its improvements in mining lagged far behind England and Germany. Yet it progressed in those industries where chemistry was crucial, and the French Academy of Sciences was the organ of change. As early as 1727, René-Antoine Ferchault de Réaumur, best known for his study of thermometry, drafted a broad research program. "Chemistry," he wrote,

> could become one of the most useful parts of the Academy. Let us not boast of the help that medicine could draw from it; let us look at it only from the standpoint of the arts, to which it could be more useful even than mechanics itself. The conversion of iron into steel, the methods of plating and whitening iron to make tin-plate, the conversion of copper into brass, three great industries which the Kingdom lacks, are in the province of chemistry. . . .

Lavoisier fell heir to such interest, even before choosing the realm of "sooty empirics" as his.

Paris had been one of the medieval *studia* and a love of learning was part of the culture; now French architecture, fashion, and furniture dominated Continental tastes and educated Europe adopted its language. By 1767, the Enlightenment as a movement had coalesced. Persecution by the Catholic Church had given Voltaire, Montesquieu, Rousseau, and other writers a sense of common purpose; the seventeen-volume *Encyclopédie*, launched in 1751 by Denis Diderot and d'Alembert, was complete by 1765. Rousseau's two best-selling novels, *Julie* and *Emile*, and the *Social Contract*, his most enduring work, were published in 1761–62, moving readers to tears at the triumph of innocence and introducing into the political language the sovereignty of the "general will."

But the general will was still fragmented between the classes. The clergy made up the first powerful order, or "estate"; the nobles, the second. The other four-fifths of the population made up the lowly Third Estate, which consisted of everyone from peasants to the bourgeoisie. While the clergy and nobility were exempt from taxation, the Third Estate was not: the peasants especially could not afford taxes, yet every year half to three-quarters of their meager incomes went to support a lavish court and huge military machine.

Poverty was the nation's Achilles heel. England had its working poor and gin houses, but France's poverty was medieval. "All the country girls and women," noted the English traveler Arthur Young, "are without shoes and stockings; and the ploughmen at their work have neither sabots nor stockings for their feet. This is a poverty that strikes at the root of national prosperity." Peasants comprised 80 percent of the King's subjects, and only a fifth lived in communities of more than 2,000 people. A good 250,000 people did not live anywhere, a floating population of vagabonds that terrorized isolated farmers and filled city streets with armies of beggars. In the best of times, the poor numbered nearly 8 million, or a third of the population; in times of famine or other disasters, another 2 or 3 million joined the ranks. As the century progressed, more peasants straggled to the cities, where survival meant begging, smuggling, prostitution, and violent crime. In the 1760s, Paris boasted 25,000 prostitutes; by century's end, its population of new orphans and foundlings numbered 40,000 a year.

Education was a route up and out, but by 1767 the French educational system was in shambles, caused by the dissolution of the Jesuits, who had dominated the education of Catholic elites since the sixteenth century. Although not disbanded by the Pope until 1773, they'd been expelled from France nine years earlier and their 113 colleges sequestered. The British alternative of the Dissenting academy was unknown in France; the curriculum fell apart, and by 1789, only one boy in fifty-two would attend college.

That Antoine got as far as he did can be attributed to his family's riches and the influence of their friend Guettard. It is uncertain when Jean Lavoisier met the famous geologist: until Antoine graduated from the Collège Mazarin in June 1761, he was still under the wing of Lacaille. From the famous astronomer he absorbed logic and a preference for numbers: "I was accustomed to the rigorous reasoning of mathematicians," he later said. "They never take up a proposition until the one preceding it has been determined." Lacaille advised him to study meteorology, and he started recording barometric readings three or four times daily in hopes that someday he might discover the laws governing the atmosphere.

But when Lacaille died in March 1762, Antoine was left without a mentor. His father pushed him toward the law, a profession inclined to justify decisions by appealing to precedent, and although Antoine entered the School of Law, his heart lay elsewhere. When he should have pored over texts on disputation and oratory, he took "philosophic walks" with the botanist Bernard de Jussieu. Instead of hanging out in court, he learned the rudiments of anatomy at the School of Medicine and witnessed the abbé Jean-Antoine Nollet's experiments with electricity.

Then Guettard took him in hand. The geologist would be to the young Lavoisier what Ben Franklin was to Priestley—a mentor, model, and sponsor, but without Franklin's charm. In 1762, Guettard was forty-seven, but he already seemed old. Described as a "fighting animal," a misanthrope, and, by his own estimation, a "Hottentot," the philosopher marquis Marie-Jean de Condorcet would say that "few men have had more quarrels." Guettard spoke his mind with brutal frankness and was rude and pious, and most conversations disintegrated into diatribes about the expulsion of the Jesuits, about whom he was passionate. "What a man this is," he wrote once in declining an invitation to a *salon*. "He is a monster in society. . . . What can one do with this savage?"

He was also brilliant, in his way. His mind was encyclopedic regarding the sciences of mineralogy, geology, and chemistry; he was an academician, curator of the Museum of Natural History, and in 1750 solved the imperial problem of porcelain manufacture by discovering sources near Alençon for two prime ingredients, kaolin and feldspar. He proposed his mineralogical atlas as early as 1746, but it was not until 1767 that it finally received state funding. The minister of state ordered him to begin with a survey of the Vosges, and Alsace and Lorraine, an area rich in minerals and mines.

Guettard may have begun preparing Lavoisier as his assistant as early as 1762 or 1763. He seemed to recognize a kindred spirit in the young man. By age nineteen, Antoine was a handsome and eligible bachelor who detested the social whirl. To escape the obligations of social life, he would claim illness, at one point stopping all nourishment but milk. In response to his fast, a family friend sent wheat flour, butter, and advice: "Your health, my dear mathematician, is like that of almost all educated people whose minds are stronger than their bodies. So spend less time on your studies and remember that one more year on earth is better than a hundred in men's memories."

Lavoisier definitely preferred work to pleasure. Work *was* his pleasure, and Guettard tapped into that. If Lavoisier was too ambitious, as later claimed, then ambition is nothing without discipline and his discipline in pursuit of a problem was as stern as a monk's. In 1764, for example, when the Academy of Sciences sponsored a 2,000 *livres* prize to determine the best way of lighting city streets, Lavoisier disappeared from view. The problem was a vexing one: a British traveler, William Cole, who visited Paris in 1760, said that the streets were "most abominably lighted up," strung with drooping ropes of hanging lamps—on windy nights the lamps blew out, and one either hired lantern bearers to light the way home or fell prey to thugs. Lavoisier covered the walls of his study with black cloth and shut himself up in darkness for six weeks to sensitize his retinas to minor differences in illumination. Although he did not win, his 70-page paper so impressed the judges that the King gave him a special consolation prize.

From 1763 to 1767, Guettard set the young Lavoisier on a course of study in chemistry and mineralogy; he regaled his protégé with tales of his travels and inspired the young man. Notes from these years show Lavoisier proposing "a fine course of experiments to run." What they might be was yet to be determined, but he was already dreaming of breaking new ground.

Nevertheless, he pursued his legal studies, and on July 26, 1764, qualified as Licentiate of Law and was admitted to the *Parlement de Paris*. His identity papers describe him as five feet four, tall for the era; fair-complexioned, with chestnut hair and brown eyes. His forehead was prominent, nose long and hooked, mouth small and thin. Those who knew him said his eyes were gentle, though his bearing and the set of his face struck many as severe. To strangers he seemed aloof, while friends insisted he was timid, more at ease in his lab than dancing a minuet in society.

But the law was a social world, and perhaps he voiced his trepidations to Guettard. For all his infamous misanthropy, Guettard could be kind to his few friends. At this critical point, Guettard performed for Antoine his greatest favor, encouraging him to enroll in the private and public classes of France's premier chemist, Guillaume-François Rouelle.

▪ ▪ ▪

At 1742, the year before Antoine's birth, chemical instruction in France was at a low ebb. The nation's seat of chemical instruction was the Jardin du Roi, founded in 1626 by Louis XIII as a center for medical education (which then included botany, anatomy, and chemistry). But the lectures were largely pharmaceutical in nature since the field was considered an adjunct of medicine. Even as France began to recognize the importance of chemistry if it wished to remain a world power, there were no influential chemists on par with England's Robert Boyle, Scotland's Joseph Black, or the several German scientists whose theories formed a battleground for all that would come.

Then Rouelle arrived. Although the tall and lanky "demonstrator's" theories descended from the alchemists, and he never published a textbook, Rouelle achieved fame through his wildly popular courses and inspired an entire generation of French chemists. "When Rouelle spoke, he inspired, he overwhelmed," wrote Louis-Sébastien Mercier, the chronicler of Paris who was a pupil of Rouelle. "He made me love an art about which I had not the least notion . . . it is he who made me a supporter of that science which would regenerate all the arts, one after the other."

Rouelle apparently overwhelmed Lavoisier in a similar way. Before Antoine started attending Rouelle's public lectures in the Jardin du Roi's 600-seat amphitheater—and later in his private course in the rue Jacob during 1762–63 or 1763–64—chemistry had been for him just another tool in the study of minerals. After Rouelle, it was the whole world.

By the time Lavoisier discovered the famous chemist, Rouelle was sixty, nearing his retirement in 1768 and his death two years later. The years never mellowed the man. As *"démonstrateur en chimie,"* his role was officially subordinate to the *professeur,* Louis-Claude Bourdelin, since in theory he was supposed to perform experiments that illustrated the professor's lectures. In practice, Rouelle would begin his demonstrations by announcing, "Gentlemen, all that *monsieur le professeur* has just told you is absurd and false, as I will prove to you." He came to his lecture hall in full dress—velvet morning coat, well-powdered wig, small hat under his arm—but as he spoke, the coat and waistcoat were tossed to the floor, the wig hung upon a handy hook; in short sleeves, his red hair unruly, he strode about the hall, performing experiments. Mild and even-tempered in most respects, he grew insulting when challenged on anything touching chemistry. Passed over once by the Academy in favor of a man he considered a lesser chemist, Rouelle never forgave the slight; though later elected, he would snipe, "So you are going to choose a chemist. . . . In such matters, the Academy has sometimes chosen the worst." He was obsessed that his lectures were being plagiarized, but forgot that he revealed his secrets to his students almost every day. Once, when a spectator chided him for his foul language, he retorted: "Where are we? At the Academy of Proper Speech?"

The public lectures, which began in June and lasted through November, were free and open, which meant the auditorium was filled with every strata of French society. On the lecture desk stood rows of beakers, flasks, and glassware, all checked by his assistant, a belabored soul he called his "damned nephew." Aristocrats, dipping at snuffboxes, sat near the top. In the middle tiers sat academicians and the bourgeoisie. Up front sat the students, notebooks ready.

He was so famously absent-minded that sitting too close could be dangerous in Rouelle's lab. One day Rouelle put the fear of God in his audience when he pointed to the mixture he stirred and said, "You can see very well, Gentlemen, the cauldron over the flames. If I were to stop stirring it for one moment, there would be an explosion that would blow us to the rooftops." Unfortunately, while making his point, he stopped stirring. The promised explosion blew out two windows, demolished the fire hood, incinerated Rouelle's wig, and sent the terrified audience scrambling for the garden.

But it was in his private course that hard questions of chemistry were explored. The course began in December and lasted from 11 a.m. to 1 p.m., Monday, Wednesday, Friday, and Saturday. By the time Lavoisier enrolled, Rouelle was sick: he had been charged in 1753 by the minister of war to develop a new method for refining saltpeter, and he'd been slowly poisoned by breathing the phosphates contained in the nitre. Acute phosphate poisoning could be horrific: in the early 1800s, when English farmers noticed that crop yields in those fields treated with phosphate-laden guano jumped from 30 to 300 percent, more than one tasted the guano to assure it was the real article. In one case, a farmer swallowed some and it burned him internally; his throat swelled and constricted, followed by a hemorrhaging so acute that he died within forty-eight hours. After his death, the coroner wrote that "the blood continued to exude for several hours, till the entire sanguineous system . . . had become emptied of its contents."

Although saltpeter may have affected Rouelle's health, it did not diminish the force of his mind. High on the wall of his private lab was posted in large letters the peripatetic motto

Nihil est in intellectu quod non prius fuerit in sensu

or literally, "Nothing is in the understanding that was not earlier in the senses." This was the central doctrine of empiricists and a catch phrase for thinkers like John Locke and John Stuart Mill; its presence was meant to impress on Rouelle's students the lesson that the only way to conduct labwork, and science, was through a strict empiricism. Though Lavoisier had learned this lesson early under Lacaille, it took on the force of dogma under the tutelage of Rouelle.

Rouelle's other catechism was the doctrine of phlogiston, developed in the early eighteenth century by a German chemist, Georg Ernst Stahl. Stahl's theories were taken from the ideas of his predecessor, Johann Joachim Becher, but more elegantly explained. Until the end of the seventeenth century, the ancient theory that treated fire as one of the four basic elements—a substance rather than a form of energy—was still firmly entrenched. Stahl, court physician to the King of Prussia, was fascinated by burning, the central question of alchemy. According to alchemists, something burned because it contained the "sulphurous principle" of combustibility, so in 1716 Stahl updated the principle by giving it a new name. Stahl's new "phlogiston theory," as he called it, was the first unified, systematic approach to understanding chemical reactions, and for all the flaws that accumulated over the next half century, it was still the only working explanation chemists had at the time. Today, phlogiston theory is a historical curiosity, a transition between alchemy and chemistry, but in the eighteenth century it was a reality as accepted as gravity is today. As the race to discover and comprehend oxygen heated up, it became a war zone.

According to Stahl, all bodies contained phlogiston, derived from the Greek *phologizein,* or "set on fire." Phlogiston was the active "principle of combustibility"; although, like gravity, it could not be weighed or measured, it was a primary component of all things. Some substances contained more phlogiston, others less, and good fuels like natural gas, anthracite coal, and whale oil (all of which burned and left little or no residue) were essentially pure phlogiston. Fuels that did leave a residue, like peat, wood, and bituminous coal, were made of phlogiston and ash. Although phlogiston itself could not be isolated, it did become visible when escaping a substance—and it did this as fire. Thus, during a fiery chemical reaction, phlogiston changed from an inert combined state to a free one. Although there were still four elements in Stahl's worldview, they were changed slightly from the ancient Aristotelian standard to water and three kinds of earth, one flammable. Fire was no longer an element but a sign of phlogiston's presence; air, no longer deemed chemically active, was needed as a place and means for absorbing freed phlogiston.

Today, such reasoning seems circular and forced, but belief in the theory was deep-seated. It was the flat-earth theory of its time, the geocentric universe, both accepted so strongly because, like phlogiston, they conformed to common sense. The Earth *does* look flat when viewed from the surface; the stars, planets, and Sun *do* seem to revolve around the Earth in great circles like the Moon. Something *does* seem to escape from a substance when burning occurs: Look at the smoke, the fumes, the flame. The theory seemed to link many isolated phenomena, and was seen as a working hypothesis for what happened in chemical reactions.

The doctrine of phlogiston had the added advantage of explaining parallels between combustion and respiration. When a candle burned in a limited supply of air, it went out. When an animal was placed in a closed vessel, it died. Respiration, like combustion, must be a transfer of phlogiston from the body to the air, and both processes ceased in a closed container because the air became saturated. Too much was poisonous.

Thus, phlogiston seemed to make perfect sense. It was the first theory to give order to a large number of physical *and* chemical reactions. It explained why bodies burned—they were rich in phlogiston. It explained why different metals had properties in common—they were all composed of phlogiston and one of the three elementary earths—sulfur, or *terra pinguis* ("fatty earth"), which was the essence of inflammability; mercury, or *terra mercurialis,* the essence of fluidity; and salt, or *terra lapida,* the essence of inertness and fixity. Stahl admired his predecessors Becher and van Helmont, but was convinced he'd surpassed them in theory. Phlogiston not only explained the subtleties of combustion; it was the common link between all bodies, be they animal, vegetable, or mineral, he said. A global circulation of phlogiston constantly renewed the world: when substances decomposed or eroded, they released phlogiston into the atmosphere; this was then reabsorbed by substances in a variety of ways. To Stahl, Rouelle, and other phlogistonists, the theory explained the entire cycle of life and death. Phlogiston was the first principle; it *was* the vital flame.

Unfortunately, there were problems. For example, phlogiston theory could not explain the discoveries of Joseph Black, John Mayow, and other British pneumatic chemists. If the air was as inert as phlogistonists claimed, then what explained all these amazing new gases? Even more worrisome was the problem of weight. According to the theory, a metal should lose weight when burned or roasted, since phlogiston was released into the air. But in fact the opposite occurred. Metals gained weight, a problem first noted by medieval Arabic chemists and then, in increasing numbers, by their seventeenth- and eighteenth-century European heirs. In 1688, John Mayow had developed the idea of a "spiritus igneo-aureous," a more "active and subtle" part of air that combined with a metal and increased its weight; his "spiritus," which sounds today like oxygen, would also explain why a candle went out in a closed flask even though air remained inside. Breathing and burning must both exhaust the "spiritus," Mayow said; but this was an inconvenient idea, throwing all other tenets of the theory into question, and so he was ignored. After all, if a chemical reaction could alter the volume, texture, color, and smell of a substance, why not its weight, too? In addition, metal's behavior ran counter to the rest of the physical world: most bodies, like wood and bone, lost weight during burning, just as the theory said.

By the time Lavoisier studied under Rouelle, most phlogistonists were explaining this weight loss through backward reasoning. Since a flame soared up when leaving a substance, it possesses the quality of "levity," of being lighter than air. It must have "negative mass," since when it was gone, the substance weighed more. The lightest substances might well contain the most phlogiston.

Such "blurring of a paradigm" and "loosening of the rules for normal research" is, according to the historian Thomas Kuhn, the hallmark of a science in a state of crisis: a worldview which can no longer adequately state why the world behaves as it does. Insecurity rattles the faithful and the crisis spreads. Although one of the West's dominant myths is that science heralds truth, more often than not that "truth" is a servant of power. Science proscribes ways of behavior and belief, thus defining the status quo. An explained world may not be perfect, but at least it makes sense. No one asks questions; no one seeks change without an alternative.

But what happens to a science when a crisis evolves? As more flaws are discovered in existing theory, they are either ignored or twisted until the sheer volume makes accommodation impossible. The power of the theory to order the mind and explain life, to predict and effect change, comes to a halt. The field must either reconstruct itself from "new fundamentals" that change "some of the field's most fundamental generalizations," or it slips into an intellectual coma, its adherents continuing aimlessly, playing by the old rules while knowing full well the game is useless. Today, such a state would be dubbed existential despair.

Which raises the ageless question, older than the Greeks: What is truth? What is its physical form? Does it exist as a separate state, waiting to be discovered, or is it formed and shaped in the mind? The ancient question informed everything that drove and surrounded Priestley and Lavoisier: science, revolution, the madness of crowds, the nature of God. Perhaps hardest to accept was the idea that so much depended upon modes of perception and ways of viewing the world, yet nothing was more threatening, destabilizing, or revolutionary than a shift in basic reality. The British historian Herbert Butterfield compared such a basic reorientation in science to someone "picking up the other end of the stick"—one approaches the same set of data from the other end. Others have compared such a revolution to "a change in visual gestalt: the marks on paper that were first seen as a bird are now seen as an antelope, or vice versa." Kuhn disagrees with this metaphor: "Scientists do not see something *as* something else; instead they simply see it," he says. Yet at the very moment that Lavoisier and Priestley were entering the scene, a rabid debate on the fundamentals of perception filled the halls of philosophy.

It boiled down to this: the eighteenth century was divided between Isaac Newton's followers and those of his rival intellect, Gottfried Wilhelm Leibniz. Their themes were huge: How does one see the world around one? How does one perceive the limits, or even the ground rules, of knowledge? Were reality and space absolute, as Newton suggested, or relative and multifaceted, scattered into independently developing "monads" of thought, as Leibniz said? Even their careers illustrated these opposite poles of thought, for both men developed the system of calculus independently and within years of one another, a fact that apparently supported Leibniz's structure. Yet Newton could as easily say that the laws of discovery, like his laws of gravitation, were part of a larger, if still undiscovered, scheme.

One could argue endlessly about such notions yet come no closer to clarity than theologians arguing about the number of angels that could dance on the head of a pin. Such disputations were immensely frustrating to Antoine Lavoisier. They seemed to have no purpose, yet filled the halls of learning. Science should be *exact,* but it wasn't. No wonder the world was in such a state . . . something was badly amiss. There had to be a better way.

▪ ▪ ▪

Lavoisier and Guettard left for the Vosges Mountains on June 14, 1767. Antoine was accompanied by a servant, Joseph, and took a good horse, 50 *louis* in his pocket, and his equipment—three thermometers, a barometer, chemical reagents, a mortar, and a silver hydrometer for measuring the density of mineral water. The men carried pistols as protection against highwaymen. His father told him to write; Aunt Constance wept, associating mines and quarries with deadly gas and cave-ins. "Antoine," she told him, "I will await the post like I do the coming of the Lord."

Antoine promised to write, but asked his family to check his barometers and record the readings religiously.

He felt on top of the world. He was on his first great scientific expedition—a wonderland of undiscovered fact stretched before him. His letters home were frequent throughout the trip, and that first evening he sent his first from Brie-Comte-Robert, seventeen miles away. "Animals and people are delighted and in good health," he wrote. "The horses seem especially gay and have good appetites. So there is every reason to hope that all will go well."

But country life proved more difficult than he'd imagined. His first inkling of rural conditions would be the isolation of those with whom he stayed. Luxuries that he took for granted were unattainable less than fifty miles from the capital. Near Chaumont, he asked his father to send a case of ratafia, a cordial flavored with peach and cherry kernels and bitter almonds; some roman candles and pinwheels; and a dozen small goldfish for his hostess to put in her moat. His father complained about the "nasty and extremely awkward errand" assigned to him by his son, since he would have to travel the entire distance with the bowl of goldfish on his lap, but he did so anyway. The weather ruined Antoine's suit and he asked for a new one. Aunt Constance handled the details, sending back one of green wool with small pockets for pistols.

The two great surprises were weather and poverty. They were

drenched, then roasted, then drenched again. "One would hardly put a dog outside in this weather," Guettard complained. "One day you are utterly soaked with rain, the next with perspiration. . . . Then a plague of burning rays grills you like a smoked herring. . . . Is this a life for a Christian?" More alarming was the poverty. In one village, they lodged in a drafty loft that smelled of onions; they scoured the entire village before blankets could be found. The peasants suffered from malnutrition: "The dairy farmers, who are ordinarily very well off, eat neither bread nor meat," Antoine observed. They did not have wine, and subsisted almost entirely off a dry, low-fat cheese, which they shaped into loaves. All could not be well in a country with such wretched conditions: his firsthand knowledge of such conditions would influence his later attitudes on revolution and reform.

But this was no sightseeing trip, and Guettard worked him hard. They reached Basel, in Switzerland, on July 25, where they met Daniel Bernoulli, the famous mathematician, then proceeded to Gérardmer, where they climbed the Grand Ballon, the highest point in the Vosges. The steep, wooded ridges cut off the Franche-Comté and Alsace-Lorraine from the rest of the nation: Franche-Comté slumbered through the eighteenth century, missing the worst of the Revolution but preoccupied with bitter infighting between families. In Alsace-Lorraine, to the north, the inhabitants spoke German and all economic life swirled around the Rhine. At every stop, the two men recorded the nature of the soil, vegetation, mines, and local industries; they inspected and inventoried the collections of local naturalists; they took temperature and density readings of the lakes and rivers; and every night, in true family tradition, Antoine made note of the day's expenses. By September 3, on the long loop back, they visited Strasbourg, where Antoine spent the equivalent of $20,000 on 118 books, including texts by electricians, "aerial" philosophers like Mayow and van Helmont, and the phlogistonists Becher and Stahl. He returned home at eleven-thirty on October 19, dirty and tired. His aunt and father hugged him, then he checked his barometers before washing up, and went to bed.

Their four months over plain and mountain did little to advance the theory of the Earth's formation, but Lavoisier learned other lessons. More crucially, he developed a working method. For years he had watched Guettard fritter away time with the gathering of hundreds of bits of information without any cohesive theory to guide his steps. It was the same deductive approach espoused by Newton, even if his greatest works were leaps of inductive imagination. Collect the facts, wrote Newton, and the patterns would emerge.

Beginning with this journey, Lavoisier instinctively seemed to disagree. He would proceed in the opposite direction, even though the accepted myth of objectivity made such a course suspect to colleagues. "There are two ways to present the objects and subject matter of science," he wrote. "The first consists in making observations and tracing them to the causes that have produced them. The second consists of hypothesizing a cause and then seeing if the observed phenomena can validate the hypothesis." Although the second method was rarely used in those days "in the search for new truths," it was used often in teaching, "for it spares students from difficulties and boredom." He borrowed the method from his teachers—either Lacaille or Rouelle, if not both—then expanded it for his geological research. The method was so efficient that he turned it into personal dogma and adhered to it religiously for the rest of his days.

Inductive reasoning was not new—it had merely fallen out of favor due to Newton's hold on the scientific imagination. Science follows fashion, as does art and literature. Both Francis Bacon in 1620 and Christian Huygens in 1690 set down the intellectual basis for induction that became what is accepted today as the "scientific method." They saw that it was not possible to explain the laws of Nature by deduction: in the end, there was simply too much for us to know. Every explanation reached beyond our immediate experience, said Huygens, to become a speculation. Such speculation is a choice, and we live out our days as best we can based on such educated guesses.

But there are ways to test these guesses, and that is the scientific method. The elegance and irony involves the artifice used to probe the natural world. Every person who designs an experiment based upon inductive reasoning puts it in a box, separating the idea from the rest of the world. To paraphrase the Talmud, we put a fence around the law, fencing in the laws of Nature that we are trying to determine. Everything inside the fence has meaning while everything outside is irrelevant: it is a lie, of course, for the limitations of the mind are such that individuals, entire groups, and nations can accept erroneous concepts as truths, based on what comfortably dwells within the fence—truths that seem "self-evident." But every induction guesses at a unity, and it is in unity that we ultimately find meaning.

Of more immediate advantage to Lavoisier was the influence of the *Atlas* in his election to the French Academy. On March 10, 1768, Théodore d'Enouville, known for his research on borax and potassium chloride, died, thus freeing a chair for an assistant chemist. Lavoisier had applied in 1766 and been turned down; now, between his early paper on lighting, an analysis of gypsum, and his service to the state in the mineralogical atlas, he was a stronger contender. In addition, the Academy, so chronically short of money, tended to act with an eye on the future. To advance "a young man with knowledge, brains, and energy, with sufficient private means to enable him to live without practicing a profession, would be extremely useful to science" as well as to the Academy, said Joseph Lalande, the astronomer who'd predicted the end of the world.

Out of seven candidates for the seat, Lavoisier had one serious competitor—the metallurgist Gabriel Jars. Eleven years Antoine's elder, Jars was essentially France's first industrial spy of record, and at a crucial time. After the end of the Seven Years' War, the state—worried by Great Britain's industrial superiority and its progress in metallurgy—sent Jars to England to learn what he could. He was there from June 1764 to September 1765, and he did not disappoint. He visited the coal mines of Newcastle and the Scottish border, where, according to his marching orders, he studied the "various uses made of the different types of coal." He visited mines where sulfuric acid was produced; he reported that only Swedish iron was suitable for conversion into steel; he observed the widespread use of machinery and steam engines, commented upon the division of labor in complicated industries, and speculated that Birmingham was thriving because of its free competition and spirit of *laissez-faire.* Most important, he sent back to France a detailed report on the use of coke in smelting. Upon his return, he was despatched to the mines and ironworks of Liège and Holland; to Hanover, the Hartz Mountains, and

Saxony; to Sweden and Norway. Sixteen detailed reports came from this mission, and when he returned, he traveled throughout France to disseminate the techniques he had learned.

Clearly, Jars could not be ignored, especially by a body that depended on the state for money and whose every appointment was ratified by the King. The election was held on May 18, 1768. Lavoisier had the highest count of votes, Jars next; the final choice was left to the minister of state and the King. Jars was chosen in consideration of his service to France, but an additional temporary chair was created for Lavoisier, with the understanding that he would be named officially without election when the next vacancy occurred.

▪ ▪ ▪

With that, Antoine's star began to ascend. "How splendid that at so early an age, when other young men are occupied with amusing themselves," a friend wrote Aunt Constance, "he should have made contributions to the progress of science, and have obtained a position which is usually won, with great difficulty, by men past their fiftieth year." By age twenty-five, he was a recognized prodigy.

But the good fortune of 1768 didn't end there. On May 2, little more than two weeks before his election to the Academy, Lavoisier accepted a position with the *Ferme générale,* or General Farm, a decision that would affect every sphere of his life: reputation, romance, and mentality. He'd reached the age where he had to choose a profession, but the law was no longer appealing and the 2,000 *livres* stipend he could eventually expect from the Academy would not allow him to live in the manner to which he was accustomed. Then in March of that year he learned that a Farmer, seventy-four-year-old François Baudon, hoped to sell his holdings and retire.

The General Farm was the tax collector of France, a private company of sixty stockholders charged with collecting the "indirect" taxes of the King. Through a wide-ranging network of agents, they collected the salt tax, or *gabelle;* the taxes on tobacco, alcohol, playing cards, meat, oil, and soap, called *aides;* the custom duties, or *traites,* collected at the 1,600 toll gates at the nation's borders and entrances into cities; and the tax on goods entering Paris, or *entrées.* The practice of auctioning tax collection to private individuals or companies was a practice as old as Rome; the opportunity for enormous fraud and extortion was ancient, too. In seventeenth-century France, "indirect revenue" had been entrusted to a web of official and semi-official agents who received a percentage of the collection, but the difficulties involved in controlling such an operation led the state in 1681 to turn the entire concern over to a single outfit of financiers. The annual sum paid by the Farm for this privilege was enormous: 75 million *livres* in 1700, more than doubling by 1774, to 152 million. This annual payment was based on an estimate of the taxes to be collected, and the Farm's profit came from the difference between this payment and the taxes actually taken in. Since the state's finances were in chaos, this difference could be huge and hard to anticipate from year to year. The Farmers themselves tried to stay discreet: each six-year contract was drawn in the name of a "man of straw," while the identities of the controlling financiers remained in the background.

Not surprisingly, a Farmer's income could be vast, second only to the King's. By one estimate, real income, which included a fixed salary, travel allowances, and a percentage of the taxes collected, amounted to 60,000–120,000 *livres* a year, or $2.7–$5.5 million in current dollars. Other estimates placed annual incomes as high as 200,000 *livres.* For entry into such a life, Lavoisier was required to invest an astronomical sum. The Farm's assets in 1768 totaled 93.6 million *livres,* and each member's share cost him 1.56 million. On May 2, he purchased a third of Baudon's share for 520,000 *livres,* or nearly $24 million in current dollars; he supplied 340,000 *livres,* which was the lion's share of his inheritance from his mother, and his father loaned him the rest. During the next six-year contract, Lavoisier advanced his share to a half, and when Baudon died in 1779, he became a full Farmer with a complete share.

Lavoisier's decision to join the Farm worried those supporters in the Academy who'd hoped that by electing a young man of wealth they'd bring a full-time scientist into the fold. In truth, the Farm would divide his energies immensely, and it was later estimated that he only gave one-fifth of his time to science during his years as a *fermier.* Since the General Farm did not have a director, jobs were parceled among members; they served on committees that made decisions, on commissions that carried them out, or as inspectors sent into the field. Lavoisier's first job was that of a *tourneur,* or regional inspector, for the Tobacco Commission, and his first two years in this post drew him away from Paris and the Academy. Still, the problem wasn't considered great enough to block his election: as one academician said of his admittance to the Farm, "That's just fine! The dinners he will serve us will be all the better!"

Of greater concern was Lavoisier's alignment with a system that was hated and despised. In the seventeenth century the Farmers were either considered the King's lackeys or the sons of lackeys; by the eighteenth century, the membership had changed, the ranks now filled with professional financiers, but old hatreds remained. In addition to the dislike of the taxman, this was a royally chartered, unaudited private company in which a few secretive financiers were rumored to pocket most of the revenue. There were stories of *fermiers* furnishing their *salons* for 400,000 *livres;* of others whose horses ate from silver mangers. The Farm's agents were notoriously insolent; the Farm employed over twenty thousand men to suppress the huge black market, and in one year alone they arrested more than one thousand smugglers, most of them sent to the galleys.

By law, these agents had the right to enter any house, and tales of injustice and abuse of power were rife. The most famous would occur eight years after Lavoisier joined the company; sadly, it was fairly typical of justice under the Farm. In 1776, a merchant named Monnerat was suspected of smuggling, and one of the agents obtained a *lettre de cachet,* or special royal order by which individuals could be arrested and imprisoned on suspicion alone. Monnerat was thrown in a dank cell in Bicêtre prison for three months with a 50-pound chain around his neck, then transferred to a slightly better cell, where he spent another seventeen months. When his friends intervened, it was discovered that someone had made a bureaucratic error: the Farm had actually wanted a man called La Feuillade. Meanwhile, Monnerat had contracted scurvy. He claimed damages but the Farm refused to pay; when he appealed to the more sympathetic *Cour des Aides,* the Farm used its influence and had the case transferred to the Royal Council, where the unlucky Monnerat's claims were ignored. Of the Farm and its abuses, Adam Smith would say, "Those who consider the blood of the people as nothing in comparison with the revenues of a prince may perhaps approve of this method of levying taxes."

Lavoisier would grow rich as a Farmer, becoming in time one of the richest men in the nation; he would try to reform the corruptions and even ease some of the taxes, such as his lifting the toll against Jews who entered Metz, an action for which the Jewish community later gave him a testament of gratitude. Nevertheless, later biographers proved as uncomfortable with his membership as they were with Priestley's blindness in defense of phlogiston. Both seemed out of character, and hard to reconcile. Some of Lavoisier's defenders would claim that his motive was to secure a large income to serve science, yet his fortune was already more than enough to let him spend his days in the lab uninterrupted. Scientific equipment was as expensive then as now, but the Academy provided grants; Lavoisier's income without the Farm would still have been greater than most of his colleagues'. Finance and science would seem to work in opposition: if nothing else, both were full-time jobs.

If anything, Lavoisier's decision to join the Farm made and broke him; his life became "one long rush from the counting house to the laboratory," serving science and Mammon. It brought him opportunities and influence such as he had never imagined, and a host of enemies such as he had never dreamed of. He came from a family of social climbers, and for many in the bourgeoisie the social ladder became the purpose of life itself. He came from a lineage of means and ambition, where ambition was a kind of discipline and the purpose of money was to make more.

Yet before the countinghouse swept him up, he would have time for one more experiment, his first since joining the Academy. As always, he was neither conservative in testing new methods nor modest in his ambitions. This time, he took on the ancients themselves.

▪ ▪ ▪

For the twenty centuries after Thales of Miletus worshipped the Nile and attributed the origin of all things to water, science accepted as fact that earth came from fluid. Leave water sitting long enough and, through evaporation, it turned into dirt and stone. Once again, as with phlogiston, the belief was a triumph of the senses. Alchemists took flasks of water and heated them over a flame until all the water boiled out; inside they found a dull, earthy substance that must have come from the liquid. More recently, Robert Boyle had observed that the same quantity of water, subjected to the same number of repeated distillations, still left a solid residue. Even more convincingly, van Helmont planted a 5-pound willow tree in a container with 200 pounds of earth, then covered the tree with a perforated lid and nourished it for five years on nothing but rainwater. When the experiment ended, the willow weighed 169 pounds 3 ounces. Since the weight of the soil had decreased by just 2 ounces, van Helmont believed the additional weight could only come from the transformation of water to a solid. Similar experiments with squash, pumpkins, mint, and other plants seemed to yield the same results.

Lavoisier thought they were wrong, but at this point it appears he proceeded on little more than a hunch. But hunches should never be discredited. A 1931 study by W. Platt and R. A. Baker has suggested that the unexplained hunch is more important in research than generally admitted: in a survey of 232 chemists, 33 percent said that assistance from some revelation or hunch in solving a problem was not unusual; another 50 percent said they had such hunches occasionally; only 17 percent said never. Walter Bradford Cannon, famous for his research on the "flight-or-fight" response in humans, says that "in typical cases, a hunch appears after long study and springs into consciousness at a time when the investigator is not working on his problem. It arises from a wide knowledge of facts, but it is essentially a leap of the imagination, for it reaches into the range of possibilities." Cannon's own hunches came during the period of wakefulness before sleep; his mind leapt over the day's events, when suddenly the light snapped on.

Lavoisier's hunch about the ancient doctrine of transmutation seems to have arisen after his analysis of 128 water samples from across France with Guettard. He'd worked so often with distilled water that he knew the fluid did not behave according to accepted wisdom. Yet to act upon his hunch, he had to break with tradition in a couple of difficult ways.

The first break was one he'd already contemplated while compiling the mineralogical atlas—he started with a hypothesis, rather than be led in a million different directions by disparate phenomena. Today, this is the accepted method of science: good or bad, you need a hypothesis to organize your thoughts, and from this you design an experiment based on your anticipation of results. You run your experiment, and your idea is confirmed or shot down. If confirmed and replicated, the idea is on the way to becoming accepted theory; if not, try again. Of course, it's not always that cut and dried: Stephen Jay Gould observed that "theory and observation intersect in subtle and mutually supporting ways." In his earlier analysis, Lavoisier often switched between objectivity and subjectivity, using preliminary observations to build his model, then returning to the field or lab for a more extensive testing of his ideas.

Thus, it is probable that he proceeded on instinct at this stage. The second break was with a tradition that was even more firmly entrenched—the prejudice in chemistry against the empirical method. The idea that Nature was balanced, known today as "the conservation of matter," was an old one, as shown by van Helmont's careful weighing of his willow. The debate, however, lay in the discipline's acceptance of the weight and measure. As Gabriel Venel wrote of chemists in the third volume of Diderot's *Encyclopédie*, instruments and "artificial measurements" had no place in their work: while the physicist investigated the gross properties of bodies by calculation and measurement, the chemist sought the "elusive inner properties" of matter, accessible only by methods that were indirect and intuitive.

Lavoisier didn't buy this line of reasoning. Borrowing the French Mint's most sensitive balance, he weighed a glass vessel called a pelican—a flask shaped so that boiling liquid would drip from the head back down through tubes into the round body. He heated 3 pounds of rainwater, distilled eight times, in the pelican to a temperature just under boiling; the experiment started on October 24, 1768, and continued for 101 days. For nearly a month, no change occurred. Then, on November 20, Lavoisier noticed some very fine particles floating in the fluid. By then, he'd nearly given up all hope of success: when he spotted these specks, he thought his eyes were playing tricks, and used a hand lens. They appeared to be thin, earthy flakes, and for three weeks they slowly increased in size. Their size and number grew until December 15, after which no more flakes appeared. Soon, they began to settle to the bottom of the pelican, and by February 1769 none of the flakes was suspended in the fluid. The experiment seemed to be at an end.

On February 1, Antoine put out the lamp and let the pelican cool. The binding was removed from around the stopper: he weighed the pelican before opening it, and noted that there was no appreciable change in weight from when the experiment began. Next, he took the stopper out; there was a whistle as air entered the flask, satisfying him that the seal had been complete so that nothing entered from outside. Earth had been produced: the question was whether it came from the water or the flask. He poured the water and earth into a glass flask, and found that the weight of the water had remained constant during the long boil. Now he weighed the pelican, and found that it had lost 17.4 grains. The total weight of the earth in the pelican and water came to 20.4 grains, 3 more grains than the weight loss of the pelican, but he attributed this to matter that had dissolved from the pelican walls. He concluded that the "earth" had been stripped from the glass vessel during the long distillation and not transmuted from the water itself.

Not everyone believed his conclusions, but Antoine had proved something to himself. For the first time, he'd used a scientific method and defined a chemical composition entirely on the balance beam. He'd relied on the doctrine of conservation: nothing new was created, nothing was formed. It was precision, not tradition, that determined truth. In the balance lay reality.

But this led him to a greater question. If earth, fire, air, and water did not transmute one into the other, what was the basis of chemical change? Was there something more basic than the old elements? Could the building blocks of Nature be shuffled in ways never dreamed?

He had to know.

CHAPTER 5

The Goodness of Air

THE PATH FROM SPARKLING WATER TO OXYGEN WAS NOT STRAIGHT, but tended to curve and wind at a leisurely pace through the English countryside. Priestley's thoughts were winding also, pausing at purlieus of poetic similarity. Liberty was for Joseph what the sea had been for the Greeks: a drifting aesthetic that joined God and man. The sea had been the basis of Greek civilization, a *pabulum* for food, trade, and colonization, its hundreds of islands afloat in a medium as transitory and necessary as air. Empedocles preached that Love and Strife were the counterbalances of the universe: a true balance was sought, but never achieved. One prevailed, then the other, changing as quickly as light and color across the surface of the waves.

These too were times of change, when tyranny and liberty struggled like passing shades. The year 1768 was seven years away from the War of American Independence, and everyone—except for George III—felt the seismic rumbles. The efforts by Dissenters to repeal the Test Acts and by colonists to ease crown control sprang from the same source: a need for breathing room. "So long as we continue Dissenters," Joseph preached, "it is hardly possible that we should be other than friends to civil liberty."

Resistance to executive power became reflexive on the British street, and the "Wilkes and Liberty" riots of London occurred as the last links between the goodness of air and of liberty were being forged in Priestley's mind. In 1764, the parliamentarian John Wilkes was found guilty of sedition after the forty-fifth issue of his broadsheet, *The North Briton*, described the King as an idiot controlled by vipers: if such men stayed in power, he warned, all liberal institutions in the nation would be destroyed. He was jailed, then released on a technicality, then hounded again on related charges. When he was finally thrown into the Tower in 1768, riots burst forth that were the worst London had seen. Rioters pulled nobles from their sedans and chalked "45" on the soles of their shoes; the King himself was pelted with rotten fruit whenever he appeared. From that day forward, George III took cries for liberty as a personal affront and acted accordingly. When Wilkes was finally released in late 1769, his image was that of St. George rescuing English freedoms, and whenever rights seemed threatened, crowds would cry: "Wilkes and Liberty!"

It was during this chaos that Priestley released his first political salvo: the *Essay on the First Principles of Government*, published in 1768, argued for a new political system maximizing civil liberty. True liberty existed only when each member of society had the freedom to pursue his nation's greatest privileges, including its highest offices; since "all people live in society for their mutual advantage," it followed that the happiness of the majority "is the great standard by which every thing . . . must finally be determined." Yet why had this first principle been so commonly ignored? Priestley reminded readers that leaders were servants of the people and thus accountable for their actions; if that trust was violated, "principles of revolutionary behavior" existed in the world. Such talk was seditious—more dangerous in its reasoned approach than anything said by Wilkes—but Priestley tempered his argument by adding that he did not think revolt in England was justified since the Glorious Revolution of 1688–89 already laid the foundation for peaceful change. Nevertheless, the present government, filled with "tools of a court party" and "narrow minded bigots," was not fit to control matters of private conscience, and he called again for the reduction of articles in the Test Acts from thirty-nine to one. All that was needed for full participation in English public life was profession as a Christian, no matter what the denomination or creed.

A union was forming in Joseph's mind between political and scientific revolt: as of yet, its details were still indistinct to him, like the blurred and foggy silhouettes he'd seen on the heath as a child. But once the outline took form, it would inform much of his thinking. He was too prolix to boil his ideas down to a syllogism, but in this case his evolving thoughts took that form. Knowledge of the world, discovered and unleashed by science, defined truth and dispelled superstition; the old edifices of power, the Church and Crown, kept men in thrall through the suppression of knowledge and superstitious fear. Priestley was taking to heart Francis Bacon's dictum that knowledge was power. To him, knowledge meant freedom. Science spelled liberty.

He was not alone in his views. Those writers and thinkers who would coalesce around Priestley to form the English Radical Movement were finding their thoughts drawn in the same direction. A sense of malfunction lay everywhere, caused by outdated systems and faulty worldviews. When the malfunction reaches such a critical point, says Thomas Kuhn, society divides into competing camps, "one seeking to defend the old institutional constellation, the others seeking to institute some new one." At such a polarized point, diplomacy is useless. Mass persuasion and the use of force are the only means of change.

With the publication of the *First Principles*, the attacks against Priestley began. What he'd experienced from Timothy and Aunt Sarah were mere walk-ups to what he encountered now. Early in 1769, the Archdeacon of Winchester told believers that if the Church relaxed its Articles of Faith, the time would come when every sect, no matter how seditious, would call for recognition. In the fourth volume of his *Commentaries on the Laws of England*, William Blackstone wrote that change in one part of the law upset the whole structure. Priestley didn't back off, but released his *Institutes of Natural and Revealed Religion*, which made him the main spokesman for Unitarianism. He attacked the "idolatrous" rituals of Christianity, urged the use of reason in belief, and told readers to "have nothing to do with a parliamentary religion, or a parliamentary God." Unitarianism was the most radical of the Dissenting sects, professing belief in one God, not the Trinity, and preaching universal toleration—

even of the Catholics, which meant "the French" in the English mind. In 1769, he founded the *Theological Repository*, the first British journal for free religious enquiry, and three years later helped draft a petition to relax the Thirty-nine Articles, a proposal that died in the House of Commons. One of the drafters, the liberal Anglican pastor Theophilus Lindsey, was appointed minister soon afterward of the first formal Unitarian church in England, on Essex Street off the Strand. With that, the religious wars had been joined.

At every point in the *Institutes*, Priestley courted controversy, and thirty years of vituperation started with its release. In the opening line he maintained that Paul's arguments in his epistles to the early Christians were faulty, and that the early Church lied by deifying Jesus. Many Christian "truths" were incompatible with other kinds of truth, especially those discovered by science. These early corruptions must be eradicated or the Church would flounder.

Critics saw the book as a threat to the Church and an attempt to lure away the young, while friends recognized that these were fighting words. Priestley had told the young to examine the faith of their fathers, to look closely at links between the Church and State, and to examine the Constitution and rights of even the King. This enraged not only the Anglicans but many Dissenters who lived in a "Don't ask, don't tell" state of tolerance and resented too close a scrutiny. Nor did they welcome anything likely to rouse the lethargic Church and Crown, since in this inertia they'd found a measure of security. There were still penal laws on the books that made it possible for the crown to punish anyone not subscribing to the Thirty-nine Articles, and any number of suppressions could result if Priestley awoke the sleeping giant.

Joseph was shocked by the rancor rising against him. Scientific friends ridiculed his faith; Christians damned his science. The Methodists in Leeds composed a hymn begging God to

The Unitarian fiend expel
And chase his doctrine back to Hell.

The first of his recurring ailments began during this onslaught, painful bouts of kidney stones and rheumatism. In May 1770, he wrote to a friend that "I have frequently written until I could hardly hold the pen; for writing by hand is irksome, and indeed painful." Three months later, in July, he confided that his theological writing caused offense even among his friends: "By one means or another I believe I have more enemies among Dissenters than in the Church [of England]. I shall soon be obliged to court the Papists and Quakers in order to have any friends at all, except a few philosophical people." Science, in addition to its other delights, was a relief from controversy. He was tired of others' hatred; that same year he told a friend that he had "quite done with controversy." He'd been away too long from science. From henceforth, his search for truth would be conducted in the lab.

▪ ▪ ▪

Once Priestley returned to his experiments, discoveries came quickly; when one considers everything else in his life, they unfolded rapidly indeed. During his six years in Leeds, he published twenty-eight non-scientific works in first edition, another ten in second edition, three each in third and fourth edition, plus four new books of science, two revisions, and an addendum to one. People were astonished by his output, but criticized him for popularizing science, criticism that didn't make the least impression on him.

Nevertheless, when Priestley returned to his lab, he was initially drawn in three directions. He could continue his electrical studies; write a history of light and optics as a sequel to his electrical treatise; or give in to temptation and revisit his brief fling with unknown airs. Pneumatic chemistry had a real allure. It was pursued by people who did not think of themselves as chemists. *Real* chemistry, as practiced in France and Germany, focused on salts and acids and was steeped in theory—and Joseph never saw himself as a theories man. Even later, when he embraced phlogiston, he always seemed happiest with research alone. "Though I have made discoveries in some branches of chemistry, I never gave much attention to the common practice of it, and knew but little of the common processes," he contended, and was perfectly at ease with Newton's deductive method, rolling blithely from point to point as inspiration moved him. "If we could content ourselves with the bare knowledge of new facts, and suspend our judgment with respect to their causes, till, by their analogy, we were led to the discovery of more facts of a similar nature, we should be in a much surer way to the attainment of real knowledge," he said.

Pneumatic chemistry was also British, which meant his friends were part of the same grand game. And fortunately, it was cheap. In 1770–71, Priestley was overwhelmed by monetary woes: he was now supporting three children and a wife on £100 per annum, aiding Tim financially, and had spent £100 on texts for his proposed history of optics and vision. The cost of platinum used for crucibles and minerals for research could be dear: "Phosphorus is too expensive for me to have much to do with," he noted. But the equipment used for pneumatic chemistry was "exceedingly simple and cheap," considerations of thrift that never plagued Lavoisier.

Most of his equipment came from Mary's kitchen, the tools of man's advancement serving double duty with more daily household chores. Just as the furnace was the center of an alchemist's workshop, the "pneumatic trough" was the centerpiece of Priestley's lab. This was simply a large basin containing some liquid over which gases could be collected, and he began his research with an oblong earthen trough used by Mary for washing linen. His glassware was from her kitchen, and drawings show a 2–3 ounce beer glass with a mouse inside. He bought glass vials and jars with ground-glass stoppers since corks were corroded by acid vapors, and used leather and glass tubing to connect his flasks. For a sealant, he used cement made of equal parts turpentine and beeswax, and for reactions requiring high temperatures, he used a gun barrel rather than glass or earthenware containers since the latter tended to crack. This worked well, but he worried that the heated metal might affect his experiments in ways he didn't yet understand.

Part of Priestley's strength as a scientist lay in his inventiveness in the lab. One early innovation transformed the way that researchers investigated the air. Until Priestley started, gases were collected in balloonlike bladders, a time-consuming and frustrating method: since the bladders weren't transparent, a gas could not be seen. Gases were collected over water, and scientists did not know that the water absorbed great quantities of these new "airs." Exactly when the revelation hit him is uncertain, but he soon invented a method of collection still in use today. He substituted liquid mercury for water after discovering that mercury did not absorb the gas as readily; he filled a glass bottle with mercury, then inverted it over a larger container of mercury so that the mouth of the smaller bottle was below the level of the quicksilver in the larger tub. A tube was inserted through a cork and sealed to the source of the gas; the other end of the tube was placed beneath the mouth of the small mercury-filled bottle. When he heated or agitated the substance being tested, a gas was produced: this flowed through the tube and bubbled to the top of the small bottle, displacing an equal volume of mercury inside. The gas was now trapped in a transparent glass container, through which Priestley could measure and note its properties.

Much has been said about Priestley's mix of science and religion, but one secret to his success was his decision to investigate something less expensive. "Good scientists study the most important problems they think they can solve," Peter Medawar has said; a corollary to this is that they work in a field they can *afford*. Priestley was gifted with fine powers of observation and a certain manual inventiveness, but pneumatic chemistry did not require advanced theory or high-priced equipment. Plus, the field was ripe for exploitation. Only three distinct gases had been identified by then: the "fixed air" of van Helmont and Joseph Black; the "inflammable air," or hydrogen, of the wealthy and reclusive Henry Cavendish; and common air.

In the wings, the Scottish chemist Daniel Rutherford was just about to discover nitrogen, in 1773. His experiment was an exercise in exhaustion: he kept a mouse in a closed container of air until it died, then burned a candle in what was left until the candle went out. He then burned phosphorus in *that* until even the phosphorus would not burn. This was passed through a solution that measured trace amounts of "fixed air," or carbon dioxide. What remained was nitrogen, but Rutherford called it "phlogisticated air"—air saturated with phlogiston.

That left the thing in the air that kept mice and men alive: the distant "oxygen" that was dimly sensed but as yet didn't even have a common name. No one knew how to isolate it; no one knew its properties. The slavish fealty to phlogiston masked its importance and worked against its discovery. Instinctively, many chemists knew the race for the ultimate secret of life was at the starting block. "The game's afoot," as another famous Englishman would one day say.

▪ ▪ ▪

When Priestley returned to research, he did so by revisiting first obsessions: the "goodness" of air. This was a moral point: Air "injured" by human and animal respiration must somehow be restored. The secret had theological import since it showed that God would not allow man to be suffocated by continual exhalation. But what was the missing piece? What renewed the bad air?

He would not see success until the summer of 1771. Everyone knew that animals could not be kept in sealed spaces without changing the air at intervals; it was assumed that the same was true for plants, but no one had tried this experiment before and reported what they'd seen. On August 17, 1771, on a whim, Priestley placed a sprig of garden mint into a container whose air had been "exhausted" by a candle, then went about his business, leaving the mint inside for ten days. Although he may not have been aware of the fact, he was working well within a philosophical frame. "Nature" was a controlling idea in Western thought: it was the alternative to all that man created on earth. If man was the great destroyer, Nature healed. All politics, art, religion, law, and Utopian dreams must be based upon Nature for man to have any chance to survive. But how did you build such a structure if you didn't know the basic design?

When Priestley returned to his mint on August 27, he placed a burning candle in the glass and *it burned perfectly well.* He didn't believe it; he *must* have done something wrong. He repeated the experiment ten times in what remained of summer: some sprigs died, but most thrived "in a most surprizing manner." The discovery was one of Priestley's most historically important, for even though he did not recognize the importance of light in the photosynthetic process, his observation was the starting point for understanding the effects of plants on the atmosphere. Still, Priestley knew he was on to something important, and wrote: "Plants, instead of affecting the air in the same manner with animal respiration, reverse the effects of breathing, and tend to keep the atmosphere sweet and wholesome."

In October, he wrote to Ben Franklin and Richard Price about his successes, as well as of some "unsuccessful trials" in restoring spent air. These included letting "putrid" air stand in sunlight for months, letting it stand unshaken in fresh or salt water, and introducing to it the "fumes of burning brimstone." But nothing worked like the sprig of mint, and, in later tests, groundsel and spinach. He wondered if something in nitre might act as a restorative, so he heated saltpeter until fumes appeared. Candles *did* burn in it, but he didn't know why. Unwittingly, he had prepared oxygen, but he did not yet see the significance and passed on to other things.

If Priestley was preoccupied by his tests, others knew that something momentous was taking place. The word, most likely, was spread by Franklin and Price to friends in the Royal Society. Priestley had opened a door into an understanding of biological processes as profound as those opened in the physical realm by Newton, but he was isolated at Leeds and did not comprehend the stir he'd caused. The first hint came at the end of 1771, when Sir Joseph Banks invited Priestley to serve as ship's scientist and astronomer during James Cook's second voyage to the Pacific. The offer included a "handsome provision" for Joseph and his family.

At first Priestley hesitated, thinking his absence might bring hardship to Mary and the children. "No man can have a more domestic turn than myself, or be more happy in his family connection," he responded, still wavering. "I shall leave an affectionate wife, and three children, at an age which is, of all others, the most engaging." Why was he being chosen, he wondered, since most of the duties would consist of comparing new techniques for measuring longitude? Still, the idea of adding to the storehouse of knowledge was appealing, and at the end of his *History of Electricity* he'd referred to the excitement of such expeditions without ever thinking he might be involved. He wrote again to accept, full of enthusiasm. It was time to see the world.

And then the rug was pulled out from under him. His letter of acceptance was barely in the post when Banks wrote again. "Some clergymen in the Board of Longitude, who had the direction of this business," objected to Priestley's appointment because of his Dissenting principles and past publications, Banks explained. Biographers have said that Priestley took the rejection in good humor, but he did feel bitter, though his rancor was not directed at Banks. In a letter dated December 10, 1771, he told the botanist:

> I thought this had been a business of philosophy, and not of divinity. If, however, this be the case, I shall hold the Board of Longitude in extreme contempt, and make no scruple of speaking of them accordingly. . . . I am surprised that the persons who have the chief influence in this expedition, having . . . minds so despicably illiberal, should give any countenance to so noble an undertaking. I am truly sorry that a person of your disposition should be subject to a choice restricted by such narrow considerations.

The rejection was a sign of even greater forces aligning against him. Priestley had the bad luck to be alive when a Hanoverian was King, and George III was no fan of Dissenters. "I own myself . . . a great enemy to any innovations, for, in this mixed government it is highly necessary to avoid novelties," he once complained. Each time the Dissenters tried to repeal the Test Acts, George made sure the Houses of Parliament stood firm. "I think the test was never so necessary as at present for obliging [various Dissenters] to prove themselves Christians," he said.

In 1772, George III had been on the throne for twelve years; although he was more popular than his father or grandfather, they all had common traits. Their minds were dull, their tastes drab, and their family feuds (especially between father and son) vindictive. George was jealous of royal power and anxious, during this age of creeping democracy, to keep the court at the center of all things English. Like Louis XIV before him, his philosophy of state was "One law, one King, one God." The existence of Dissenters like Priestley seemed a challenge to the royal will.

George's strength was that he was a traditional Protestant in a Protestant land. This would haunt Priestley; even as he grew eminent, he could not expect friendship, patronage, or protection from the King. By 1772, George had suffered his first mild attack of porphyria: the genetic blood disease had tormented Mary Queen of Scots and her son James I, and would attack George's son, George IV, and his granddaughter, Queen Victoria. In addition to the physical symptoms, the behavioral changes would account for the confusion, delusion, and paranoia of his famous "madness." When the disease returned, George saw enemies everywhere.

Priestley belonged to this category. As a Unitarian, he was an enemy of the state. In a letter to his prime minister, Lord North, King George suggested that another minister should end his friendship with Priestley due to the pernicious influence the scientist might have on the man. George remained Priestley's foe for decades, and as did the King, so did England.

But there were triumphs, too. During four successive meetings at the Royal Society in March 1772, Priestley presented his findings on "fixed air," "inflammable air," "air infested with animal respiration, or putrefaction," "air in which iron filings and brimstone has stood," and other observations. Word got out about these experiments, including the curious episode of the beer, and in early spring 1772 he was invited to dine with the Duke of Northumberland, a powerful patron of science and friend of the party of the King.

The duke was a curious man, interested in all the invisible wonders Priestley had discovered. He also was connected to the Admiralty, and had the ear of Lord Sandwich, head of the Royal Navy. The ships of the line were England's greatest hope for empire, but their greatest threat was neither French guns nor the Spanish navy, but something more insidious: scurvy. The Admiralty was besieged with reports of crews halved by disease, where survivors dragged themselves around the deck with putrid gums and bleeding skin. It was known by then that a sufficient supply of fresh vegetables prevented scurvy while a diet high in the salt meat of the regular rations was a leading cause, yet vegetables rotted quickly on a long voyage.

"What about fixed air?" the duke wondered aloud. A philosopher at the College of Physicians had come up with a novel (if erroneous) theory: since a diet of vegetables resulted in greater flatulence than one of meat and potatoes, maybe all that was lacking to combat scurvy was a healthy dose of fixed air. The learned doctors had confused the methane of solid English burps with carbon dioxide, but scurvy was so horrendous that they were willing to try anything. The duke produced a bottle of distilled water made by Naval Surgeon Charles

Irving as suitable for long voyages. Priestley took a sip—it was wholesome but flat, and he doubted seamen would like it. Then he remembered his experiments with sparkling water and told the duke it might be easy to make a pleasant drink that could double as a weapon in the navy's war against disease.

The reasoning, of course, was flawed: it was lack of vitamin C that caused the scurvy, not carbon dioxide or methane. But Priestley's "impregnated water" caused excitement anyway. The next morning, he and Mary gathered flasks and troughs from her kitchen. Perhaps here was a way from their financial bind, he chattered, while Mary assured him she'd always had faith in his schemes. It took thirty minutes to impregnate a bottle of water. He made several, and they made their way up the line, from Northumberland to his friend Sir George Savile, then from Savile to the notoriously rakish but powerful Lord Sandwich.

Priestley's sparkling water was loaded aboard a number of vessels, including Captain Cook's *Resolution* and *Adventure*—the very ships he'd hoped to join. Both lay in anchor at Longreach, taking in supplies. In addition to exploration and testing the newest schemes for longitude, the voyage's added purpose was to test special foods that might fight scurvy. These included "marmalade of carrots," "rob of oranges and lemons," "inspissated juice of malt," and now Priestley's "windy water," as it came to be called. Soon afterward, Priestley published his pamphlet, *Directions for Impregnating Water with Fixed Air*: it was written as a how-to manual, and the home brewing of "windy water" became as popular as that of bathtub gin in later years. A French translation was published soon afterwards, spreading his name overseas.

By July 1772, Priestley was able to generate "new airs" with every known metal but zinc, and for the next few years he would go on to discover more new gases during this short time than any other man before. The first of these discoveries apparently occurred on June 4, 1772, with his discovery of nitric oxide: he called it "nitrous air," produced by heating brass with spirit of nitre. The same gas was given off by iron, copper, silver, tin, mercury, bismuth, and nickel; more remarkably, one measure of the colorless nitric oxide mixed with two measures of common air suddenly generated heat. In addition, turbid red or deep orange fumes spread through the vessel. Finally, there was considerable shrinkage in the volume of the gas: if left to sit a day or more, the remaining quantity of common air seemed devoured by the nitric oxide.

This discovery, above all others, seemed to suggest a practical use, largely because the change was so dramatic and visible. Joseph thought he'd found a way to measure "the fitness of air for respiration": this fitness must somehow be proportional to the rate of diminishment in the vessel. He'd always felt guilty killing mice in his tests, and the new procedure made him sit back and smile. He showed Mary, the children, his friends. The man who'd nearly died from bad air had discovered a non-lethal method of testing the "goodness of air."

It worked this way: When he'd first prepared nitric oxide, some properties were immediately evident. The new gas did not dissolve in water. It reacted easily with common air, giving off those beautiful red fumes that *did* dissolve in water—and in the process formed Spirit of nitre. Why not put these reactions to use? He'd noticed that if he mixed 1 measure of nitrous air with 2 of common air, the red fumes burst forth and only 1.8 measures of gas remained. He analyzed the remaining gas—it was comprised of all the nitrous air and 40 percent of the common. But when he used common air that had been "spoiled" by a burning candle or the labored breathing of a mouse, a volume greater than 1.8 measures, and sometimes greater than 2, remained in the vial. The worse the air's quality, the less would be absorbed. Thoroughly "bad" air should not diminish at all, he thought: as much as 3 volumes might remain.

He realized to his delight that he'd invented the first quantitative test of air quality ever devised by man. A shrinkage to 1.8 volumes represented the "common air" we all breathed, while a measure greater than 1.8 indicated a measure of "spoilage," his term for air pollution. The red color varied, too. He could measure air quality throughout the country, and that's exactly what he did, rushing around England with his little vials like a one-man EPA. He collected air from the country, from the city, from Birmingham, Manchester, London, and Leeds. Air was not "good" in crowded rooms, poor slums near manufactories, and great cities; but in the country, the air was "sweet." The very best air in England was found in Wiltshire, while the air in a crowded room was "much contaminated" with a "goodness" index of 1:31. By comparison, the index of sweet country air was 1:25, and of air which smothered a candle 1:43.

The red fumes were fascinating, too. Red was the color of the philosopher's stone. The alchemist Berigard of Pisa claimed the stone was the color of a wild poppy; Paracelsus compared it to a ruby, as transparent and brittle as glass. Others called it a red powder, or a shimmering fluid as red as blood. As late as 1777, Priestley told Franklin that he "did not quite despair of the philosopher's stone." Franklin advised that if he found it, "take care to lose it again."

It was still easy to be tempted by the claims of alchemy, and for a man like Joseph, the ancient goal of joining every bit of chemical knowledge into a deeper understanding of the cosmos was no less presumptuous than that of Newton and the natural philosophers, who hoped to tie "another knot in the 'net of the World.'" There were plenty of famous "alchymists" still drifting through the capitals of Europe, evoking wonder and scandal. The blacksmith-turned-"famous alchymist" Jean Delisle had taken France by storm in 1705–11 with his transmutations of lead to gold "by means of his wonderful oil and powder." Delisle claimed his miracles were made possible because he had, indeed, discovered the stone; today it is presumed he did so with gold powder concealed in a double-bottomed crucible or hollow wand, but the exact mechanics of his trickery were never discovered. Even Louis XIV and the Church were intrigued. When Delisle was arrested and died in the Bastille in 1711, it was more due to his continual reluctance to answer the King's summons to demonstrate his art than to any exposure of fraud. Currently, the Count de Cagliostro and his beautiful wife Lorenza were traveling through the capitals of Europe "transmuting metals, telling fortunes, reviving spirits, and selling the *elixir vitae.*"

If there was a false aesthetic driving the lives of such adventurers, there was a true one implicit in Joseph's experiments. With each new discovery, he experienced anew an overpowering feeling of beauty or awe. His writings are filled with sudden exclamations of joy: "the sweetness of air," "a white cloud of delicate beauty." He described experiments as "charming," "delightful," "striking"; his results were "amazing," "startling," leaving him "utterly at a loss to account" for the things he observed. He bubbled his red vapor through a series of bottles with a little water in each; the water in the first bottle sparkled, emitted gas, and turned blue. It continued to give off gas until the water changed from green, to black, then yellow. By adjusting the quantities of water and flow of gas, he could isolate each change in a separate bottle. There was no practical reason for this: it was merely a parlor trick, but he loved its beauty.

Such wonder seemed necessary to maintain the reactive level of his mind. Another English scientist, a century later, would show a similar worship and inspiration flecked with awe. In his *Beagle Diary*, the young Charles Darwin, then twenty-three, would note that "The world is a chaos of delight out of which a world of further and more quiet pleasure will arise." As he watches the flight of a "gaudy" butterfly, his eye is arrested by some strange fruit or tree, and in contemplating this the eye retracts to take in all the landscape: fecundity and sensory overload spring from the same source, he said. A quarter century later, when writing *The Origin of Species*, he visited the theme again in his closing image:

> It is interesting to contemplate an entangled bank, clothed with many plants of many kinds, with birds singing on the bushes, with various insects flitting about, and with worms crawling through the damp earth, and to reflect that these elaborately constructed forms, so different from each other, and dependent on each other in so complex a manner, have all been produced by [simple and straightforward] laws acting around us.

Throughout his writing, there was a dualism like Priestley's: complexity and simplicity, wild and tame. Years later, in his *Autobiography*, Darwin lamented the loss of this aesthetic sense. Science was no longer fun after his brain had atrophied into a mere theorizing machine.

In quick order, Priestley discovered a host of today's most common and useful gases, identifying them by their properties and method of preparation, leaving only the modern names to later chemists.

He discovered hydrogen chlorine gas in 1772, ammonia gas in 1773–74, sulfur dioxide on November 26, 1774, and silicon fluoride in late 1774. He heated common salt with sulfuric acid, then dissolved the gas with water—the water greedily sucked up hundreds of volumes to create the hydrochloric acid used today for cleaning metals and making glue and gelatine. He prepared ammonia, driving Mary and the children out of the house with the fumes. He brought together the two disagreeable gases, hydrogen chloride and ammonia, thinking they might cancel each other out. Instead, he watched in awe as the gases vanished and formed a "beautiful white cloud" that filled the glass vessel, then settled out as a white powder. He called it "sal ammonia," known today as ammonium chloride, used as an electrolyte in the dry battery.

If ever a man embodied an entire branch of a science, Priestley was the face of the "doctrine of airs." In 1772, he issued his paper on carbonated water; in 1773, he issued one on his experiments in air. And on November 30, 1773, at the Royal Society's one hundredth anniversary meeting, the honor was made official: Priestley was awarded the Copley Medal for his work, the Nobel Prize of his day. Single-handedly, he'd shifted the focus of chemistry from mineral to gas, yet the grandest prize of all, the secret of the flame and the breath, was just beyond his reach.

After centuries of false starts, the race for oxygen was on.

▪ ▪ ▪

Plato, who saw much, once observed that inspiration and creativity can be as infectious as a disease. The muse "first makes man inspired, and then through these inspired ones others share in the enthusiasm, and a chain is formed." The excellence of the epic poets was not an inborn quality but a form of possession that blazed through a clutch of like-minded men. Possession was creativity's vital flame, the fuel that drove men forward.

The same possession gripped the great chemists of Europe in the first half of the 1770s, and Priestley was the cause. In addition to his tract on carbonating water and the groundbreaking *Experiments and Observations on different Kinds of Airs*, there were other papers, talks, and lectures. His total output during these few years rekindled a fire that had lain dormant. He demonstrated the possibilities of labwork, but more important, suggested what was left undone.

Oxygen—the unnamed thing in the air that controlled life and burning—was the grail. Yet in an age before instant communication, the urgency to find it was more instinctively sensed than consciously realized. Near misses and hits, unacknowledged and unrealized, were arising in all of Europe's scientific centers, goaded by this strange phenomenon of new gases that rose from heated metal and the unexplained fact that the residue *gained* weight instead of losing it, as it should if this were indeed a smoothly running phlogistonic world. Such near simultaneity in effort was not unusual: Copernicus complained about problems in Ptolemy's cosmology even as Arabic scholars sweated through the same problems; the system of calculus was developed independently by Newton and Leibniz; Charles Darwin was not the only biologist puzzled by evolution. There would be eight simultaneous discoveries of the cellular basis of organic life, at least three independent demonstrations of artificial immunity following inoculation of cultures of the anthrax bacillus, five demonstrations of the value of cowpox vaccinations, five discoveries of heart block, three discoveries of the vaso-constrictive nerves, and five independent introductions of ether as a surgical anesthesia. While the old order fell apart, the new one struggled to find form. This was the "crisis phase" of "paradigm change" that Thomas Kuhn would later argue so persuasively; the local disturbances predating general revolution; the small incendiary particles floating in advance of wildfire. A wind was blowing and the *savants* felt it. The emergence of a world-altering discovery was right on the horizon, driven by the confusions of combustion, respiration, oxidation, and formation of acids.

Before Priestley and Lavoisier came near the truth, two other chemists edged perilously close to being named the first discoverers. Even today, both have their fans. The first to approach the answer—in fact, to discover it independently, but because of his delay in publishing to be cast from the running—was Carl Wilhelm Scheele, an obscure apothecary who lived a solitary life in the small town of Köping on a lake in Sweden. Scheele was hobbled by debt, suffered depression, and died at age forty-three in 1786 most probably of poisoning from three toxic gases (hydrogen fluoride, hydrogen sulfide, and hydrogen cyanide) that he regularly tested and tasted in his lab. But like Priestley, he had a genius for experimentation, and even exceeded his more famous contemporary in the number of discoveries.

From Scheele's notes, it is now believed that he discovered chlorine and manganese; hydrofluoric, nitrosulfonic, molybdic, tungstic, and arsenic acids among the inorganic acids; and lactic, gallic, pyrogallic, oxalic, citric, tartaric, malic, mucic, and uric among the organic acids. He isolated glycerin and milk-sugar; determined the nature of borax, Prussian blue, and microcosmic salt; and prepared hydrocyanic acid. He invented new processes for preparing ether, phosphorus, algaroth, calomel, and magnesia alba; discovered ferrous ammonium sulfate, the analytical separation of iron and manganese, and the decomposition of mineral silicates by fusion with alkaline carbonates. But he was painfully slow in publishing his results—his greatest book, *Air and Fire*, which detailed these findings, was not released until 1777. By then it was too late for any claim to be seriously recognized.

Sometime between 1771 and November 16, 1772, Scheele discovered that manganese oxide, when heated white-hot, discharged a gas he called "fire air." He went on to obtain oxygen by heating other sources—mercury oxide, silver carbonate, magnesium nitrate, and potassium nitrate—and each time powdered charcoal glowed brightly and burned rapidly in the presence of the released gas. "I have often taken great pleasure in observing the extraordinary effects produced by heat alone," he wrote in his solitary lab. He believed that common air was a mixture of two fluids, "foul" and "fire" air, later identified as nitrogen and oxygen; but he explained everything in terms of phlogiston theory and did not demonstrate the importance of "fire air" in respiration or combustion, as Priestley and Lavoisier did.

The second researcher to come close was Lavoisier's Parisian contemporary Pierre Bayen, a master apothecary of the French army. In the early to mid-1770s, the "race" to discover this unknown oxygen was centered in Paris; Priestley and Scheele worked on the fringes, driven more by their solitary whims and scientific insights than by any awareness of the most recent developments in the field. But Paris was an entirely different world: there, the concentration of chemists was so great that they nearly tripped over one another. All sensed the vulnerability of phlogiston theory; many suspected that its overthrow lay in the study of gases; each thirsted to be the Newton of the chemical world. The newest discoveries were quickly published and replicated. In February 1774, Bayen entered the race when he released a report describing, among other things, his experiments with mercury oxide. Like other metals, mercury gained weight when heated in oxygen: we know now that it combines with oxygen to form its oxide, but under prevailing phlogiston theory this weight gain could only be explained if phlogiston had negative weight. Bayen also noted the fact that red mercury oxide could be converted directly into metallic mercury without being heated by charcoal. In the past, heating substances in this manner was believed an essential lab technique since charcoal was thought to be almost pure phlogiston. When heating the red oxide, Bayen released a gas: once again, the gas was oxygen, but Bayen mistakenly identified it as Joseph Black's carbon dioxide, or "fixed air."

Red mercury oxide (HgO) would be the key to everything. To alchemists, mercury was basic to research since they thought it was a component of *all* metals and thus must hold the key to transmuting base metals into gold. The Chinese alchemist Ko Hung wrote in the fourth century AD of his wonder when bright red cinnabar turned into silver mercury simply through heating; unknown to him, the sulfur in the cinnabar was oxidized by contact with the air, a reaction that formed sulfur dioxide gas and left metallic mercury behind. Greek and Roman writers like Aristotle, Pliny the Elder, and Vitruvius also knew that metallic mercury could be obtained in this fashion, but ignored and unknown was the oxygen released during the reaction.

Today we know what Priestley and Lavoisier did not: air is primarily a mixture of two gases, oxygen and nitrogen. We must jump ahead to understand the intellectual barriers the two needed to overcome. Combustion and respiration involve chemical reactions between carbon compounds and oxygen: the products are water and CO_2, *except* when heated by charcoal, during which CO_2 alone is formed. When a metal is heated in air, it forms an *oxide* by combining with oxygen, a product that Priestley, Lavoisier, and others called a "calx";

the process of forming that calx was called "calcination." When most oxides of metals are heated with charcoal, oxygen combines with the carbon in the charcoal to form the original metal and CO_2, or "fixed air."

But mercury oxide acts differently than the rest, and this difference makes it important. Then called "red precipitate of mercury" or *mercuris calcinatus per se,* the oxide had the unusual property of being converted directly into the mercury and oxygen when heated without the use of charcoal, a property that baffled chemists but also opened a door to understanding. An alternate method for heating an oxide—one that was just catching on in the 1770s—was to focus sunlight on it with a glass lens. Three reactions important for the history of chemistry proceeded from this technique, and can be illustrated in equations:

Oxidation (then called "calcination")

$2\,Hg$	+	O_2	*heated yields*	$2HgO$
(metallic mercury)		(oxygen gas)		(oxide of mercury with one oxygen molecule thrown off)

Decomposition of oxide

$2\,HgO$	*heated very hot yields*	$2Hg$	+	O_2
(mercuric oxide)		(metallic mercury)		(oxygen gas)

Reduction with charcoal (also called "reduction with phlogiston")

$2HgO$	+	C	*heated yields*	$2Hg$	+	CO_2
(mercuric oxide)		(carbon, as charcoal)		(metallic mercury)		(carbon dioxide)

Thus, one can start with liquid mercury, heat it in air to obtain the red oxide, then heat it again and liquid mercury reappears. Early chemists were stunned: no other substance then known transmuted back and forth so readily. What no one understood was the difference in the last two equations. When mercuric oxide was heated with charcoal (instead of by a focused beam of sunlight), the reaction yielded two gases, a detail that caused no end of confusion in the birth of the new chemistry. The carbon in charcoal was not yet seen as an element that could react with other substances, but as an inviolate inflammable agent: pure phlogiston.

Today, these equations are known as examples of basic *redox reactions*—reactions in which a simultaneous transfer of electrons occurs from one chemical species to another. In turn, these are composed of two processes: *oxidation,* or a loss of electrons, and *reduction,* a gain. The reactions are always coupled since electrons lost in oxidation are picked up in reduction: one substance's loss is another's gain. In this modern image of chemical bonding we see Lavoisier's balancing act—nothing is actually lost in a reaction, merely traded. Reduction and oxidation are commonly termed *half-reactions,* because it takes the two halves to make the whole. Even the name is a marriage: *red*uction + *ox*idation = *redox.* The confusion caused by phlogiston theory is now more evident, since in effect only one-half of a redox reaction was being observed.

Only one other substance had caused as much recent excitement among chemists, and this was the preparation of phosphorus during the previous century. The process was long held secret by German chemists, but in 1692 the French learned the secret: a long and rather disgusting procedure that entailed evaporating the residue of putrefied and distilled human urine through three times its weight in fine white sand. People were intrigued by phosphorus—the white vapors and strange jet of blue light, and even its discovery was strange. In 1670, the German chemist Hennig Brand boiled down fifty buckets of human urine to a wormy, rotting paste, then left it in a cellar for months. Some accounts say he forgot it, though the smell must have been hard to ignore. Eventually he distilled the substance until it glowed in the dark, a property that ensured its fame. Robert Boyle used the same process in England: he got the secret from Johann Daniel Krafft, a German from Dresden, who bought it for 200 *thalers* from Brand. In 1737, a simpler method perfected by Rouelle and fellow French chemist Pierre-Joseph Macquer was presented to the French Academy. Such was the wonder of phosphorus that both were immediately established as the nation's authorities in chemistry.

From the example of phosphorus, European chemists learned a valuable lesson—investigating the right mystery could make one's career. For awhile, phosphorus was believed to be the repository of that "*flamula vitalis* which animates the blood, and is, for aught we know, the animal life it selfe of all things living." Although research would disprove the idea that the spark of life was nourished by phosphorus, it seemed for a moment as if the secret of life were made visible. Now, with this invisible gas shed by mercuric oxide, the hope was reborn.

▪ ▪ ▪

On August 1, 1774, it was Priestley's turn to approach the truth, and at first he was as stumped as his predecessors. In fact, by 1774, the man who'd acted as a muse for fellow chemists was slightly behind the curve. He'd encountered too many delays.

They started in the summer of 1773. His friend Richard Price, concerned about Joseph's continuing financial straits, put a bug in the ear of a powerful political patron: William FitzMaurice Petty, the second Earl of Shelburne. An Irishman, Shelburne had been George III's aide-de-camp when he ascended the throne in 1760; since then he had fallen from the King's favor for his liberal views, but such dislike would not prevent him from becoming prime minister in 1782, at the end of the American Revolution. When Price mentioned to him that Priestley needed a patron, Shelburne offered the post of librarian and tutor to his two sons at the annual salary of £250, a 150 percent rise in Priestley's earnings.

Despite the temptation, Joseph hesitated. Working for Shelburne would bring advantages to his family and lift him from the financial pinch; he'd be able to send his children to any school they wanted. Granted, he'd have to consider Shelburne's interests before launching into controversy, but cultivating the aristocracy could have advantages. The first was obvious—for someone like Priestley who lived in anxiety over the prospects of supporting his family, a stable, steady income was hard to ignore. There was also the larger social good, which Priestley framed to himself as a classic quid pro quo. Natural philosophy sustained the "arts" upon which cultivated life depended, making it fundamental for the refinement enjoyed by the aristocracy. "Without that knowledge . . . rank and fortune would be of little value"; in other words, the "elegant enjoyment of life" depended on ingenious people like him. At the same time, natural philosophy *deserved* patronage because of its utility to all society. The aristocracy and gentry were in a position to support research simply because they had money; they were not caught in the mad economic scramble that seemed to define this age. "Being free from most of the cares particular to individuals, they may embrace the interests of the whole species," Priestley claimed.

To the modern ear this sounds self-serving, but it's not that different from the argument made by any artist, scientist, or charity writing a grant proposal or begging money from the rich today. As Joseph always did when uncertain, he sought the advice of friends. Price was euphoric; Josiah Wedgwood thought working for Shelburne a capital idea. Others said Lord Shelburne might be a difficult employer. Ben Franklin, ever the pragmatic Yank, suggested that Joseph make use of his "Prudential Algebra": Make two columns, one "pro," one "con," and cancel out items on both sides that carried equal weight. The surviving column was the best choice.

In addition, Shelburne showed promise of being an interesting employer. Though descended on his father's side from the Irish earls of Kerry, Shelburne hailed on his mother's side from Sir William Petty, the renowned financier and scientist of the seventeenth century. Perhaps because of this, he was already interested in science and an admirer of Priestley's work when Price confided in him. In addition, their politics were *simpatico.* In 1766, Shelburne had been appointed secretary of state in the cabinet of the Earl of Chatham, but was dismissed two years later when he opposed the King's American policies. Now out of office, he'd supported the Dissenters in their failed 1772 attempt to repeal the Test Acts, a big plus for Priestley. Shelburne was considered so liberal that he was detested by the King and court; George III called him Malegrida, or the "Jesuit of Berkeley Square."

What was there not to like about Shelburne? He was as liberal a patron as Joseph was likely to find in this day and age. According to Jeremy Bentham, he was the only minister at that time who did not fear the people, and the first, according to Benjamin Disraeli, to appreciate the importance of the rising middle class. He was also personally appealing, especially to Mary, and anything that made Mary happy carried a lot of weight in the Priestley household. Shelburne's wife Sophia, daughter of the Earl of Granville, had died in 1771, leaving the lordship with two sons to raise alone. Who could not sympathize with such a personable man?

On a Saturday in August 1772, Shelburne arrived at Priestley's house in Leeds and offered his terms, something *not done* by an aristocrat; after all, commoners came to them. He met every request made by the Priestleys. They'd live in a town house near Shelburne's in Berkeley Square, and have a separate house on his Bowood Estate at Calne in Wiltshire. They'd spend the six winter months in London and the summer in Calne. Priestley would supervise the education of his sons, catalogue the contents of his extensive library, and act as a kind of political researcher, collecting materials under discussion in Parliament. The Priestleys were convinced: Joseph turned in his resignation at Mill-Hill Chapel in December 1772, and preached his farewell sermon on May 16, 1773. He thanked his congregation for their tolerance and friendship, two qualities he'd learned to value the hard way.

Hence the delay in returning to his "airs." It took awhile to settle in their new home, and by July 18, 1773, everything was still in confusion. All of Joseph's books somehow ended up in London. "I cannot be said to be settled till I have got to work at my experiments," he snapped, but if anything, he contributed to the disorder. On their arrival, Mary found in each box a medley of minerals and chemicals. Joseph begged her not to worry if the clothes "were a little injured," for everything else had stood the journey well.

The stay with Shelburne was long, productive, and friendly; their years there were happy ones, and a fourth child was born, named Harry at Shelburne's request. In some ways, Shelburne seemed to value Mary as much, if not more, than Joseph; at the least it can be said that she amused the widower. On the day after their arrival, Mary stood on a high pair of steps papering the walls herself while Priestley was lost amidst his boxes. When Shelburne arrived to welcome them to the estate, Priestley hastily apologized for the state of their house, while Mary laughed and said, "Lord Shelburne is a statesman and knows that people are best employed in doing their duty." Shelburne immediately saw who was the boss of the family.

Another time, he gave her tickets to see the trial on charges of bigamy of the Duchess of Kingston in the House of Lords. When she went, she found there was a scrabble for the limited seats and watched as people's clothes were torn in the mêlée. When Shelburne asked how she liked the spectacle, Mary replied, "Indeed, my Lord, I find the conduct of the upper so exactly like that of the lower classes that I am thankful I was born in middle."

Mary Priestley, like oxygen itself, is something of a mystery in this scientific saga. She is seen but unseen; her importance is felt, but remains unrecognized. There is no documentary evidence that she worked in the lab beside her husband, as there was with Marie Lavoisier, yet Joseph occasionally let slip that he saw her as an intellectual partner. He must have discussed his research with her, using Mary as a sounding board as his thoughts and hunches came together. What we know of her comes to us piecemeal, through chance observations of family, friends, or acquaintances, and unfortunately this seems the rule in eighteenth-century England. The era's stereotyping of women was so prevalent that today it is hard to give a good picture of their lives. There were no female authority figures—no parliamentarians, explorers, lawyers, magistrates, or industrialists. To Dr. Johnson, the idea of a woman preacher was "like a dog walking on his hind legs." Women were so tightly laced into the roles of mother, wife, housekeeper, domestic servant, or maiden aunt, that few escaped being pigeon-holed and, as one commentator noted, the stereotype "created a kind of invisibility: women were to be men's shadows." Anatomy defined destiny; according to Blackstone, "In marriage husband and wife are one person, and that person is the husband," adding that "the very being, or legal existence, of the woman is suspended during marriage."

But Mary was no shrinking violet and she made an impression wherever she went. The sole portrait remaining of her is as an elderly woman, with long face and chin, wearing a cap and with a hand cupped to her ear as a hearing aid. This was such an unusual pose that it suggests both self-deprecating humor and the confidence to consign an unflattering image to posterity. If anything, she aided Joseph in a way that gave him breathing space, and this was probably the greatest gift she could give him. One tradition is that she sent him once to the Leeds town market with a large basket, and he acquitted himself so badly that she never sent him again. She was known to free him of all domestic work save lighting the fire in the morning, which meant he was first up in the cold; she helped him capture mice for experiments, and kept them warm and comfortable in their cages. No doubt she felt sorry when they died, which contributed to Joseph's unease about their fates. She divested him of the time-consuming chore of transcribing his shorthand notes, and readied manuscripts for publication: "I (or rather Mrs. Priestley) am transcribing the third volume of the *Institutes*," he wrote in 1773 to Anna Laetitia Aiken Barbauld, of the Warrington "lassies," now married and one of the most popular poets of her day.

With Shelburne, Priestley had more freedom than he'd ever experienced. His new duties were not time-consuming; he was not the boys' personal tutor, but helped arrange their course of study. He took up gardening for exercise and fresh vegetables. He had time to play three daily games of chess with Mary after lunch, and another three after supper. During the winters in London, his duties included demonstrating new experiments for Shelburne's guests, often from overseas. He helped establish a Unitarian Hall at Essex Street in London, and attended its first service on April 17, 1774, with Franklin and other notables. Shelburne gave him £40 per annum for scientific expenses, and encouraged him in his chemical pursuits.

The other real concern was the coming American Revolution. That year was momentous in many ways. It saw the death of Louis XV in France and the coronation on May 10 of Louis XVI. *The Newgate Calendar, or Malefactor's Bloody Register*, with biographies of criminals, their crimes, and executions, was released to brisk sales. Nothing focused the mind like the possibility of a hanging, Dr. Johnson quipped, but revolution in America and elsewhere had such a grip on the popular imagination. Priestley's tenure with Shelburne covered the first seven years of the American Revolution, and when he arrived in Calne in June 1773, the discontent of the colonies had already led to riots and harsh taxation. "Having taken under our most serious deliberation the state of the whole continent," wrote the members of the First Continental Congress, we "find that the present unhappy situation of our affairs is . . . evidently calculated for enslaving these colones." That year Priestley wrote *An Address to the Dissenters of the Approaching Rupture with America*, a pamphlet that circulated in great numbers. On December 13, four months before shots were fired in Lexington, colonists captured Fort William & Mary in New Hampshire. The news so embittered King George that all hopes of concession ended. War was coming soon.

In such unrest, Priestley shut himself up in his laboratory at Calne and once again attacked the problem of the vital flame. It was a sunny Sunday in the country; he donned his heavy leather apron with deep pockets and prepared for his chemical trials.

A couple of essential ingredients had changed this time. The first was his method of generating heat. When he'd started his experiments in Leeds he'd burned his metals in a charcoal furnace, but with charcoal there was always the risk of introducing fumes or ash into the vessel and thus contaminating the reaction; in addition, glass beakers melted. He'd graduated to the use of a small concave mirror to bring the sun's rays into focus, producing higher temperatures than ever before. Then, in June 1774, he purchased for £6 6*s* from a London glass manufacturer, William Parker, a glass convex lens, larger than his mirror. A "burning glass," he called it, measuring a foot in diameter with a 21-inch focal length, supported by a wooden frame. Burning glasses of this sort were created for the generation of heat and were not very good optically, but what they did, they did exceptionally well. Priestley had even heard tales that a team of French academicians had melted a diamond with a giant burning glass on wheels just a year or two ago. On a clear, sunny day such as this, heat could be produced of such intensity that wood caught fire and fumes leapt suddenly out of various metals. He loved his new toy.

The other change to his procedure was his very first use of mercuric oxide. He'd become fascinated with this red powder, as had others before him: Scheele and Bayen; Robert Boyle, a hundred years earlier; the German alchemist Eck three centuries ago. All noted the release of a gas, and except for Scheele (who theorized the production of "fire air,") all left it at that—an interesting, if unexplained phenomenon. The world, like them, had not been particularly aroused.

So Joseph made his test: he placed the red powder in a bell jar, arranged so that any gas thrown off would pass out and collect in a bottle placed over a trough of mercury. He set the lens outside the bell jar and focused the ray on the powder. A gas leapt out, with a volume three or four times greater than that of the original material. In the chance that the gas might be "fixed air," he admitted water into the collection flask, but the new air was not absorbed.

What if he'd produced Rutherford's "mephitic" air? He looked around the room—a candle burned beside him on the table. A flame would be extinguished in mephitic air.

This candle changed everything, one of those accidents that forever alter our perceptions. Joseph thrust the burning candle into the trap, expecting it to go out. Instead it leapt up, "a remarkably vigorous flame, very much like that enlarged flame with which a candle burns" in nitric oxide, or laughing gas, obtained when iron was left in nitrogen. But no nitrous material had been used in this preparation, and the candle burned with more splendor and heat than he'd ever seen. He was at a loss to explain—what was happening seemed impossible.

Then he confused himself more. He thrust a chip of red-hot wood into the trap. It sparked like paper dipped in a solution of saltpeter. He inserted a hot wire, and the metal glowed white as the sun. He heated red lead, or "minium," in the same manner, thinking perhaps that since the two substances looked so similar they might produce the same gas. He was not disappointed. Again, the candle flame leapt; again, the sparks flew off the glowing wood.

He'd made a startling discovery, but was unsure of what he'd found. In his beaker hovered a pool of the gas that so many had sought, a pure sample of oxygen. Bayen, Scheele, and Boyle had created, but missed it—perhaps Priestley too, in November 1771, when he'd heated saltpeter but was unprepared psychologically to notice what he had. Now, three years later, he was confused, but at least he'd traveled down enough mental paths to know that something momentous was taking place. Yet for the life of him, he couldn't say what it was.

Later he would marvel at chance, and how the course of science depends finally on "the observation of *events arising from unknown causes.*" How easy it was to miss the obvious. In the end, the only real difference between this test and the one in 1771 was an everyday candle.

Man himself was the most important research instrument, and the most flawed. "The force of prejudice," Joseph wrote,

> biases not only our *judgments* . . . but even the perceptions of our senses. For we may take a maxim so strongly for granted, that the plainest evidence of sense will not entirely change, and often hardly modify our persuasions; and the more ingenious a man is, the more . . . he is entangled in his errors; his ingenuity only helping him to deceive himself, by evading the force of truth.

He'd made a discovery that would change all life on earth and every science known to man. But he was still entangled in the old modes of thought and couldn't find his way.

What *had* he found?

CHAPTER 6

The Problem of Burning

AS PRIESTLEY INCHED TOWARD DISCOVERING THE SECRET OF FIRE, it seemed to Lavoisier that he was always putting them out.

Fire suppression was the order of the day during his apprentice year as regional tobacco inspector for the General Farm. The Farm held a monopoly on the production, import, and sale of tobacco, and levied taxes on its sale; tobacco duties now brought in nearly 30 million *livres* each year. The weed was as important in France as in England. In 1558, the naturalist and monk André Thévet brought back from Brazil some grains of *pétun* that he called *herbe angoumoisine;* Jean Nicot, ambassador of Catherine de Medici, launched the fashion of its use, and this spread quickly. The crown quickly grasped that the exclusive privilege of selling tobacco would be an important source of revenue; the General Farm obtained the right to collect the tax in 1727. But revenues had fallen more than 20 percent since 1759, and the reason was fraud.

Lavoisier's main duty was to combat smuggling and fraud among tobacco retailers. Contraband tobacco, ten times cheaper than the Farm's, was smuggled into France by land and sea; the contraband was ground, shredded, or pulverized, then mixed with the legal product and diluted further with ash and water before being sold as plugs, pipe tobacco, or snuff. A small amount of *mouillade,* or moistening, was necessary to start tobacco's fermentation, but ash was a dead giveaway for tampering and Lavoisier's skills as a chemist stood him in good stead. "Chemistry provides a sure, unequivocal means of recognizing ash in almost all substances," he reported. "When a spirit of vitriol, *aqua fortis* or some other acid solution is poured on ash, there is an immediate, very intense effervescent reaction, accompanied by an easily detected noise."

But analyzing people was more complicated. The acceptable amount of water per hundredweight of grated tobacco was 6 pounds 5 ounces, or 6.3 percent; dealers who added more increased their profits but risked hastening fermentation, which "putrefied" the tobacco. The problem lay in identifying the culprit. Who added the water—the dealer, or his supplier? In the field, Antoine Lavoisier was judge and jury, with full power to fine and fire, but such decisions rarely came easy. In Vittry, for example, a young female dealer was found with moistened tobacco, but she had such a reputation for "sincerity and frankness" that he believed her protestations of innocence. In Châlons, he fired a "Sieur Martin" when more than half his tobacco tested impure. A Widow Ducloir sold tainted tobacco, yet Antoine urged leniency: she was the widow of a military officer, he wrote, and "is responsible for a very large family."

Lavoisier sent his reports to his supervisor, Jacques Paulze de Chasteignalles, and soon Paulze took especial interest in his gifted young subordinate. When the older Paulze joined the Farm in 1768, he was already influential: he was a lawyer at the *Parlement de Paris*, a Royal Prosecutor, and director of the East India Company. He was also related by marriage to abbé Joseph-Marie Terray, Comptroller General of France and one of the most powerful men in the nation. Paulze soon advised Antoine as he would a son: be honest and considerate, yet wary of confiding in strangers; have irreproachable morals; and do not be tempted by gambling.

Antoine was now caught up in a different kind of conflagration—an affair of the heart, though at first he simply acted with the affable distance that characterized his activities as an inspector. Jacques Paulze had a daughter—a very attractive daughter in every sense, but Marie-Anne Pierrette Paulze was only thirteen when Antoine first laid eyes on her. Even so, she was already besieged with marriage offers; she had very blue eyes and brown hair, a petite girl who would obviously mature into a beautiful woman with, as extra incentive, a considerable dowry.

After her mother's death, Marie was brought back to Paris from a convent school in the country to be her father's hostess, a role she filled with authority and charm. But such good graces brought unwanted attention. In 1771, the abbé Terray decided to press a marriage proposal to Marie from the comte d'Amerval, a fifty-year-old penurious rake who was also the brother of an important friend. Terray was Marie's great-uncle; he'd watched her grow up, knew her wealth, and calculated the political capital of arranging such an alliance. But Marie refused, even if it was to a nobleman, and her father backed her wishes. "My daughter has a definite aversion" to d'Amerval, Paulze wrote to Terray. "I shall certainly not force her."

Terray was neither pleased nor used to having his wishes ignored. A contemporary described him as "hard, sinister, even frightening. He was very tall but stooped, and had a gloomy face with wild eyes that stared downwards." Marie was his only daughter and he missed his deceased wife, Paulze explained; but Terray was not moved. He threatened to oust Paulze from his post as director of the Tobacco Commission unless he relented. Only the intercession of powerful friends in the Farm mollified Terray, and he backed down. But Paulze knew his dangerous relation too well, and decided to head off a renewed offensive.

Only another suitor would do, and he thought of Antoine. At twenty-eight, Lavoisier was tall and good-natured, had a ready smile around Marie, and was said to carry his intellect lightly, at least in those days. Marie, for her part, was not displeased with the suggestion. When sounded out, Antoine saw all the advantages to the union and accepted without apparent hesitation.

The arrangement, though expedient and rushed, was not unusual. Marriage in the eighteenth century was a financial arrangement which, it was hoped, would blossom into love. On both sides of the Channel, dowries were published in the papers: a 1735 London broadsheet ran the following notices, which seem similar to today's listings in the Real Estate section:

MARRIAGE

* 25 March, 1735, John Parry, Esq., of Carmarthenshire, to a daughter of Walter Lloyd, Esq., member of that county, a fortune of £8,000.
* The Lord Bishop of St. Asaph to Miss Orell, with £30,000.
* Married, the Rev. Roger Waind, of York, about twenty-six years of age, to a Lincolnshire lady, upwards of eighty, with whom he is to have £8,000 in money, £300 per annum and a coach-and-four during life only.

By midcentury, this began to change. Warmth, and even desire, in couples was no longer considered scandalous: couples whose parents had called each other "Sir" and "Madam" now used more familiar forms of address, and children were allowed to call their parents "Mama" and "Papa." By the late 1750s, the change was expressed as "sentiment"; the mark of a civilized man or woman was not only the capacity for reason but also for feeling. Edmund Burke's *Philosophical Enquiry into the Origin of our Ideas on the Sublime and the Beautiful* (1756) stressed the power of passion in aesthetic response, embracing terror and awe as well as delight and harmony, while Laurence Sterne's *Tristram Shandy*, stressing feeling above all else, was the surprise hit of 1759. In France, Rousseau's *Emile* and *Julie* did well. The marriages of both Lavoisier and Priestley would be noted for affection and respect, and both were wed near the cusp of this change. Although the financial circumstances of Mary Priestley and Marie Paulze Lavoisier were radically different, their approaches to love and romance were remarkably similar. Mary chose her suitor, like Marie; although there is no record of a dowry, Mary's father Isaac Wilkinson was as much a businessman in his own right as Jacques Paulze, and there are indications that in later years the Wilkinsons considered Priestley as much an informal research associate in their ironmaking ventures as Paulze considered Antoine a partner on the Farm.

Even Terray backed off when Lavoisier gave his consent, but only when he proved to himself that the marriage was irreversible. The vulturous Comptroller General was present at the signing of the marriage contract on December 4, 1771, a fact that says volumes about the man. Marie brought a dowry of 80,000 *livres* to the union—20,000 on signing, and the balance in six years; while Antoine brought 170,000 *livres* from his private fortune, 250,000 *livres* as an advance from his father, and an additional 50,000 *livres* promised by Aunt Constance as an inheritance. Thus, he was far wealthier than his bride, but also 1 million *livres* in debt due to his half-share of the Farm. His annual income, after paying interest, came to 20,000 *livres,* or $5.4 million in today's dollars. The actual ceremony was held on December 16, and attended by a *Who's Who* of Paris society. They moved into a house purchased by Antoine's father, on the rue Neuve-des-Bons-Enfants, near the grounds of the Palais Royal.

▪ ▪ ▪

By 1772, all the factors were in place to put Lavoisier and Priestley on their trajectories toward eventual collision; their nations had collided for centuries already. So far, both were blissfully unaware of this course, each caught up in his own religious, romantic, or business affairs. All that was needed was a link, in this case a tall, bony and amiable link—a former Portuguese monk whose family traced its lineage back to the Great Navigator Ferdinand Magellan, and who at the behest of the French government played the role of spy.

Though he dubbed himself John Hyacinth Magellan for his work in London, João Jacinto de Magalhães had not originally planned a life of espionage. Born in 1722 in Aviero, Portugal, he entered an order of Augustinian monks when quite young. He studied science, and in 1754 was appointed guide to the French naval officer and astronomer Gabriel de Bary when he visited Portugal to view a solar eclipse. Magellan seemed more inspired by the faith of the scientist than by that of the Jesuits; the next year, he asked to leave the order, and at the urging of de Bary moved to a new life and identity in Paris. He traveled around Europe, making the acquaintance of scientists and their instrument makers; then in 1764 he took up residence in London, where he stayed with brief returns to the Continent until his death in 1790.

By the time of his move, the French system of espionage was a well-oiled machine. Daniel Trudaine, director of the Bureau of Commerce, had sent Gabriel Jars to Britain with roaring success; now his son and successor, Jean-Charles-Philibert Trudaine de Montigny, continued the legacy with Magellan. The ex-monk may have been de Montigny's first confidential agent when he assumed the directorship in 1764; he was certainly the most valuable. Soon after arriving, Magellan sent Trudaine pamphlets and books describing the newest English discoveries and inventions: on one occasion, he tried to smuggle to France a newly developed loom that the British had decided not to export; another time, he sent back samples of flint glass so French astronomers could craft achromatic refractors as delicate and sharp as Britain's. His cover was perfect. He collected and manufactured scientific instruments for sale abroad, particularly to the rulers and nobility of Portugal and Spain, but his more secret efforts were so successful that he was made, at de Montigny's behest, a corresponding member of the French Academy on September 4, 1771, and assigned to correspond with his friend de Bary.

He seems today a perfect spy—amiable, intelligent, and slightly absurd, at least absurd enough to be thought harmless and so pass beneath the radar. Described in a 1790 issue of *The Gentleman's Magazine* as tall, bony, "plain of dress, unaffectedly mild, and decent in his whole demeanor," he seemed to fit in everywhere. He was a tireless letter writer, and because of his dogged correspondence with Continental men of science was considered by the British as a "scientific intelligencer," which meant a purveyor of scientific gossip. Above all else, he was a clubbable man. He belonged to an informal group of manufacturers and scientists that met at London's Chapter Coffee House (a group that included James Watt, Matthew Boulton, Josiah Wedgwood, and Priestley); he was a member of Ben Franklin's Honest Whigs, as shown by a letter from Franklin to Priestley dated May 4, 1772. Franklin was more worldly in politics and probably suspected that Magellan was a spy, but he saw his use and promoted him. Some, like James Watt, were suspicious, warning Joseph Black that Magellan had "an extensive correspondence and may circulate erroneous doctrines to the prejudice of your honour," yet Watt used him also, sending through him drawings and specifications to the engineers building his improved steam engine in France. Only Priestley seemed to genuinely like him, enjoying Magellan's conversations and worldly knowledge, speaking of him in his *Memoirs* as "very friendly, especially in everything that is related to my philosophical pursuits."

It is uncertain when Priestley and Magellan became friends. It may have been through the coffeehouses, or even earlier, when Magellan was involved in an incident important to proponents of religious freedom. In 1768, he and three Catholic priests were indicted for saying mass: the regulation, finally overturned in 1778, made it a felony for a foreigner, and high treason for a native, to celebrate mass on the grounds that doing so was subversive to the Protestant state. The suit, one of the last of its kind in England, was overturned by the King's Bench; afterward, the cautious Magellan converted. On March 14, 1771, Magellan attended a meeting of the Royal Society as Priestley's guest, and on May 4, he sent a sample of Priestley's "rubber" back to France, as well as an English translation of the *Dictionary of Chemistry* by the French chemist Macquer. The translator criticized Macquer for his lack of knowledge about the discoveries of English chemists, particularly the "fixed air" of Joseph Black and the "inflammable air" of Henry Cavendish. Rather offended, Macquer wrote back for more information on this "fixed air."

But it was not until March 19, 1772, that Magellan's role in the discovery of oxygen becomes clear. That day, Priestley presented his *Experiments and Observations on different Kinds of Airs* to the Royal Society; the next day, Magellan wrote to his French handlers. Ten days later, on April 1, Macquer spoke of Magellan's information to the French Academy.

Lavoisier was present, but paid little heed. His reaction was typical of his colleagues, but that would soon change. On July 5, Magellan sent de Montigny a copy of Priestley's recently published *Directions for Impregnating Water with Fixed Air*. This modest pamphlet on soda water woke the French from their slumbers concerning the possibilities of the British pneumatic chemistry. Priestley's proposal to the Lords of the Admiralty that his soda water might cure scurvy suddenly assumed the importance of a stolen military secret. Another war was looming, this time between Britain and its American colonies, and even now it looked as if France might be drawn in. The sailors of both navies were still vulnerable to scurvy and their effectiveness in battle suffered. Priestley's new therapy must be known and analyzed, and on July 14, 1772, Lavoisier received a summons.

Up to this point, Lavoisier had shown little or no interest in the mysteries of the air. In spring 1771, a note he wrote to himself outlined subjects he hoped to pursue: research on nitre and indigo, studies in the causes of barometric variation, the improvement of hydrometers, and a reevaluation of his earlier paper on lighting the streets of Paris. Many interests, but no frame. De Montigny's note seemed to change everything; the two men were close friends from childhood, but de Montigny's letter, while acknowledging Lavoisier's expertise by asking him to translate and analyze Priestley's pamphlet, had more the tone of a polite but unmistakable command. "May I ask you to repeat these experiments and add your own observations," he said. "Since the value of these new discoveries depends on the rapidity with which they can be applied, I hope that you will not be long in getting out this little work."

From that day forward, in one form or another, Lavoisier's path was set. He would become the nation's prime researcher and theorist on the gases in the air.

▪ ▪ ▪

National defense has a remarkable focusing effect on science, just as the prospect of hanging does on the individual. Unlike England, where research still remained at the mercy of private funding and enterprise, science in France was largely state-funded—and without these government grants, a scientist was left hanging. Science has historically accommodated itself to power, be it the Church or the State, whether through secrecy (as did Copernicus), capitulation (Galileo), or implication (countless arms makers and alchemists seeking a better gunpowder). By 1772, the French were learning a lesson the British had yet to grasp: one key to a nation's primacy lay in the wonders that emerged from its labs.

Unfortunately for Antoine, he could not comply with his friend's first request simply because he could not translate Priestley's work. His English, if he had any command of it at all, was not equal to the challenge, and so the task went to François Rozier, a chemist and

Parisian scientific publisher. Antoine realized he'd made a misstep, and in the future tasks of translation would go immediately to Marie. Yet by this juncture she was still taking classes in English and Latin, and when de Montigny's request was made, they'd barely been married half a year.

But Antoine was not sitting by idly. His experiment disproving the transmutation of water had convinced him that the Greeks were wrong. If water was composed of even more basic substances, why shouldn't it be true of air, the other great "fluid" in the ancients' equation? And as he chewed on this, here came Priestley with his experiments releasing a host of unknown "airs" from metals simply by heating them.

Priestley was approaching the problem from one end of the stick: What was it in the atmosphere that sustained life? Now Lavoisier picked up the other end: What was in the air that aided burning? The Continental chemists, and even Rouelle, were more steeped in phlogiston theory than the British pneumaticists, yet this intimacy always brought them back to the problem of burning. If phlogiston were released, as theory demanded, then all substances should lose weight during combustion. Instead, some metals *gained* weight when burned in the air.

"What if," Lavoisier asked himself, "something in the air *combined* with metals and minerals when burning occurred?"

The first hint of these thoughts is contained in his lab notes dated August 8, 1772. For the first time he expressed on paper his dissatisfaction with phlogiston theory and asked himself whether air might combine with other substances. He was not alone in such speculations. By the summer and fall of 1772, a number of Frenchmen had become fascinated with the oxidation problem. In August 1771, the economist Turgot had advanced similar thoughts in a private letter, and at least five Paris chemists—Condorcet, Sagé, Búcquet, Bayen, and the younger Rouelle—were duplicating British experiments, especially those regarding release of "fixed air." François Rozier was collecting for publication in his journal the principal foreign studies on pneumatic chemistry. If Antoine wanted to get a jump on the pack, he would have to act now.

During this same period, when Priestley was inventing and refining his test for the "goodness" of air, Lavoisier began a series of oxidation experiments with phosphorus and sulfur, experiments which, unknown to him at first, repeated the tests of the British chemist John Mayow a century earlier. Sometime between September 10 and October 20, 1772, he bought an ounce of "good German phosphorus for forty-five *louis,*" then placed the phosphorus in a bottle and heated it over a fire. White fumes appeared, then stopped abruptly. A phlogistonist would say that the air in the bottle was steeped in phlogiston, thus extinguishing combustion, but Lavoisier had a hunch, just like Priestley when he thrust the candle in the jar of oxygen. Since Lavoisier was steeped in quantification, his hunch had to do with numbers. He weighed the phosphorus residue and found that it weighed more than before it had burned.

But that was ridiculous—the measurement seemed to defy common sense. Phosphorus before burning was a solid white mass; the residue afterward was light and ashen. If phosphorus gained weight during burning, what about a similar flammable substance, like sulfur? Around October 24, he repeated the test with sulfur, and it too gained weight. Between October 24 and November 1, he heated lead oxide and confirmed his suspicions again. The lead oxide (which he called *"litharge"*) gave off a considerable amount of "air" the moment it changed back to metal.

During the same period, Lavoisier was involved in a series of dramatic, if failed, experiments involving the burning of diamonds with a huge burning lens, the tests Priestley had heard rumored in England. For centuries, diamonds were associated with mystical power—they supposedly protected the owner from the plague, warned their wearers of poison, banished ghosts, and exorcised demons. Now, through a series of experiments, they disappeared in the focused rays of the sun. In 1694–95, Cosimo III, Grand Duke of Tuscany, caused a stir when he took the diamonds of his guests, placed them in the reflected beams of a large mirror, and made them disappear forever. In 1751, the German emperor Francis I placed diamonds and rubies worth 6,000 florins in a furnace and watched the diamonds vanish but the rubies remain unchanged. Obviously, these tests were not for the faint in spirit or poor in capital. Yet they mystified everyone. How did the hardest substance on earth simply vanish? French scientists like Rouelle concluded they had evaporated in the intense heat. But would this happen, Lavoisier wondered, if they were heated in a vacuum? Was air necessary for the drama?

From August 14 to October 13, 1772, Lavoisier and his fellow Academy chemists conducted 190 experiments on diamonds and other metals and minerals with the largest "burning glass" ever constructed. A glass of such magnitude could only be built with a subsidy from the state: two large convex glasses with a radius of 8 feet were joined at the edges, the space between the lenses then filled with 140 pints of alcohol. This "liquor lens" was set up outside the Louvre in the Jardin de l'Infant, a vast terrace facing south that stretched between the palace and the Seine. The monstrous contraption, lifted and lowered by cranks and riding on a wooden platform of six wheels, caught the rays of the sun in the larger lens, concentrated them in the smaller lens, and focused the intense heat on the substance being tested. The academicians and their assistants, garbed in white wigs and dark glasses, rode on the wooden platform as if on the deck of a ship, while elegant women and their escorts strolled by and stared.

The tests were a success, and a failure. True, the diamonds disappeared, but Nature did not reveal the reason. In order to prove that air was necessary, Lavoisier had to trap the gas that was released, but the intense heat generated by the lenses cracked the container and allowed the gas to escape. Other times, a haze of clouds obscured the sun. There was still no proof that air was necessary for combustion and involved in the weight gain of metals, but between these tests with diamonds and his experiments with phosphorus and sulfur, Lavoisier grew more convinced that air was a necessary chemical agent whenever burning occurred.

There was just one problem: What accounted for the flame? It was no small problem, and one that would not be adequately explained until the mid-nineteenth century when the English physicist James Prescott Joule demonstrated the link between mechanical energy and an increase in heat, thus showing that both heat and fire were manifestations of energy. In 1772, however, the only logical explanation seemed to be Joseph Black's "caloric theory," and this was what Antoine chose to adopt in place of phlogiston. Caloric was an all-pervading fluid, Black postulated: like a fluid, heat passed from one body to another; it must have weight, since heated metals gained weight, too. *Calorique,* the *élément impondérable,* was Antoine's substitution, a subtle fluid penetrating the pores of all matter. Yet he had to admit that he had no clear conception of its nature: "Since there are no vessels which are capable of retaining it, we can only come at the knowledge of its properties by effects which are fleeting."

But this was merely a diversion, not the main thrust of his coalescing theory. He was excited beyond expression: he didn't know the exact process yet, but he knew he was pointed at the heart of the mystery. He must have time to investigate the matter further, and couldn't stand the thought that someone else might see the truth and claim the laurels before him.

On November 1, Antoine wrote down a note and sealed it, only to be opened when he had more evidence. He deposited the now famous *pli cacheté* the next day with the astronomer Grandjean de Fouchy, perpetual secretary of the French Academy. A week earlier, he said, he had discovered that sulfur gained weight when heated; it was the same with phosphorus, too.

> This increase in weight comes from an immense quantity of air. This discovery which I have established by experiments that I regard as decisive, has led me to think that what is observed . . . may well take place in the case of *all* substances that gain weight by calcination or combustion. Since this discovery appears to me as one of the most interesting of those that have been made since the time of Stahl, I felt that I ought to secure my right to it by depositing this note in the hands of the Secretary of the Academy to remain sealed until the time I shall make my experiments known.

Shrewdly, he claimed provenance. If there was to be a revolution in chemistry, he would be the chief revolutionary. Yet why such caution when he wasn't even sure he might succeed?

Years later, around 1792, Lavoisier wrote a paper with the lengthy title *Détails historiques sur la cause de l'augmentation de pords qu'acquierent les substances métalliques, lorsqu'on les chauffe pendant leur exposition à l'air.* It was published posthumously in the 1805 *Mémoires de chimie*, by which time Lavoisier was famous, canonized, and eleven years dead. He repeats the 1772 note, but in a modified version: he admits that the fervor of youth and a desire for scientific glory played their part, but more important was the cause of patriotism. It was a rivalry with English scientists (most likely, Priestley) that led him to such secrecy. He was saving the laurels of discovery for himself, but for France as well.

But there is some deception here. As the biographer Henry Guerlac revealed in 1961, Lavoisier's later version of the note—the one most extensively quoted in histories of science—was different from the original note of 1772. In the original, Lavoisier admitted that he feared "he might inadvertently disclose to his friends in conversation something that could lead them to the truth." By 1792, France was on the eve of war with Britain, and Lavoisier was the acknowledged leader of a new chemistry now called the New *French* Chemistry. It would hardly do to tell the truth: that his chief rivals at the time he penned the note were not the hated English, but fellow chemists whom he saw at least twice a week and who were pointed in the same direction as he.

It was only his first deception. There would be others. But right now, it gave him breathing room. Throughout 1773–74, he tested gases with limewater; repeated Priestley's experiments; placed sparrows, mice, and rats in released gases to see what would happen. He placed burning candles and red-hot charcoal in the gases to see if they would sputter or burn.

But something was going wrong. Unlike Priestley in 1774, he used charcoal for heating, and as we saw earlier, heating metals with charcoal introduced carbon dioxide into the final mix. At one point, like Priestley, he heated mercuric oxide, but the oxygen released was tainted by "fixed air"; unlike Priestley, he did not heat the mercuric oxide directly with the burning glass, thus bypassing the contaminating fumes. He thought he had done everything right, but something was holding him back. Where was the secret to burning?

The air held the answer, but all his claims meant nothing unless proven in the lab. By autumn 1774, the prodigy of whom so much was expected was in danger of being left behind.

PART II

Solution

Pilate saith unto him, What is truth?

—John 18:38

CHAPTER 7

The Sentimental Journey

JUST WHEN PRIESTLEY SEEMED ON THE VERGE OF UNLOCKING one of Nature's great secrets, duty called: Lord Shelburne required him for that peripatetic ritual of the rich, the Grand Tour.

By 1774, the idea of a Continental tour was as much a part of one's cultural education as a college road trip is today: you might learn the Great Ideas in class, but you only learned about life on a journey. The phrase itself first surfaced in Richard Lassels's *Italian Voyage* (1670); the Tour was originally seen as a young nobleman's exercise, and in addition to visiting the great cities of Europe with their high altars to history, art, and culture, the tourist was expected to shake off family discipline and learn firsthand of politics, language, and love. The tourists changed with technology: improved lines of coaches, with their new springs and soft upholstery, made travel more comfortable, thus opening it up for women and the older traveler. By Priestley's day, a Tour for a learned gentleman was de rigueur. The standard itinerary included Paris, Florence, Rome, Naples, and Venice, with frequent side trips; political circumstance barely affected travelers, even during war. Laurence Sterne's "sentimental traveler" exclaimed in 1768 that "I left London with so much precipitation, that it never entered my mind that we were at war with France." The fact that he left without papers was rectified by authorities in Versailles. The very novelty of a visit by a foreigner excused most gaffes and opened many doors. "Once you are known to be a foreigner," said one visitor, "you are received everywhere; and the fact that you are a Protestant minister, incommodes no one."

Travel was considered the main means for spreading knowledge. Thomas Nugent's *The Grand Tour* (1756), the pre-Baedeker guide for travelers, exclaimed that "it is by means of such traveling philosophers, that the sciences were first diffused through several parts of the inhabited globe." A change in scenery inspired genius, an argument bolstered by Edward Gibbon's idea for the *Rise and Fall* arriving as he sat amid the ruins of Rome. Another moment of inspiration concerned the "goodness" of air. In January 1729, when Montesquieu arrived in Rome, he wondered whether the Eternal City's infamous air pollution was the cause of the decadence of the modern Roman. Did air influence character? Although he never addressed the Roman question, he conceived his theory on the influence of air on health and history after journey's end.

Despite the interruption in his research, Priestley had good reason to go. He'd never ventured any farther from his birthplace than London, and he'd been elected to an honorary chair in the French Academy in 1772. The names of the Paris academicians were familiar to him from Magellan's gossip: he'd like to share ideas with men like Macquer and the famous Condorcet.

And there was a new name that intrigued him: Antoine Lavoisier, an ambitious young man who had sent him a book of his own experiments earlier in the year. He wrote in the text that he was captivated by Priestley, "who was the most brilliant . . . chemist of the age," but in the same breath described his work as "a web of experiments almost uninterrupted by reasoning." What kind of praise was that? If nothing else, Priestley recognized damnation by faint praise. Inherent in the comment was a jockeying for ascendancy that Priestley didn't appreciate, and even worse, the young man's experiments were flawed. Lavoisier was still confused by the role of carbon dioxide in combustion: he seemed to think *it* was key to the process, not Joseph's mysterious air that ignited the candle.

Lavoisier's book, *Opuscules physiques et chimiques*, is interesting today for its style and the differences illustrated between the minds of Priestley and Lavoisier. The Frenchman described Priestley's experiments in detail, then repeated them himself. "No piece of recent work has made me appreciate more strongly how many new paths in physics and chemistry still remain to be trodden," he wrote, calling Priestley's discoveries "an assembly of facts, most new to us, either in themselves or because of the novel circumstances which accompany them." Yet praise was spiked with condescension. None of Priestley's sense of play or wonder at novel and pretty experiments peeked through the writing; the tone was serious, that of a professional who did not have time for an amateur's idea of fun. Although Lavoisier admired the scope of Priestley's work, he also disparaged the failure to theorize. More than anything else, Lavoisier wanted to know how a new discovery fit into chemical knowledge. How did it further the sciences?

One success of the book was the drawings, signed "Paulze Lavoisier." By 1773–74, when the book was sent out, Marie had transformed herself into a valuable partner. She was a competent student of language and spent hours translating English manuscripts; she took drawing lessons to aid her husband's work from the famous Jacques-Louis David. She was his lab assistant and confidante through frustration, and proved as ambitious and strong-willed as he.

It was good he had her for support, because there was plenty to frustrate him that year. Lavoisier had started his book with every intention that it would be groundbreaking, yet all it did was confirm the works of Joseph Black and Priestley. Worse, a pernicious trait was coming to light, one fueled by competitiveness and ambition. He ignored in the text the work of fellow countrymen: their work was insignificant to him since they were couched in the language of phlogiston. The Academy spotted the omissions when he submitted *Opuscules* for approval and demanded that the work of his colleagues be included. He added a few hasty chapters, including a summary of work by his old professor Rouelle. But word got out. Failure to give credit to the findings of predecessors was a major sin among Enlightenment thinkers—the ethos was that knowledge should be shared, not hoarded. Though understandably human, Lavoisier's omissions went against the dogma of the new scientific faith, a heresy the young chemist would commit with increasing frequency.

Priestley and Lord Shelburne left Dover in late August on the boat trip to Calais. The passage itself was uneventful, though Calais "was the first fortified town I had ever seen" with a moat and strong walls, Priestley wrote in letters to Shelburne's sons. Throughout the trip, Joseph's constant emotion seemed one of surprise: he was either awed or disapproving of the Continent's very un-British ways. In Belgium, then called "Austrian Flanders," he was amazed by the "prodigious quantity of tall, fine beans"; puzzled by the bountiful poppy crop (until told that oil pressed from the seeds was used to light lamps); amused by the little dog-drawn carts and the "slenderest and leanest pigs we had ever seen. They might almost have been taken for greyhounds." But the Puritan in him was easily scandalized, especially by the women, who did not wear hats and walked the cobblestone lanes "in slippers, without any thing to cover the heel."

It was classic culture shock, a malady suffered by many neophyte travelers. Generally, however, Priestley made an amiable traveling companion. Their route took them north from Calais through Belgium, and on to Amsterdam. Shelburne, as a former cabinet minister, was dined and fêted as befitted a man of importance; at one point, they stood in review as a French regiment performed new drills designed by the King of Prussia. "The pain that I felt on this occasion did not arise from any consideration of the mischief that this new discipline might enable the French to do us in any future war," Priestley wrote, "but from a cold that I got at the time, which affected my teeth very much." Belgium was "highly cultivated and populous." Although he admired the harbors and architecture of Rotterdam and Amsterdam, where he'd been headed as a young man, he was glad now he didn't go. The Dutch seemed an industrious but selfish people, and they smoked so much tobacco that they threatened to "poison everybody that comes near them." He considered Holland a dreary place: "I can hardly express how very low, beastly, and sordid, the manners of the common people in this country are," very different from "the roughness and brutality of some of the low-bred people of England."

By the 1770s, tourism and its parasites had developed symbiotically. Street vendors did headstands for pennies; guides tended to cheat on the duration of their sightseeing tours. The most prevalent scam was the trafficking in "Old Masters," especially in Belgium. The only rule in this counterfeit art market was "Let the buyer beware."

In Ghent, they fended off an overzealous painter; in Antwerp, a dealer who swore he owned an "original" Rubens. "Our guide, who, no doubt, was in league with him, avouched it," noted Priestley, "but going immediately from thence to the house of a rich and whimsical canon, we saw the real original of the very same picture."

From Amsterdam they turned southeast into Germany, where they followed the Rhine to Alsace-Lorraine. It was a leisurely journey, and Priestley's spirits soared. Although the roads were sloppy from the autumn weather, he found the Rhine Valley "exceedingly romantic," so well cultivated in places that he was reminded of the lavish gardens of royalty and wrote:

> We had the Rhine to the left, with hills and rocks covered with vines and woods close to it, the vines intermixed with kidney beans and pumpkins. Sometimes the road was cut in a rock almost perpendicular, the river being below us, and the houses above us, with chapels neatly cut in the rock.

Shelburne and he got along exceedingly well. They often strolled in front of the coach, commenting upon the soil and the crops: England and the Continent were still agrarian societies, whose upper tiers were filled with landed gentry like Shelburne. At least twice they got lost by missing their carriage, a feat "which threw me, who was the occasion of it, into great consternation," Priestley lamented. But Shelburne always calmed his excitable companion down.

By September 22, 1774, they reached Strasbourg and turned west into France, taking the same route as that of Guettard and Lavoisier in their return from the Vosges Mountains. As with many English travelers, Priestley arrived with a preconceived set of stereotypes and expectations, many of which he had to reevaluate. "In the art of living the French have generally been esteemed by the rest of Europe . . . and their manners have been accordingly more imitated," wrote Arthur Young a decade later, and Priestley seemed to concur. Young went on:

> On comparison with the English, I looked for great talkativeness, volatile spirits, and universal politeness. I think, on the contrary,

> that they are not as talkative as the English, have not equally good spirits, and are not a jot more polite: nor do I speak of certain classes . . . but of the general mass. I think them, however, incomparably better tempered.

They reached Paris on September 30, 1774. The city sprawled beyond its ancient boundaries, forming new and outlying suburbs. As they approached from the east, they could see twin domes rising from a gray jumble—the leaden dome of the new Panthéon, and the golden dome of the old Invalides. They saw this juxtaposition of the old and new everywhere. In the seventeenth century, only a few public eating houses existed in Paris; by 1774, there were an estimated six hundred cafés where citizens could read their journals or broadsides and discuss the news. The theater was booming, and although actors and actresses were still officially denied the benefit of sacraments, the old ecclesiastical ban was virtually ignored. The Italian opera had recently been introduced and was a hit; the avenues rang with street musicians belting out the latest tunes.

In other ways, though, Paris seemed medieval. Cabriolets sped down narrow streets in which no pavements were built for a quick escape. A foul gutter carrying away sewage ran down the center. The streets were jammed with water carts since there was no pipe system. Nearly twenty thousand people were employed in the business of carrying water to private residences, and just as many in carrying away "night soil." The slaughter of cattle, allowed in the city, meant the streets were often blocked by bellowing herds. Though disease was rampant, admission to a hospital meant a death sentence from sepsis or infection. Though Priestley was impressed by the city's size and magnificence, he was sickened by the narrow streets, dirt, and smell.

The two visitors arrived at an important moment of change. On May 12, 1774, smallpox had ended the misrule of Louis XV; his remains, covered in quicklime, were quickly carried to the Saint-Denis burial ground. Gone were the royal mistresses, Mesdames du Barry and de Pompadour; gone the abbé Terray, cadaverous uncle of Marie Lavoisier, his effigy hung on the gallows at Sainte-Geneviève. The funeral march was barely over when twenty-year-old Louis Capet ascended the throne as Louis XVI, his nineteen-year-old Austrian bride Marie Antoinette by his side. Both were received in a wave of goodwill. The Most Christian King of France and Navarre swore to uphold the peace of the Church, impose justice, exterminate heretics, prevent disorder, and pardon no duelist. A year later, when he was officially consecrated at Rheims Cathedral, men would weep. Louis said he was a liberal, and many wanted to believe.

Yet even with such hope, an event occurred as Priestley and Shelburne rode into Paris that took France a step closer to the Revolution. The exiled *parlements,* or royal courts of law, one of the few surviving holdovers of the medieval French constitution, were recalled to Paris; in addition, the "honest" Swiss commoner Anne-Robert-Jacques Turgot was installed in place of Terray as Comptroller General. The trappings of absolutism seemed to pass, yet the King could override the judgments of Turgot and the *parlements* should they ever resist his will.

The two spent a month in Paris, and from the beginning Joseph was uncomfortable. "A fish out of water" was not yet a cliché, yet one wonders if he gave rise to the phrase. He was a man of black ministerial dress in a world of soft pastels and rich brocade; his was a habit of personal fastidiousness in a city where the inhabitants of all classes were "exceedingly deficient in cleanliness," he said. Again it takes the more exacting Arthur Young to provide the explanations. "The French are clean in their persons, and the English in their houses," he observed. A French *bidet* was as common as an English washbasin, a statement of personal hygiene Young wished more popular in England, yet he found their "necessary rooms . . . temples of abomination." But it was the practice of spitting that shocked Young the most; the fact that it was common in all ranks left him dumbfounded. Marveled Young, "I have seen a gentleman spit so near the clothes of a duchess that I have stared at his unconcern."

Shelburne and Priestley decamped in a grand hotel; since his employer was an important personage, they were swept into the social whirl of assemblies, dinners, theatrical entertainments, "seeing and conversing with every person of eminence, where-ever we came; the political characters by his lordship's connexions, and the literary ones by my own." One gets the image of Joseph being strangled by chatter—and Joseph, who liked to talk, couldn't get a word in edgewise. "As far as I can judge," he wrote in exasperation, "the French are too much taken up with themselves. . . . This appears in nothing more than their continually interrupting one another in discourse, which they do with the least apology; so that one half of the persons in company are heard talking at the same time." By mid-October, he begged off Shelburne's lionizing circuit and stayed in the hotel. He was tired. He missed Mary, the children, his tidy lab with the neat rows of bottles and chemicals glinting in the sun. He wanted to go home.

But at this low ebb, the *philosophes* arrived, and he sprang back to life. Paris was not only the heart of the European Enlightenment, it was the center of a brilliant scientific circle, and Joseph seemed to suddenly wake to the fact that he was in the presence of the French Academy, which had honored him two years ago. This was the home of the *savants* Cadet, Condorcet, Macquer, Laplace, and Lagrange. And there was the young upstart, Lavoisier. The academicians, for their part, were charming. "They could not possibly . . . show more respect to anybody than they did to me, especially on account of my *Observations on Air* which have engaged the attention of almost all the philosophers on the continent." Soon, Priestley was off on his own whirl of labs and experiments, visits to the Academy and dinners at the homes of academicians.

Two men in particular took Joseph in hand. The first was the new Comptroller, Turgot. An Encyclopédiste, disciple of Voltaire, economist, and administrator, Turgot assembled the best minds in Paris to entertain Priestley and in turn be illuminated. The British natural philosopher met the French *philosophes,* and the difference stunned Priestley as much as it confused the Parisians. To an eighteenth-century Englishman like Priestley, a natural philosopher was a man of science whose inquiries revealed the workings of God the Clock Maker, the vision of Nature begun by Descartes and perfected by Newton. Uncovering God's plan was a calling worthy of Protestant theology, but the French *philosophes* didn't see things that way. Although Voltaire had taught them to regard themselves as heirs of Newton, the *philosophes* did not believe that science revealed God's plan or served its purposes. "Reason is to the *philosophe* what grace is to the Christian," noted César Chesneau Dumarsaid, the French grammarian who wrote the entry for "Philosopher" in the *Encyclopédie*. "Grace causes the Christian to act, reason the philosopher." Their business was not of God, but man; their purpose, as Diderot urged, the utilization of scientific knowledge to liberate man from despotism and superstition. This was why Diderot's *Encyclopédie* was initially banned: such a synthesis of empirical data and liberal thought had a political agenda, not a godly one.

Priestley was watching the birth of a new religion, and it troubled him. Enlightened humanism had grown out of Greek rationalism in the same broad way that Christianity grew from Judaism or Buddhism from the Vedic status quo. A dis-ease with life's misery and pointlessness spawned new faiths that sought ways to transcend human weakness and live in peace; their answer was to seek reality within themselves. The process he witnessed was similar to that during the Axial Age, the period stretching from 800 to 200 BCE that gave rise to Siddhārtha Gautama, best known as Buddha; to the great Hebrew prophets of the eighth through sixth centuries; to Lao Tzu and Confucius, who reformed Chinese religious traditions; to the sixth-century Iranian sage Zoroaster; and to Socrates and Plato, who spurred the Greeks to challenge even the most self-evident truths. In the same way, the *philosophes* sensed fresh possibilities that could be achieved only by breaking away from old traditions. Yet where Axial sages sought this truth embedded deep within the self and called it God, Enlightenment thinkers sought the truth through mind and reason—and called it science, Nature, or Universal Law.

The revelation apparently hit Priestley at a dinner hosted by Turgot. "Who are the two men across the table from us?" he asked M. de Chatellux, the man seated beside him.

"They are the Bishop of Aix and Archbishop of Toulouse," his neighbor said, and chuckled, "but they are no more believers than you or I."

"But I am a believer," Priestley said, surprised.

At that Chatellux laughed the harder, and Turgot, within hearing range, jumped in. *"C'est impossible,"* he said. "No man of science can truly believe in God."

As Priestley reiterated his belief, the French listened to this English conundrum with greater wonderment than ever. "If that is true," offered an academician seated nearby, "then you are the only man of science I know that is a Christian."

Therein lay the controversy that has since bedeviled science. A century before the supporters of Charles Darwin opened the subject to public debate, Priestley privately discovered the growth of the idea that God had no place in scientific inquiry. This was very disconcerting, and a proposition it seems he hadn't considered previously. That men of science, so united in their interests, should diverge so widely in their belief in the deity so bewildered him that his best response at the moment was to act confused.

Priestley may also have become aware for the first time of his own paradox. He had no higher regard than his hosts for Catholicism (even though he advocated toleration of Catholics in England), and, like his hosts, saw science as a way to banish superstition and improve the lot of man. In his *Experiments and Observations on different Kinds of Airs*, he'd said the papacy itself was threatened by the growth of reason; in the *Disquisitions*, he'd rejected the "soul." Yet he was not an atheist, and when he protested again that he was Christian, the Encyclopédiste looked at him oddly. "Then you are the first believer I have ever met of whose understanding I had any opinion," Turgot said.

As usual, Priestley refused to be mollified. He attributed such atheism to the corruptions of the state-sanctioned Church, a point he'd made regarding Anglicans. Men disbelieve because the Church conceals the truth for its own ends. He'd argued this point long and hard with Ben Franklin, and worried about his mentor's soul in much the same way that Timothy Priestley worried about his brother's. But now the position was reversed and he became an underground hero for those Parisians who'd thought they were the last Christians in town. One young man embraced him when he learned Priestley was a believer. A scientifically minded priest "embraced me with tears." Priestley cautioned that although he was a Christian, he was also what Catholics might call a heretic. "No matter," said the priest. "You are a Christian."

But in many ways, the real issue was skirted: Did a belief in science cancel out a belief in God? It is a continuing division today. There was one *philosophe* in attendance who took the issue very much to heart, and this was the astronomer Pierre-Simon de Laplace. In 1774, Laplace was a young man; he'd graduated five years earlier from the University of Caen with a master's degree in arts, and was on his way to ordination. Jean d'Alembert, the Encyclopédiste who would be his greatest sponsor, mentioned that Laplace was an *abbé* in August 1769. Soon afterward, he abandoned the priesthood to follow mathematics and astronomy; but by 1774 he still believed, like Voltaire, that religion maintained the fabric of society.

Yet all this would change, and Laplace would symbolize the shift in thinking. To see it more clearly, we must jump ahead to a famous story told of him in the following century. By 1827, Laplace was considered the greatest physical scientist of his time, dubbed the "Newton of France" for his extensive writing on the system of the world and the mathematics of probability. That year, an anecdote involving Laplace, Napoleon Bonaparte, and God's disappearance began to make the rounds. Napoleon was talking to Laplace about his most recent book when he supposedly commented, "Newton spoke of God in *his* book. I have perused yours, but failed to find His name even once. How come?"

"Sir," replied Laplace with a combination of reverence and arrogance, "I have no need of that hypothesis."

God is a superfluous hypothesis: God had vanished from the physical universe, and Laplace, the scientist, didn't need Him anymore. What a difference a century made between Isaac Newton, who saw God in all laws of Nature, and the Newton of France, who did not see Him anywhere. But Laplace's answer had more to do with method than theory. Laplace used his history of astronomy to show how the advancement of science came about through the gradual abandonment of superstitious belief, much as Priestley had done in his own *History of Electricity*. Science progressed through disagreements between experts who still agreed on a rational experimental method; religion, on the other hand, was mired in debate that could not be objectively tested or proven. To Laplace, interminable metaphysical arguments were a waste of time. The business of science had everything to do with well-tested knowledge; the business of religion misled otherwise well-meaning natural philosophers. Why return to a system that impeded progress? he asked, and felt justified to exclude theology from the world equation.

Laplace's philosophical stance was an extension of Lavoisier's research method; the existence of God, phlogiston, and other immeasurable forces had no meaning in the lab if they could not be measured, weighed, or quantified. Such deliberate refusal to engage in abstruse debate was a result of the growing autonomy of science from other human endeavors, but it was also an admission that human understanding was flawed. We are limited observers, and so never truly know the nature of life. We can never really *know,* objectively, the force behind gravitation, molecular attraction, or light. We can only know what lies within the narrow and artificial space defined by experiments and measurements . . . what lies within the fence we put around the law.

In replacing God the Intervening Power with God the Clock Maker, who sits back and watches after establishing eternal laws like gravitation, Isaac Newton planted the seeds of God's vanishing role. Once created, it was only a small step to the idea of a universe without God. The world could run by itself. As Jacques-André Naigea commented in his 1790 address to the National Assembly during the height of the Revolution: "God is like an excess wheel."

▪ ▪ ▪

Priestley's second guide through the scientific *salons* of Paris had a more pragmatic reason for taking the Englishman in hand. Jean-Charles Philibert Trudaine de Montigny was the director of the Bureau of Commerce, spymaster for Priestley's friend John Hyacinth Magellan (who just happened to be in Paris), and Lavoisier's best friend. Since he was the one to pass Priestley's manuscript on gases to Lavoisier for translation, he now wanted to get a close look at this remarkable Englishman. He took Joseph from the Academy of Sciences, where the latter blushed at a standing ovation, to his own home lab, where the two tried to make some "volatile vitriolic acid" by burning oil of vitriol in a cracked pan over charcoal. Nothing happened, so they added turpentine : this time a sudden rush of gas overturned the receiver, releasing flames and fumes. They ended the experiment quickly before burning the house down.

Now Montigny tried something just as volatile. He introduced Priestley to Lavoisier.

There is a direction to Trudaine's handling of the guileless Priestley, and if one looks closely, its purpose can be seen. Priestley was surrounded while in Paris by those interested in his secrets: by Trudaine himself, who'd been convinced that the Doctor's soda water had military import; by Magellan, who just happened to be in Paris and took Joseph to the best pharmacists with the purist chemicals; and now, by Lavoisier. Everywhere Priestley went, he'd been beside himself with excitement about this new gas he'd discovered . . . something about heating the "red calx of mercury" with a burning glass. Trudaine, a chemist in his own right, had three years of life remaining before a sudden illness struck him down at the age of forty-four; he knew enough to recognize Priestley's ramblings as touching on the same obsessions of his friend Antoine. Yet somehow, Priestley's procedure was different. Was this the answer? He didn't know, and probably only considered it on an intuitive level. Yet by handing Priestley over to Lavoisier, he handed a gift to his friend and glory to his nation, both on a silver platter.

In October 1774, Lavoisier was thirty-one; in his prime, no longer a prodigy or rising young star. He'd worked on the problem of burning for two years without results; he was still certain that something in the air was added during combustion, but had no proof. In May 1773, he'd revealed the contents of his sealed note to the Academy in a series of lectures: these made public his suspicions that phlogiston was a dying theory, but in the end proved nothing. *Opuscules* was released that same year; it was politely received, but that was all.

Among scientists there is a process analogous to realizing one's mortality. Lawrence Kubie of the Yale School of Medicine has compared it to the self-deceiving fantasy of personal invulnerability young soldiers take with them into battle, one that quickly vanishes when friends around them begin to die:

> The man who went into battle squarely facing the fact that he might be mutilated or killed was less likely to break down under the stress of combat than was the soldier who went into battle with a serene but unrealistic fantasy of his personal invulnerability. Among young scientists we find a similar self-deceiving fantasy: that is, that a life of science may be tough for everyone else, but that it will not be for him.

Until he'd tackled combustion, everything had been easy for Lavoisier. True, he'd lost his mother and his sister, but who in this age of disease and violence was not acquainted with loss? It was a tragic fact of life. But in the realms of science and finance, his paths seemed paved with gold. His was a fantastic future, a charming and lovely wife, a calculating mind for business, myriad prospects for gain. He'd befriended France's most famous *savants,* was honored by the King, studied under France's premier chemist, and apprenticed under a brilliant geologist. He was the youngest person in any field ever to be elected to the French Academy, the greatest scientific institution in the civilized world. Yet for all this, his ultimate goal of overthrowing ancient doctrine had come to naught and he was feeling his mortality. What if . . . and here was the question he would not have considered two years earlier . . . What if he should fail?

"This is the ultimate gamble which the scientist takes," says Kubie, "when he stakes his all on professional achievement and recognition." Success and failure are ruled by chance, dependent not on what is discovered, but *when,* and by whom. Antoine had finally realized the awful truth, and as Kubie would see in modern patients, the vision shook him deeply:

> The rewards of a career in science are slow and also uncertain; bad luck can frustrate a lifetime of sacrifice and ability. Every successful scientific career is an unmarked gravestone over the lives of hundreds of equally able and devoted, but obscure and less fortunate, anonymous investigators. Science too has its "expendables"; but these do not earn security or tenure, or veteran's pensions—not even the honors which accrue to the expendable soldiers of war.

Perhaps Lavoisier had already started down that dead end. By mid-October, Priestley was a frequent guest at the Lavoisier mansion on the rue Neuve-des-Bons-Enfants. He witnessed a spectacular experiment of "the rapid production of, I believe, near two gallons of air from a mixture of spirits of nitre and spirits of wine." When his host applied a candle, "it burned with a blue flame" that streamed from the source for "a considerable distance." Privately, however, Priestley did not seem impressed. In a letter to Trudaine soon after his return to England, he spoke of Lavoisier in encouraging fashion, but not as one would a peer: "The world has great expectations of him and I doubt not that he will abundantly answer them." In his *Opuscules*, Lavoisier had made mistakes in interpreting Priestley's experimental results and attributed to him things he had never done. In the preface to his second volume of *Experiments and Observations on Airs*, Priestley complained that the mistakes made by "several foreign philosophers" were so egregious that he felt required to insert a new chapter correcting their errors. He listed four pages of mistakes and then added, "I shall not even think it worthwhile to note all the mistakes of Mr. Lavoisier."

Mistakes and sins of misinterpretation were not Lavoisier's only failings, for in his despair he'd turned truly deceptive, something that would have seemed impossible when he was still the bright young prodigy. Earlier, he'd failed to list the research of fellow French chemists; he was gently chided by the Academy's editors, and forced to add a summary chapter to his *Opuscules*. But in 1774, before he met Priestley, worse omissions were beginning to occur.

The first came quickly after publication of *Opuscules*, in February 1774. That month, Pierre Bayen published his report that various mercury oxides could be "reduced" to mercury without the use of charcoal. He added an essential fact that Lavoisier had missed: that when the oxides returned to mercury, "they lose one eighth of their weight." He was describing a redox reaction, though he didn't know it, and two months later he struck again, specifying that this weight gain was due to the addition of some "elastic fluid." This challenged the existence of phlogiston, and Bayen wondered if the change could be produced by the burning rays of the sun.

When Lavoisier later claimed the laurels for oxygen, he failed to note Bayen's contribution. Yet the army chemist's speculations had been public, and so nothing was hidden. On September 30, 1774, however, the lonely Swedish apothecary Carl Scheele wrote a letter to Lavoisier from his lab in Uppsala; it arrived fifteen to seventeen days later, about the time of the famous dinner with Priestley. Scheele thanked Lavoisier for a copy of *Opuscules*, which had come to him via the secretary of the Academy of Sciences in Stockholm; more important, he called attention to the preparation of a new gas he would call "fire air," or *Feuerluft,* obtained by heating silver carbonate and then absorbing the carbon dioxide in alkali. Scheele asked Lavoisier to repeat the experiment with his superior burning glass and tell him the results. "It is by this means that I hope you will see how much air is produced during the reduction," he wrote hopefully, "and whether a lighted candle will maintain its flame and animals survive in it."

Lavoisier pocketed the letter, never replied, and never mentioned receiving it to anyone.

▪ ▪ ▪

There are several versions of the conversation that shook up chemistry as it was practiced and dislodged one of the last holds of the ancients upon man. Some locate it in Lavoisier's laboratory, others in old Macquer's house, but most evidence seems to point to a supper hosted by Antoine and Marie sometime between October 15 and October 17, 1774.

In all of Antoine's despair and panic, his greatest asset was Marie. By October 1774, though still not quite seventeen, she was beautiful, charming, and fitted in easily with the older men. In addition to her drawing lessons and instruction in English, she would soon learn Latin—she asked her older brother to help her learn to "decline and conjugate for my own pleasure and to make me worthy of my husband"—and she asked Antoine to teach her chemistry so she could work with him in the lab. Lavoisier happily complied, as well as arranging for a private tutor; she also attended lectures at one of the Paris colleges. In a word, her chemical education was taking a similar form to her husband's, with particular emphasis on filling Antoine's deficiencies.

Seated across the table, Joseph could not help but be impressed. French meals were still served *à la française,* which meant many plates were brought to the table, but you helped yourself and were limited to that within your reach. The number of dishes, ornaments, and silverware would have been overpowering to a member of the English middle class. The sheer scope of the dishes could be breathtaking, too: for example, a supper given for Marie Antoinette on Thursday, January 24, 1788, had four soups, two main entrées (rump of beef with cabbage and loin of veal on a spit), sixteen entrées, four *hors d'oeuvres,* six roast dishes, and sixteen smaller dishes. A dinner served in 1793 at the height of the Revolution featured twenty-four desserts. The Enlightenment delighted in uniting the pleasures of the table with those of the mind, and Lavoisier probably employed a special cook and may have printed menus for the occasion. Caviar, beefsteak, and curry were available by then. A 1766 painting by Michel-Barthélemy Ollivier showed a huge table with guests seated in high-backed chairs, their faces glowing in the gold light of candles. A harpist played nearby. Even the Christmas dinners held at the Royal Society could not compare in lavishness to these late-night Parisian meals.

As Priestley watched the Lavoisiers, seated beside each other across the table from him, he could not help but reflect upon the differences between Marie's role in society and that of Mary's back home. Mary never expressed dissatisfaction at her lot, but she was so intelligent, and could have accomplished so much here. In England, a woman was automatically excluded from philosophical discussions; she belonged to the "soft sex," existing to serve men's needs. The idea of a female scientist was as alien as that of a female minister; although there was one known female scientist, Carolyn Herschel, sister of the astronomer William, she disparaged her work, saying that what she did was nothing more than the tricks of a well-trained puppy dog. This included tracking the cosmos and discovering eight comets and several nebulae independently of her brother. When she was elected to an honorary membership to the Royal Astronomical Society in 1835, it was long overdue.

Such was not the case in Paris now. According to Enlightenment thinking, respect for women's intellectual accomplishments was the mark of an advanced civilization. A woman's role was to eliminate brutality and promote intellectual exchange. Marie was not yet a *salonnière,* the guiding spirits of the *salons;* women who assumed that role always apprenticed before breaking out, and first mention of Marie's *salon* does not occur for another decade. Rousseau, who despised *salon* society, divided the world into the calming nature of women and the "active, feral nature" of men. He may not have been far off the mark: such conversational styles seem to bridge the centuries, at least according to the present-day linguist Deborah Tannen. As she observes in *Gender and Discourse*: "That the girls and women showed less discomfort finding a topic, elaborated topics at greater length, physically squirmed less, and generally looked more physically relaxed, seems to indicate that they find it easier to fulfill the assigned task of sitting in a room and talking to each other than did the boys and men."

This was the France that an Englishman expected, the France that was the apex of Continental culture and cuisine. But tonight Priestley was the focus of the affair. He had spoken to others during his stay in Paris of the mysterious air that he'd released on August 1 of that year, but this was apparently the first time that "most of the philosophical people of the city were present," he later recalled. So he described the experiment again. He narrated the heating of *mercuris calcinatus per se* in the rays of the burning glass, the prodigious release of gas, the fact that the gas was not soluble in water. Lavoisier sat across the table, Marie by his side, their faces glowing in the flame of the candles. Old Macquer seemed to be doing the questioning, while Priestley struggled to make himself understood in his halting tongue. His French was imperfect. He labored to explain how he'd tested both red lead *(plumb rouge)* and red calx of mercury *(minium),* but kept getting the terms confused. He smiled as he envisioned revealing this surprise, but his translation got in the way in the same way that a poor storyteller garbles a punch line.

"I heated *plumb rouge* in the burning glass—"

"Plumb rouge?" asked Macquer. *"Mais qu'est ce que c'est plumb rouge? Vous voulez dire minium?"*

"Oh yes, *mais oui, minium,*" Priestley corrected. "I heated the *minium* in the burning glass, then applied a candle to the gas that was released." He smiled mischievously. "I don't know what to call it yet, but the candle burned much better than in common air."

There was a moment of stunned silence at this last disclosure, then the entire table seemed to erupt in questions. As late as 1800, Priestley would remember the effect of his words: "At this, all the company, and Mr. and Mrs. Lavoisier as much as any, expressed great surprise."

This is an understatement. Antoine awoke as if from a trance, checked the English with Marie, then started firing questions with the other startled academicians. He'd labored so long at this one problem, but he hadn't seen the way. Now Priestley had imparted two key pieces of information: oxidized mercury contained a gas that could be released by the simple application of heat (without the use of charcoal). More important, this gas sustained combustion—the very secret he sought—much better than did atmospheric air!

This was the transparent secret. How could he have been so blind?

Yet the final puzzle piece arrives in just this way. The imaginative leap comes after much preparation, but unexpectedly. F. A. Kekule dreamed of a snake whirling in space, and this fantastic image solved for him the mystery of benzine. Perhaps, says Lennard Davis, the emotions, like the intellect, have to be surprised and engaged.

> Perhaps the proof of that observation lies in the appearance of such insights so often at fairly unintellectual moments. For me, inspiration tends to come when I've spent the morning stuck on an issue and then go running. Friends have told me that their ideas come when they are swimming, doing a hobby, foaming a latte, or soaping up in the shower. The key, of course, is not to wait for inspiration, but to invest in the hard intellectual work that evokes it, and then to relax enough to allow the subconscious to do its part. You have to be steeped in your project. It has to be rolling around your consciousness day and night. That is when the intuitive process seems to happen—when emotion combines with thought, activity, and dreaming.

Priestley and Lavoisier sized one another up, each a little dazed and not quite sure what had just occurred. *Something* had happened, they were aware of that—and they would go over this moment a thousand times in their minds. It is from such details that history is cut, the shimmer of light on a goblet, the glow of excitement in a woman's blue eyes. If all of life is chemistry, then this night the chemists were human expressions of a redox reaction—

For Priestley gave something up.

And Lavoisier took it away.

CHAPTER 8

The Mouse in the Jar

PRIESTLEY SEEMED EXHAUSTED AFTER THIS DINNER, YET HE MADE one more stop—a visit with his friend Magellan to the lab of Cadet de Gossicourt, a pharmacist of reputed skill, for a pure sample of mercuric oxide. The dinner conversation had made him restless. He'd talked too much and the luxury of time was no longer his. He had to return to his lab.

Shelburne stayed behind another ten days, the politician in him attracted by the pomp of state occasion. Priestley may have been troubled by the reinstatement of the ancient law courts, which few expected to yield real justice in modern times. Accompanied by Magellan, he took the stage to Calais and then the boat to Dover. He arrived back in London on the evening of November 2.

Despite his provincial criticism of the Continent, friends remarked that the trip had done him wonders. "I leave the Doctor to give you a long list of one thousand things his curious mind has picked up," wrote Theophilus Lindsey to a friend. "One thing I will venture to say, that he is much improved by this view of mankind at large."

Back in his lab at Bowood, Priestley turned again to the problem of the strange gas. He focused sunbeams on Cadet's pure sample of mercuric oxide: an ounce of the gas burst from the quarter ounce of red calx, a quantity that always amazed him. He was still stuck on the idea that this was "diminished nitrous gas." On November 19, he shook it vigorously with water, knowing that nitrous air would dissolve, but "a candle still burned in it with a strong flame." Two days later, he shook the same sample, left standing over water—and once again the candle burned.

Now he let the gas stand two more days, until November 23. It had decreased in volume only by one-twentieth of its original measure; he shook it violently in water for five minutes until his arm hurt, as if frustrated that the gas insisted on reacting inappropriately. *Still* the candle burned. The test became a personal contest, as if he *would* find a way to penetrate this gas's invincibility. He could not believe that he had isolated something from the common air that was better than the air itself; his religious belief simply would not allow it. A central tenet of "philosophic necessity" held that God provided the best conditions possible for man's existence and perfection. In God's benign clockwork, how could anything be better than the air provided for our benefit? And yet he'd seemed to have stumbled upon that very thing.

From December to March 1775, he displayed every hallmark of extreme distraction. After this last test, he abandoned the new gas and dove into other projects, including tests on sulfuric acid. He made mistakes and experienced health problems that were uncharacteristic of him. He burned a hand in the sulfuric tests; he suffered from painful boils. They still troubled him when he returned to the mystery gas in March, but disappeared with the resumption of the tests. He'd forgotten until March to subject the new gas to the most obvious of his tests, the one he loved most—his test for goodness. One wonders how he was so blind.

Several scholars have noted the extent to which entire societies resist new discoveries and "truths," and Joseph was repeating this resistance on a personal scale. "The mind delights in a static environment," says Wilfred Trotter in his attempt to understand why scientists themselves are often the most resistant to new ideas. Change from without "seems in its very essence to be repulsive and an object of fear," a reaction so universal that "a little self-examination tells us pretty easily how deeply rooted in the mind is the fear of the new." The reaction is now recognized as so common that it is included in the curriculum for future scientists. "There is in all of us a psychological tendency to resist new ideas," warns W. I. B. Beveridge in *The Art of Scientific Investigation*. It is one of those points where the mass mind intersects so smoothly with the individual's that questions arise about the existence of the *private* man. Such resistance is so similar in both science and religion that Sigmund Freud, for one, questioned whether the underlying dynamics were not the same.

> Probably we do not adequately appreciate the fact that we have here to do with a manifestation of mass psychology. There is no difficulty in finding a full analogy to it in the mental life of an individual. In such a case, a person would hear of something new which, on the ground of certain evidence, he is asked to accept as true; yet it contradicts many of his wishes and offends some of his highly treasured convictions. He will then hesitate, look for arguments to cast doubt on the new material, and so struggle for a while until at last he admits it to himself: "This is true after all, although I find it hard to accept and it is painful to have to believe in it."

There were other distractions, too. The march of folly was so entrenched in King George and the Parliament that by early 1775 there seemed no way to avert the American war. On January 25, the Americans dragged cannon up Gun Hill Road in the Bronx to challenge an anticipated march of the British; on February 9, Parliament declared the Massachusetts colony to be in rebellion. Edmund Burke, Priestley's friend through liberal acquaintances, addressed the Parliament in one last plea for reason: whether England liked it or not, the American spirit of liberty could not be suppressed. Their ancestors emigrated for that very reason, he said. "It cannot be removed, it cannot be suppressed, therefore the only way that remains is to comply with it, or if you please, to submit to it as a necessary evil." His words had little effect. On March 23, Patrick Henry proclaimed: "Give me liberty or give me death!" On April 18, Paul Revere rode from Charleston to Lexington. On April 19, the Revolution began on Lexington Common with the shot "heard round the world."

Long, long had Joseph thought about such convulsions, off and on since 1771. In a letter that year to Theophilus Lindsey, he feared that current events, particularly the developing crisis with America, foretold the long-prophesied Millennium. "To me every thing looks like the approach of that dismal catastrophe," he wrote. "I shall be looking for the downfall of the Church and State together. I am really expecting some very calamitous, but finally glorious, events." He expanded on the theme in his *Institutes of Natural and Revealed Religion*, which would be his most widely read theological work: the rise of democracy paired with the folly of nations pointed directly to the Second Coming of Christ and the end of the world.

So he lingered and mused, drawn to the mysteries of his new gas, repelled by the truths it might reveal. The fires of revolution, revelation, destruction; the flame of the candle renewed by his enigmatic air. He thought about such things, only half formed in his consciousness, as he sat at night with Mary and the children, staring at the warm fire in the hearth, watching the dance of the flames. Fire was like a moral choice: in one direction lay light and love; in the other, death and destruction. He remembered his lessons in Greek. "All things," said Heraclitus, "are an exchange for fire, and fire for all things, even as wares for gold and gold for wares." The Stoics believed in rebirth, a rise from the ashes; every being contained a divine spark, breathed into life by God. The world was a living entity, animated by fire.

▪ ▪ ▪

On March 1, 1775, he returned to the lab and faced his apprehensions. Problem solving was also a moral choice and an act of will. Once both were faced, his mind worked clearly again.

He started this time by subjecting the new gas to his "goodness" test, an application that should have been obvious long ago. Yet by doing so he made a mistake, one built into the process and which he took "so strongly for granted, that the plainest evidence of sense will not intirely [sic] change." He was speaking of himself when he lamented that "the more ingenious a man is, the more effectually he is entangled in his errors; his ingenuity only helping him to deceive himself, by evading the force of truth." In review, when 2 volumes of colorless nitric oxide (NO) were mixed with 1 volume of "common air," red fumes poured forth that were soluble with water. The formula is expressed like this:

$2NO$	+	O_2	→	$2NO_2$
nitric oxide	plus	oxygen	*immediately yields*	red fumes

The "best" air that he'd ever tested yielded 1.8 volumes: that is, the gas left over equaled the entire volume of the nitric oxide plus another 40 percent of the common air.

Priestley's problem at this point was his unshakable belief that nothing could be more pure than common air. This was the wall he could not see over. His test had worked with other gases, so why not now? But when applied to oxygen, a new and confusing result occurred. Instead of a result of 1.8 volumes, the final result with pure oxygen was closer to 1.6. Priestley noticed the small difference, but thought it negligible. He couldn't see what was in front of his eyes. He *did* notice that when he mixed the gases, the fumes were redder and "deeper" than usual, and this, rather than the difference in measurement, nagged at him. But once again he did not seem to pay attention. Fortunately, some insurance was built into his system.

The first redundancy was his faithful candle. Just as in August 1774, he had one handy. In most tests of "goodness," the candle flame went out when inserted in a gas being tested; this time, however, it continued to burn. He was surprised and tried the test again: "I wish my reader be not quite tired with the frequent repetition of the word *surprize*." He left the air in the flask overnight, and came back the next morning. He lit another candle and stuck it in the flask. It, too, burned.

By now, Priestley was truly taken aback. There is a pause in his narration . . . an audible gasp. A frustrated groan. A modern scientist would likely say, "What the hell is going on?" but Priestley didn't swear. He rose from his long table set with beakers, flasks, and the pneumatic trough; his laboratory at Bowood was a large, airy room, provided with big windows for ample light and ventilation. He pushed back his chair and walked to the window overlooking the sloping lawn. He'd thought that perhaps he'd discovered a new way to produce common air, yet ever since last summer he'd also suspected he might have stumbled upon something greater. One test remained—its fitness for breathing. For that he needed a mouse. Several mice. He'd have to set his wire cages in the wainscoting.

Priestley made no secret of the fact that he did not like to use living things, usually mice, in his pneumatic investigations. This was one reason for his happiness when he developed his "goodness" test: nothing would be harmed. It was painful to watch a creature die. He'd place a mouse in a bell jar full of gas and a shock would go through it with the first breath. Sometimes it would fall on its side and its lungs would heave. Sometimes, it would go into seizures. Often the fading would be gradual—minutes, rather than seconds, and the mouse simply went to sleep. But always he was reminded of his own fight for life at Aunt Sarah's and how hard it had been to draw a good breath. How darkness edged upon him from every side.

All of life was like that—a fight against encroaching darkness. A minister could not help but draw parallels between the plight of his little prisoners and man. The one to three years of life for *Mus domesticus* was not so much greater in the scope of the cosmos than man's three score and ten: the mouse's 51 heartbeats and 163 breaths per minute meant it lived faster, but did it experience more as well? A century later, the psychologist William James would have the same questions about the senses and our experience of time: "Suppose we were able, within the length of a second, to notice 10,000 events distinctly, instead of barely 10, as now: if our life were then destined to hold the same number of impressions, it might be 1,000 times as short." What if, like mice, we were merely captives in God's evolving lab? We sit uncomprehending under the sky's clear arc, just as Joseph's mice did in their transparent jars. We twitch our whiskers in anticipation, slaves to the Experimenter's whim.

Joseph was not alone in seeing the metaphor. In 1773, Anna Laetitia Barbauld wrote a poem about one of Priestley's mice trapped in a cage. The poem first ran in *The Annual Register*, then was reprinted in a collection of her works; since it was written before the oxygen experiment, it suggests that Priestley's use of mice was ongoing and well known. Barbauld's tone suggests a belief that the mice faced a cruel fate, and Priestley's softheartedness suggests that he agreed:

Oh! Hear a pensive captive's prayer
For liberty that sighs;
And never let thine heart be shut
Against the prisoner's cries.

For here forlorn and sad I sit,
Within the wiry grate;
And tremble at th' approaching morn,
Which brings impending fate.

If ever thy breast with freedom glow'd,
And spurn'd a tyrant's chain,
Let not thy strong oppressive force
A free-born mouse detain. . . .

The cheerful light, the vital air,
Are blessings widely given;
Let nature's commoners enjoy
The common gifts of heaven. . . .

So when unseen destruction lurks,
Which men like mice may share,
May some kind angel clear thy path,
And break the hidden snare.

By March 8, he'd caught enough mice in his traps and was ready. He prepared two jars of air that were inverted over the filled trough—one jar contained common air, the other this unknown oxygen. He picked up a mouse by the scruff of its neck, talking to it softly in hopes of allaying its fears. He placed it in the upended jar of ordinary air, letting it rest safely on a platform above water. He repeated the process with a second mouse, placing it in the jar with the unknown gas. He sat back and waited, playing the flute as he watched. He had no idea how long he must wait, or what he would see. This was the scientific method at its simplest: *Try X, and see what occurs.* Theory came later. All that mattered now was the experimental act, the drama of the laboratory and necessity of observation.

After awhile he stopped. Something was happening. He put the flute aside and leaned forward for a better view.

The mouse in the jar of ordinary air was showing signs of distress and fatigue. It moved on the platform as if seeking some escape; its ribs moved like rapid bellows. He glanced at his pocketwatch: little more than ten minutes had passed since he'd put it on the platform. Another five minutes and the mouse was unconscious. He plucked it out by its tail, hoping to save it, but realized it was already dead.

He glanced over to the jar containing the mystery air. The mouse in there stared back, wriggling its whiskers, as healthy as when he'd put it in there fifteen minutes earlier. It lay on the edge of its platform, breathing easily, staring at the water beneath it as if contemplating a swim.

This continued another fifteen minutes. Thirty minutes in the jar—it was impossible! For thirty minutes, this mouse had prospered in the "mystery air," while his brother, breathing the air of which we all partake, died in half the time.

After the thirty-minute mark, however, the remaining mouse began to show unmistakable signs of fatigue. Joseph perched on the edge of his chair as he watched: from outside came the sounds of spring, the yells of his and Shelburne's children on the lawn, ravens calling from tree to tree. But he did not hear them. The prisoner's movements grew slower. Sluggish. A stupor came over it; the mouse closed its eyes. Joseph rushed to pluck it from the glass, but such creatures are fragile and it seemed that life had already fled. Through his fingertips he could feel that its skin was chilled, yet he thought he still felt its heartbeat through his fingers. He rushed to the fire and held the mouse to the heat, cradled in his hands. A whisker tickled his palms. He opened his hands and watched the mouse revive.

How does this feel, standing on the brink of something you know will change everything? Dizzy. Joseph felt dizzy. He was unable to believe his senses as the surviving mouse came back to life. For thirty minutes this animal sat in the new air and thrived, while the first mouse, breathing the same air as Joseph did, had died.

That night, Joseph got very little sleep: he pondered the "extraordinary fact" of the survival of the mouse and existence of a miraculous gas latent in the air around him. He lay still "on my pillow" and thought, but could not account for what he had seen. Was it possible that he had indeed discovered something purer than the common air,

But caution was needed, too. Perhaps breathing in this air "might not be proper for us" in a healthy state, he said, predating the concern of William James about "burning out" by at least a hundred years. For as a candle burned faster in oxygen, "so we might . . . *live out too fast,* and the animal powers be too soon exhausted in this pure kind of air. A moralist, at least, may say, that the air which nature has provided for us is as good as we deserve."

Perhaps he was a moralist, but this could not stop his imagination, and now his mind veered from health to destruction. It was known that "inflammable air" (hydrogen) mixed with two measures of common air caused an explosion of "great advantage." But what if it were mixed with this pure air? He tried it: a mixture of the same proportion caused a report "as loud as that of a small pistol; being . . . not less than forty or fifty times as loud as with common air."

That led next to his air's effect on the instruments of war. If gunpowder were stored in casks filled with this gas, the result might exceed anything seen in history. Such oxygen-enriched gunpowder might yield ten times the explosive effect of powder stored in common air: "I should not . . . think it difficult to confine gunpowder in bladders, with the interstices of the grains filled with this, instead of common air; and such bladders of gunpowder might, perhaps, be used in mines, or for blowing up rocks, in digging metals. . . ." Or for war. Knowledge *was* power in ways Francis Bacon never dreamed.

Now Priestley envisioned welding: he got the idea after filling a bladder with his oxygen and puffing it through a glass tube upon a piece of lighted wood. It would be easy enough to "supply a pair of bellows with it from a large reservoir," he conjectured; "the strength and vivacity of the flame is striking, and the heat produced . . . is also remarkably good." The candle or wood itself burned with a crackling noise, "as if full of some combustible matter."

With that, he could contain his discovery no longer. On March 15, 1775, he wrote to Sir John Pringle, secretary of the Royal Society, announcing his discovery of this new pure air. His letter was read before the assembled body on March 23—a formal announcement of his discovery to the world. He didn't depend entirely on them, either, to make his excitement known. On March 19, he wrote Jeremy Bentham, apologizing for missing an appointment. "You will blame me for not having waited on you," he lamented, but some health problems (either his boils or a recurring attack of gallstones) and "a great hurry of engagements" had completely filled his day. On April 1, he wrote Richard Price, describing the discovery in detail. At some unspecified date, he showed Magellan the improvised blowtorch and no doubt talked about his ideas regarding the increased destructiveness of gunpowder that was stored in his new gas. There is no record of whether Magellan informed his spymaster about this meeting, but there is also no reason to think the agent failed to report the most momentous thing the little minister had ever done. Between 1772 and 1777, Magellan sent thirteen letters describing English scientific work, and in particular Priestley's, with whom he had the closest ties. Magellan rarely hesitated, and soon afterward Lavoisier was assigned by Montigny to improve France's own gunpowder stores. The *pabulum* of life held deep secrets for war and death, and with March 1775, we see the arms race between France and England truly begin.

Priestley rightly felt that he'd surpassed Newton, that he'd discovered one of the greatest secrets of all time. Yet in the midst of his greatest triumph, he made his greatest mistake. He needed to name his discovery, for in naming lies possession, existence, identity. This was the chance to break free of the ancients and say, "I've discovered something truly different. The new millennium begins with this."

But Priestley could not break with the past. Despite the evidence of his senses, he was still tangled up in phlogiston: he claimed his discovery was air somehow stripped of that flammable, mythical fluid. He'd broken through in the lab, but not in his mind.

Oxygen was not yet *oxygen*. Joseph stuttered in the worst way he ever had, and called his new find "dephlogisticated air."

CHAPTER 9

The Twelve Days

WHAT EXACTLY WAS THIS AIR OF PRIESTLEY'S THAT SET EUROPE'S halls of science abuzz? Was it, as he told the Royal Society, air that had been mysteriously "dephlogisticated," stripped of its toxic phlogiston as if run through a microscopic sieve? This was unsettling enough, for although it supported the ancient definition of air as a homogeneous fluid, it also suggested improvements upon Nature, allowing scientists to glimpse a world in which they could play God. Or was the threat deeper—that the air we thought we knew was not, after all, as we'd known? Could we break down the air and put it back together in ways never foreseen? This was an alien idea, one that went far beyond Priestley's claim that his pure air was merely old air that was in some way refined or filtered. Would every day hereafter be another Day of Creation?

Yes, it is quite possible, Lavoisier replied—privately at first, then more publicly as the new year dawned. Not only possible but probable that for century after century, we'd gotten it wrong. In the period from Priestley's departure from Paris in October 1774 to March 1775, Lavoisier tried to replicate his rival's experiments down to the letter. He obtained from the pharmacist Cadet the same sample of red mercuric oxide that Priestley had taken home to England; he burned a candle, as had Priestley, in the gas thrown off. The flame "threw out such a brilliant light that the eyes could hardly endure it," he said. In a public session of the Academy held on November 12, 1774, he announced: "Atmospheric air is not a simple body, but is composed of different substances." A few days later, in the Paris journal *Observations sur la physique*, he refined this statement by adding that air, "far from being a simple being, an element, as has been commonly thought, must on the contrary be placed at least in the mixed class, and perhaps even in the compound one."

Yet Lavoisier's experiments were not *exactly* like Priestley's, and in this he erred. Though Lavoisier would eventually take Priestley's discoveries farther with his insights, the younger man was not yet able, like Priestley, to pare down tests to their most simple form. His experiments were not sufficiently fenced off—he was unable to control for confusing variables. He did not heat the red oxide directly, as had Priestley, with the burning glass and without the aid of any other substance. Instead, he ran two sets of experiments in which mercuric oxide was heated in a small retort: in one test, the red oxide was mixed with powdered charcoal; in the other, the red oxide stood alone. The gas thrown off flowed through a long glass tube into a bell jar filled with water, and bubbled up into the top of the jar. The gas thrown off from the red oxide mixed with charcoal tested positive for carbon dioxide; the gas from the red oxide alone did not. This showed Lavoisier the danger of using charcoal for heating, but he still did not see as clearly as Priestley the importance of the unknown gas. He seemed to think it was common air, not something unique—he failed to subject it to Priestley's "goodness" test, or to place a mouse in the jar to see how long it lived. When he read his results to the Academy on April 26, 1775—in a paper known as the "Easter memoir"—he said the gas thrown off was common air, but in a form "more respirable, more combustible, and . . . even more pure than the air in which we live."

Worse, past sins were catching up, sometimes in embarrassing ways. Although he was never caught for covering up Scheele's discovery, he was found out for a lesser omission in his 1774 *Opuscules*. He'd failed to mention the rival discoveries of two Italian chemists, Gianbatista Cesare Beccaria and his nephew, Giovanni Francesco Cigna. Their supporters' letters to the French press forced him publicly to recant his mistake: "I have no intention of taking over somebody else's work," he wrote in the December 1774 issue of abbé

Rozier's journal, "and I am convinced that scrupulousness in literature and physics is no less essential than in morals."

As soon as he wrote this, his honesty was questioned again. This time the injured party was Pierre Bayen, still smarting over being muscled aside earlier by Lavoisier. In January 1775, Bayen took revenge. He republished a book by the Périgourdian doctor Jean Rey, which had appeared 150 years earlier; in it, Rey said that the weight gain of metals during heating must be due to the fixation of air—the same discovery Lavoisier claimed for himself in his "sealed note." Aware of the novelty of his statement, Rey considered this "something that until now has never been understood." Bayen asked Rozier to publish Rey's findings in his journal so that chemists everywhere would know that he'd guessed the secret long before Lavoisier.

Lavoisier was furious, calling Rey's speculations "a vague assertion, thrown out by chance and backed up by no experiments": it was an obscure work, virtually unknown. Although Lavoisier was probably telling the truth, he refused to acknowledge his earlier injustice to Bayen. And still he did not learn. In the "Easter memoir," Priestley's hand is seen throughout, especially in the choice of heating mercuric oxide, first mentioned over dinner. "Mercury *precipitate per se* . . . seemed to be eminently suitable for the purpose I had in mind," Lavoisier wrote, but nowhere did he mention Priestley. If he'd acknowledged this debt, he would have avoided all the charges of dishonesty that followed the rest of his life, even when he no longer deserved them.

Priestley's response to the slight turned a row among local colleagues into an international affair. Previously, he'd seen Lavoisier as simply too ambitious, a common trait in the young and talented. This time, however, his charge was weightier, and for Lavoisier to be blasted by one of the world's preeminent chemists turned previous minor complaints into a chain of evidence supporting a pattern of dishonesty and plagiarism. Priestley made his complaint seven months after release of Lavoisier's Easter memoir, a quick response given the era's slow state of scientific communication. His reply was blunt, printed in Volume II of his *Experiments and Observations on different Kinds of Airs*, released on November 1, 1775. The fact that he made his complaint in this book assured that it was aired before all the chemists of Europe; it was included in a section entitled "An Account of some Misrepresentations of the Author's Sentiments, and some Differences of Opinion with respect to the Subject of Air," which left little to the imagination. The dinner guest did not hesitate to list his host's litany of sins:

> After I left Paris, where I procured the *mercuris calcinatus* above mentioned, and had spoken of the experiments that I had made, and that I intended to make with it, he [Lavoisier] began his experiments upon the same substance, and presently found what I have called dephlogisticated air, but without investigating the nature of it, and indeed, without being fully apprised of the degree of its purity. And although he says *it seems to be* more fit for respiration than common air, he does not say that he has made any trial to determine how long an animal could live in it. He therefore inferred, as I have said that I myself had once done, that the substance had, during the process of calcination, imbibed atmospherical air, not in part, but in whole. But then he extends his conclusion, and, as it appears to me, without any evidence, to all metallic calces; saying that, very probably, they would all of them yield only common air, if, like *mercuris calcinatus,* they could be reduced without [the addition of charcoal].

The rebuke didn't end there. Priestley felt he'd been badly used by Lavoisier for failing to make any mention of the facts imparted during that lovely dinner in October 1774. Lavoisier insisted that he'd begun his research before the dinner, as early as April 1774, but few believed him. Before the dinner with Priestley, he had not used mercuric oxide in his tests. In addition, he admitted that he first used the burning glass in November of that year.

Battles between scientists over claiming rights were nothing new, and Priestley was a veteran of such conflicts. During the same period that Lavoisier was being called a plagiarist, Priestley was accused by Bryan Higgins, a scientist in the circle of Samuel Johnson, of stealing the idea of dephlogisticated air from Higgins's lectures. Priestley's famous creativity in the lab was nothing more than "rendering the phenomena, which all practical chemists have observed and understood, perfectly mysterious and surprising to others." Priestley's response became famous for his description of his own possible role as a scientist:

> It may be my fate to be a kind of comet, or flaming meteor in science, in the regions of which . . . I made my appearance very lately, and very unexpectedly; and therefore, like a meteor, it may be my destiny to move very swiftly, burn away with great heat and violence, and become as suddenly extinct.

Implicit within the passage was a warning to Higgins: Stay out of the way lest you get burned.

Yet more was at play here than a simple squabble over provenance. England was on edge, and the world was sliding into war. Johnson and his proxies despised Priestley for his "Whiggish democratical notions," made all the more relevant by events in America. In June 1775, British and American forces slugged it out in the Battle of Bunker Hill; in November, the Americans took Montreal, and in December they attacked but failed to take Quebec. French arms makers were rumored to be in secret talks with the colonials; there were pogroms in Poland; the Turkish state of Bukovina seceded from Austria. After a decade of uneasy peace, the world seemed once again headed toward turmoil and bloodshed, much of it inspired by ideas of freedom and other unwanted "democratical notions."

If anything, Priestley was upset at being used, yet once he complained of Lavoisier's actions, he seemed willing to let bygones be bygones. The search for truth should be communal, not competitive, said Priestley; like Lavoisier, he'd made mistakes, but such errors often led to greater understanding.

> It is pleasant when we can be equally amused with our own mistakes, and those of others. I have voluntarily given others many opportunities of amusing themselves with mine, when it was entirely in my power to have concealed them. But I was determined to show how little *mystery* there really is in the business of experimental

> philosophy, and with how little *sagacity,* or even *design,* discoveries (which some persons are pleased to consider as great and wonderful things) have been made.

Although conciliatory and good-natured, there was also a touch of smugness in Priestley's words. And Lavoisier had learned his lesson. Once again, Priestley pointed the way for him out of his experimental muddle. Once again, he saw the path ahead. The next time he challenged the older chemist, he'd have more than speculation in his quiver. Next time, he'd come well armed.

▪ ▪ ▪

The year 1776 would be momentous, amazing in the way so many loose threads wound tightly into one. It was the year of the American Declaration of Independence; of Thomas Paine's *Common Sense* and Adam Smith's *Wealth of Nations*; of Matthew Boulton and James Watt's first steam engine, a technical advance that set England's industries ahead of the competition. It was the year France began supporting the American rebels, first with loans and moral support, later with men and munitions. The success of that Revolution, fostered in the name of French ideals, inspired hope in Paris and the countryside. If America could be transformed, why not France? In truth, the change had already begun. The debts incurred by the war so weakened the fragile Treasury that there was no stopping the French Revolution.

The year 1776 could also be called the first year of the Chemical Revolution. It was the year of Lavoisier's most original experiment, the year he took the atmosphere apart and put it back together . . . the year that all the long days of questioning bore fruit and he finally saw the truth. After 1776, Lavoisier no longer stood in Priestley's shadow.

Yet if this was the year when everything finally started to make sense, the preceding one had been full of chaos. In addition to the personal attacks, Antoine's father was stricken with the apoplexy that would kill him in two years. While the old man lay in bed, food riots raged through Paris, protests against the rising cost of bread. As 1,500 men of the *Garde de Paris* leveled barricades and arrested 139 protesters, Antoine sat by Jean Lavoisier's side. Everything of real importance seemed not to last; he was mired in controversy fueled by his ambition, and all this elderly man had ever done was try to protect his children from an unfriendly world. "You know the tender friendship which has always existed between us," he wrote his aunt. "He has done great good and no harm. I hope that his spirit of integrity will always serve me as a guide."

It was a time of change, some good, some bad. As his father sickened and Antoine slogged through innuendo, he purchased for 65,000 *livres* a title—*conseilleur-secrétaire du roi, maison, finance et couronne de France*—which brought with it hereditary rank and made him a nobleman. Like his decision to join the Farm, entering the nobility would bring immediate advantage and long-term disaster. Records show he made little use of the title, yet position and privilege were important to him and by purchasing the title he gained the one thing that drove him in science and finance—a drive for order and quest for control.

Among the bourgeoisie, ennoblement was the ultimate indicator of social success, and privilege the hallmark of a nation without uniform laws or institutions. As a wealthy member of the bourgeoisie, Antoine already benefited from special rights and exceptions; as a nobleman, he entered an entirely different plane. Nobles formed a separate social order or estate, and everyone else—even his father and beloved aunt—were commoners. Nobles took precedence on public occasions and carried a sword; they were entitled to a coat of arms, a trial in separate courts, and even a distinctive style of execution should the need arise. They were not required to billet troops in their estates, or submit to conscription; they enjoyed considerable financial advantage, especially exemption from the hated salt tax, the *gabelle*. There were 120,000–350,000 nobles in France, who owned about a third of the nation's land and controlled feudal rights over the rest; so advantageous was the rank that during the eighteenth century members of the bourgeoisie bought their way into society's top tier at the rate of two per day.

If this wasn't enough, he was named *régisseur des poudres* to the Gunpowder and Saltpeter Administration, newly established to reverse the disastrous decline in gunpowder production since France's loss of India in the Seven Years' War. At first, the title meant little more than a name. But thirteen months after his appointment, Antoine and Marie were told to move to the grim old Arsenal of Paris, around the corner from the notorious Bastille. Gunpowder had not been manufactured there for decades, but now, as Commissioner of Powders, Lavoisier was expected to restart production and make French gunpowder the best in the world.

In April 1776, Antoine and Marie said good-bye to their home on the rue Neuve-des-Bons-Enfants, packed up the laboratory equipment, and moved across town. The Arsenal, built in 1599 by the duc de Sully, *Grand Maître de l'Artillerie*, may well have been the most lavish munitions plant ever built, an ornate group of workshops, warehouses, foundries, and residences. Henri IV, a regular visitor, was on his way there when assassinated on May 14, 1640; it was gutted by fire in 1716, then rebuilt according to the original specifications in 1716–29. The Lavoisiers' new home consisted actually of two parts: the Grand Arsenal, home of marquis de Palmy d'Argenson, the facility's governor, a labyrinthine palace of five courtyards overlooking the Seine; and the Petit Arsenal, which would be the home of the Lavoisiers for the next sixteen years.

Despite its name, the Petit Arsenal was anything but small. This was the facility's actual working space: surrounded by two courtyards, it served as the Bastille's gunpowder warehouse, with a passage leading into the infamous prison. Yet for all the symbols of oppression, the Lavoisiers' quarters were palatial. They had a large private apartment and library, and a huge laboratory. Adjoining sheds housed thousands of flasks, retorts, and other glass vessels; in the months to follow, Antoine added his own instruments—barometers, thermometers, calorimeters, precision balances, and scales. Between what was already stored at the Arsenal, what he brought with him, and what he requisitioned from the state, he soon turned his new domicile into a scientific paradise. For the time, it was probably the world's most current and well-appointed public lab, built on the research and development for war.

In this new home, Antoine and Marie transformed themselves. After 1776, there were few, if any, new accusations of plagiarism or intellectual thievery; in the future, such charges would point back to the bad years of 1774–75. After the move of 1776 and the death of his father in 1777, it was as if Antoine tried to atone for his dishonesty. A humanitarian streak arises in the 1780s: he throws himself into projects designed to improve the lives of the peasants. When government grants fell short for important scientific projects, he made up the shortfall from his own considerable fortune, knowing full well there was little hope of recompensation.

But more than anything else, he worked like a dog. He rose at five every morning and worked in his lab from six until nine. From then until seven at night, he went first to the offices of the fermiers, then spent the afternoon at the Gunpowder Administration and Academy of Sciences. After supper, he spent another three hours in his lab, 7–10 p.m. In addition to these six daily hours of scientific work, he reserved Saturdays entirely for experiments with chosen students, much as Rouelle had spent with him. "For him," Marie would later write,

> it was a blissful day. A few enlightened friends, a few young people, considering it an honor to be able to participate in his experiments, gathered in the laboratory in the morning. It was there that they ate lunch, discussed, and created the theory that immortalized its author. It was there that one had to see and hear the man whose mind was so sound, talent so pure and genius so lofty. It was from his conversation there that the nobility of his moral principles could be judged.

She was obviously in love, and like any ardent lover more than willing to overlook faults in the pursuit of a "higher good." At the Arsenal, Marie was finally made full partner. She'd finished her lessons in chemistry and had learned enough English and Latin to serve as a full-time translator. In the lab she sat at a small table placed to the side of the traffic while Antoine and his collaborators padded about; she noted down the protocols of every experiment and the figures of results as they were called to her. She recopied Antoine's scribbled notes into the lab register, putting them in necessary order if there was any hope of replicating his experiments. After these sessions, she presided over tea or the Monday dinner, creating a kind of Academy-away-from-the-Academy. Said the Swedish astronomer

Anders-Johann Lexell, a frequent visitor, "M. Lavoisier is a pleasing looking young man, a very clever and painstaking chemist. He has a beautiful wife who is fond of literature and presides over the Academicians when they go to her house for a cup of tea after the Academy meetings."

And before this all started, almost as soon as they arrived at the Arsenal, they performed together Antoine's most celebrated experiment: one that hearkened back to a more innocent and blameless time in his life, returning to the habits of his student days, rarely leaving the lab.

▪ ▪ ▪

There is no record of the planning of the twelve-day experiment of April 1776, but Lavoisier had pondered it for awhile. During the winter and early spring of 1775–76, before the move, he repeated Priestley's experiments, but this time with more care and understanding until he recognized the uniqueness of the little Englishman's "dephlogisticated air." In his notebook for February 15, 1776, he noted his preparation of "*l'air déphlogistique de M. Prisley* [*sic*]," but preparation was no longer enough. He wanted to find that ineffable substance responsible not only for life but for combustion, the unknown thing that reacted with metals when they were heated. After moving to the Arsenal, this was apparently the first test he set up in his lab. He took a glass retort and bent the neck so its end could be inserted under a bell jar in a trough of mercury. He introduced 4 ounces of very pure mercury into the jar, then lit a fire beneath it, keeping the temperature just below boiling. For twelve days straight, he heated, measured, and weighed.

On the first day, "nothing noteworthy happened," he later wrote. "The mercury, though not boiling, was continually evaporating." The fluid covered the inside of the vessel with small drops, first minute, then merging as they rolled down the glass into the pool of mercury below.

On the second day, an important change occurred. Small red particles the color of rust floated on the silver surface of the liquid. It had to be calx, the red mercuric oxide! His heart beat faster: the particles were so small, but in them might lie proof. He moved a bed and washbasin into the lab, and called for a change of clothes. During the next five days, the red spots increased in size and number, after which they ceased to form. Marie asked if he would not leave the lab, but when she saw the particles she stayed, too. On the twelfth day, when it was evident that no other particles would form, he turned off the fire and let the vessel cool.

Now Lavoisier examined the air left in the retort and found that it had diminished by 8–9 cubic inches, or one-sixth of its former volume. At the same time, the red calx floating atop the mercury weighed approximately 45 grains. Mice placed in the jar died within seconds; candles were immediately extinguished—both negative images of Priestley's trials. (The remaining gas, although he did not know it yet, was nitrogen, which he called *mofette.*) It would be easy to confuse this gas with carbon dioxide, as he had done in the past, but he'd learned his lesson. "Fixed air" turned limewater cloudy, but *mofette* did not. It was an entirely different entity.

The next step involved "reducing" the 45 grains of red powder that had formed on the surface of the liquid. This, too, was a replication of Priestley's procedure, but with a quantitative spin, something the Englishman had never considered. "I carefully collected the 45 grains of calx of mercury," he wrote.

> I put them in a very small glass retort, whose neck, doubly bent, was fixed under a bell-jar full of water, and I proceeded to reduce it without adding anything. By this operation I recovered almost the same amount of air as had been absorbed by the calcination, that is to say, 8 or 9 cubic inches, and on combining these 8 or 9 inches with the air vitiated by the calcination of mercury, I restored this air exactly enough to its state before calcination, i.e., to the state of common air.

Simply put, the loss in the weight of the oxide matched the weight of the new gas: he'd produced 41.5 grains of pure mercury and 8 cubic inches of a gas "greatly more capable of supporting both respiration and combustion than atmospherical air." It was, measure for measure, Priestley's "dephlogisticated air." Candles burned beautifully in it; mice scampered around without care. He tried Priestley's "goodness" test and it checked, too. Finally, when he mixed this pure gas with *mofette,* he again had common air.

As a result of this, Lavoisier was able to make a clear statement of what he had found:

> I have established in the foregoing memoirs that the air of the atmosphere is not a simple substance, an element, as the Ancients believed and as has been supposed until our own time; that the air we breathe is composed of respirable air to the extent of only one-quarter and that the remainder is a noxious gas (probably itself of a noxious nature) which cannot alone support the life of animals, or combustion or ignition.

He, and others, had said this before—but now he went one step further.

> I feel obliged . . . to distinguish four kinds of air.
>
> First, atmospheric air; that in which we live, which we breathe, etc.
>
> Secondly, pure air, respirable air; that which forms only a quarter of atmospheric air, and which Mr. Priestley has very wrongly called dephlogisticated air.
>
> Thirdly, the noxious air which makes up three quarters of atmospheric air and whose nature is still entirely unknown to us.
>
> Fourthly, fixed air. . . .

It was a remarkable conclusion, one he'd proved by sticking to Priestley's procedure with one major departure—his use of weights and measures. This very empiricism opened more doors in his mind. He'd taken Priestley's "dephlogisticated air" and rechristened it "pure air"—this, with time, would become "oxygen." That step was waiting, yet he'd already glimpsed chemical processes at work that Priestley had overlooked entirely. Burning, he said, was the union of the burning substance with the pure air, not the opposite, proven by the fact that the product formed during burning weighed more than the original substance itself. Where did the extra weight come from?

Simple, he answered. It equaled the weight of the air, plus that of the burning body.

Burning added weight: there was a union, shown by the most sensitive balances in Europe. There was no loss of mysterious phlogiston, nor of the fluid caloric, which was supposed to leave the substance in the guise of flame. "It can be taken as an axiom that in every operation an equal quantity of matter exists both before and after the operation," he wrote. All chemical changes obeyed the law of the indestructibility of matter. There were no ghosts in the process, no ether escaping notice of his scales. In the chemical change of burning, nothing was gained or lost, even in the vaporous air.

In one stunning experiment, Lavoisier had finally started the revolution which he'd dreamed about so long while ending the ancients' chokehold on chemistry. The air was not a primal element—there was more in it that he did not know about, substances more basic than he'd ever dreamed. He'd found two of these already: his "pure air," and its deadly brother. Everything was proven by his delicate balances, the most sensitive of which could be affected by the five-hundredth part of a gram.

Yet proof and acceptance are separate worlds. When he released his results, two competing experimental camps sprang up, where earlier there had been no rivalry. The first was in Britain, grouped around Priestley, the second in France, circling Lavoisier. Men's view of reality itself was at stake, made no less strident by envy, ambition, national pride, and, all too soon, the threat of revolution and war.

CHAPTER 10

The Language of War

WHEN LAVOISIER REVEALED HIS "NEW CHEMISTRY" TO THE academy in August 1778, half of his revolution was complete; yet he was wiser now, more prudent, willing to take time to cover his bases rather than leave himself as vulnerable as before. Because publishing delays would not allow a wider audience until November 1779, it was only the academicians who were initially enlightened, stunned, or enraged. In fact, much of the war over phlogiston would be fought between old-guard French chemists and the Young Turks with Antoine.

"The principle which unites with metals during calcination, which increases their weight," he said, was none other than Priestley's "dephlogisticated air." This could again be freed after combination to emerge as something "more suited" than the air around us "to support ignition and combustion." But this was just a preamble to greater things. Lavoisier believed his pure air contained heat (or caloric) that was released when it united with a substance; this was especially important in the body, for there it burned the carbon found in foodstuffs to form the carbon dioxide exhaled as breath, and release the heat that keeps an animal warm.

The latter statement was new, and indeed radical. That respiration was merely a slow form of combustion was a claim that had never been made before. The genesis of body heat was not a mechanical process, caused by the "pumping of blood" that somehow caused friction; neither was it due to some "inborn" matter, a "life-spirit" or unexplained vitalism. Respiration was a *chemical* process, involving the slow combustion of living tissues, and its fuel was this purer air. Just as wood burned in a fireplace, man burned food in his body: pure air drawn into the lungs united with the carbon in the blood to produce carbon dioxide, which was then exhaled.

But how could this be proved? It was a puzzle until 1783, when his friend, Pierre-Simon de Laplace, and he demonstrated the relationship of animal heat to breathing with a struggling guinea pig (hence the origin of the phrase). A vessel containing pure air was placed inside a second, well-insulated one containing ice. The guinea pig was placed in the inner vessel: the amount of carbon dioxide exhaled was determined by the limewater test, while the amount of heat used up was measured by the amount of melted ice. Next, a calculated amount of pure carbon was substituted for the guinea pig in the inner vessel, then ignited by a burning glass.

The reasoning was simple: If body heat is produced by the slow combustion of carbon, shouldn't the total amount of heat be equal to that produced by burning an equivalent amount of carbon? Actually, the guinea pig proved to be more efficient, since organic combustion spawned more heat than that generated by simply burning a lump of mineral. It was only later, after Lavoisier analyzed the organic compounds and discovered that they contained hydrogen as well as carbon, that he could explain the discrepancy. Today, we know that both elements burn in the body; the combustion of carbon produces carbon dioxide, while that of hydrogen produces water.

Later, Lavoisier would see the web of interdependence in plants and animals, all based on his processes. He peered into the laws that held life together on this planet in the same way that Newton gazed at the forces binding planets to the cosmos. Material for building plant tissue came from air, water, and minerals: although he never saw the importance of sunlight, he did grasp that the organic chemicals built up by the plant served as food for animals. Combustion, respiration, fermentation, death, decay . . . all were means of returning organic materials back to the air and soil. Priestley had shown that plants purify air, but once again had not seen the bigger picture. The plant and animal kingdoms were opposites of each other: the combustion and putrefaction that occurred in both were inverse phenomena. In one process, life was created; in the other, destroyed. The key to both was oxygen, necessary to maintain the balance in all things.

In fact, Lavoisier told his listeners, it seemed as if this purer air altered just about everything it touched, and did so by forming acids or oxides. When combined with charcoal, it made carbonic acid; with sulfur, sulfuric acid. With metals, it made oxides. Lavoisier hoped to one day unite oxygen "to animal horn, silk, animal lymph, wax, essential oils, pressed oil, manna, starch, arsenic, iron, and probably a great many other substances of the three kingdoms."

Granted, there were problems with the scheme. His theory of acids was too rigid, and *calorique,* his heat principle, was not much different from phlogiston. Nor was the explanation of combustion entirely clear. Where Priestley's dephlogisticated air lacked phlogiston, Lavoisier's purer air was somehow mixed with the matter of fire. Phlogistonists said his theory was merely an inverse of Stahl's and dubbed it the "antiphlogistic theory." In 1777–78, Lavoisier was not yet suggesting that chemists entirely abandon the phlogiston theory, but he was heading that way.

Yet even with its faults, the new theory was exciting—and threatening. Most threatening was the way in which it held up in the lab. Lavoisier's construct was not merely a theory of combustion; it was a theory of composition, too. A corollary of oxidation was the fact that an oxide could no longer be considered a simple substance, but a compound formed by its merger with this vital air. Metals were now recast in the role of simple substances, as were carbon, phosphorus, and sulfur. A list of basic elements was taking shape in Lavoisier's mind and would shape a research agenda for the second half of his revolution.

It was this second half, stretching from 1777 through the mid-1790s, that convinced a growing body of scientists—in the end, all but Priestley—and turned the tide. Lavoisier's operational definition of an element just worked too well. It was simple, elegant, and stubbornly straightforward. If a body could not be broken down to more basic constituents, then it must be an element, one of Nature's fundamental building blocks. If time and more advanced technology *did* see it broken apart, then from its rubble another element would be found. When in 1789 Lavoisier eventually set forth his list of thirty-three elements, he conceded that some might be decomposed further. Yet of this original list, twenty-three are still recognized as elements today.

The theory's implications were staggering, even if they were issues that he preferred to leave alone. The vital flame or animal heat were not the work of God, but the result of chemistry. Every thought, every emotion, every act of love or lust—the hand of God was not involved. They were chemical reactions, a burning in the blood. God and chemistry were synonymous. Might as well worship a test tube as the Host—and chemistry could always be manipulated by man.

But in the beginning, he had to devise a word. Every good parent names his newborn; every priest identifies the object of reverence. St. John had seen it long ago: "In the beginning was the Word, and the Word was with God, and the Word was God." The basis for his new faith needed a name, and since his vital air seemed to transform so many substances into acid, it was only fitting to name his new element *principe oxigène.* It was a nod to the Greeks even as he left them behind: "that which generates acids," or, from the Greek, *oxy* ("acid") *-gen* ("maker").

Oxygen, as an idea, was born.

▪ ▪ ▪

The naming of oxygen is treated reverently in the annals of science, a major waystation on the road to the modern. But there was nothing reverent about the christening. Lavoisier's "Chemical Revolution" is always cited as a prime example of the "paradigm shift" that occurs when one scientific reality gives way to another, yet at the time looked less like an intellectual debate and more like war. Scientists divided into hostile factions of English phlogistonists and French upstarts; converts were won by propaganda, proselytization, and intrigue. As France edged toward murderous revolution and England toward hysterical reaction—as Louis XVI strolled through Versailles as if unconcerned and George III became shut up in his madness—both nations inched with scarcely concealed deliberation toward war. And as these spasms built upon each other, the war over oxygen became a personal one between Priestley and Lavoisier.

These were apocalyptic days. Priestley wrote of the Millennium and the Second Coming. Lavoisier envisioned a new chemical reality. "If man could learn from history, what lessons it might teach us," bemoaned Samuel Taylor Coleridge, who among Romantic poets was the one who most admired Priestley. "But passion and party blind our eyes, and the light which experience gives us is a lantern on the stern which shines only on the waves behind us." Passion and party also enslave the brilliant, and in questions of faith, neither side looks noble once the smoke clears. Old accusations of Lavoisier's dishonesty resurfaced and new ones were added. In an exposé of the French chemist's "deceptions," Priestley supporter William Ford Stevenson called Lavoisier an "arch-magician," who "imposed upon our credulity"; phlogistonists saw deceit even in the fact that Lavoisier habitually misspelled English names.

Priestley especially seemed to see the contest in terms of war. His treatment at the hands of the antiphlogistons "was neither friendly nor agreeable to the rules of honourable war," he wrote near the end of his life. A colleague who tried to reconcile the two camps was told: "I thank you for your ingenious and well-intended attempt to promote a peace between the two present belligerent powers in chemistry; but I fear that your labour will be in vain. In my opinion, there can be no compromise. . . ." The bitterness ran so deep that when Lavoisier was executed at the hands of the Terror, Priestley never once offered condolences to his French friends. He would never mention his rival's death, except obliquely in an *Open Letter to French Chemists*, in which he said, "We hope you had rather gain us [to Lavoisier's thesis] by persuasion, than silence us by power."

Lavoisier was as callous, if not more so. Dogged in his attempt to force chemistry into the mold of his new science, he could be contemptuous of those he thought charlatans. As the war heightened, this included adherents of phlogiston. To the English, he was a pirate of Priestley's work; to the Germans, contemptuous of Stahl and Becher. In truth, Priestley had every reason to feel used: as Sidney French has noted, "Lavoisier had to use Priestley because he was building a comprehensive theory and Priestley's discovery was an integral part of it; Priestley, on the other hand, had no need to use Lavoisier because he was building nothing, merely discovering facts." In a happier world, they could have been collaborators, like Marie and Pierre Curie, but this was no amicable marriage of minds and there would be no compromise. Lavoisier sensed this, as the traveler Edmond Genet recalled during the height of the conflict in 1783:

> I also had the advantage during my stay [in England] of becoming acquainted with Dr. Priestley who had the kindness to repeat for my gratification his most interesting experiments on air and gases of which I sent an account to the Academy of Paris. At that time, Lavoisier was pursuing the same subject, and I was surprised on my return to hear him read a memorial at one of the sittings of the Academy which was simply a repetition in different words of Priestley's experiments which I had reported. He laughed, and said to me, "My friend, you know that those who start the hare do not always catch it."

One cannot help notice that during this period Lavoisier was far better organized than Priestley and his supporters. The rise of the New Chemistry was a war on many fronts; even before Napoleon's example, Lavoisier was good at striking several points at once, and had a keen eye for picking lieutenants. He started a school at the Arsenal for young chemists, as Rouelle had done; he instructed a new generation in the New Chemistry, proselytizing a young cadre of disciples. When France's older, more conservative monthly journal, abbé François Rozier's *Journal de physique*, refused to publish the antiphlogistonists' papers and published instead those supporting phlogiston, Lavoisier started *Annales de chimie*, a journal that is still published today.

Marie was perhaps the most persuasive weapon Antoine had. She corresponded with sympathetic and recalcitrant scientists across Europe, such as the prominent Genevan naturalist Horace-Bénédict de Saussure, who finally conceded, "You triumph over my doubts, Madame, at least concerning phlogiston." She kept lab notes and made translations. But most important, by the 1780s she had developed into an accomplished *salonnière.* In the eighteenth century, *salons* were one of the principal engines of the Enlightenment, and became invaluable to *philosophes* like Lavoisier. There they found a forum where ideas were carried into a society eager for new diversions. It took skill, tact, and intelligence for a woman to gain the respect of temperamental scientists and intellectuals, and not everyone thought highly of *salons.* Rousseau saw them as gilded prisons whose wardens were the *salonnières:* "Imagine what can be the temper of the soul of a man who is uniquely occupied with the important business of amusing women, and who spends his entire life doing for them what they ought to do for us." Lavoisier's mentor Guettard escaped the pull of the *salons* by pleading lack of social etiquette, calling himself a "Hottentot" and "savage." Banter and vitriol between the sexes were not unusual. When Denis Diderot visited the influential *salon* of Mme Marie-Thérèse Geoffin, she "treated me like a ninny and counseled my wife to do the same," he complained.

Marie's *salons* were not renowned for repartee but concerned with scientific matters and above all else the spread of the New Chemistry. They were organized around Lavoisier's famous dinners held each Monday night at the Petit Arsenal. She acted as a charming foil for her husband, and her guests allowed themselves the enchantment of her hospitality after being treated to the newest chemical demonstrations in Antoine's lab. The fateful dinner with Priestley in October 1774 was an early practice session for this choreography of dinner, discussion, and demonstration in the 1780s. Yet even in 1774 she was a central player, and one wonders how much of Priestley's discovery was charmed out of him by Marie.

All were invited to these affairs, scientist and non-scientist, sympathetic to Antoine's theories or skeptical. Many mentioned invitations in their correspondence, and few were critical. They included the deposed minister Turgot, still dreaming of an enlightened France; the suave mathematician Laplace; the equally famous mathematician comte Joseph-Louis Lagrange, so depressed with life that he stared "absently out of a window, his back to the guests who had come to do him honor"; Pierre-Simon du Pont, rising from the ruins of the Turgot ministry and destined, some think (from a veiled comment in his correspondence) to become Marie's lover; Benjamin Franklin, minister extraordinary to the French court during the American Revolution; Macquer, greatest of the old-guard chemists; Jean-Silvain Bailly, the astronomer, who would become mayor of Paris; and the duc de La Rochefoucauld, patron of science and an experimenter himself. The doors were always open to academicians, and to any foreign scientists who happened to be in town.

The latter received the most intense indoctrination: it is hard to tell where Antoine's conversational fire left off and Marie's gentle blandishments began. They fit together, hand-in-glove. James Hall, a young Scotsman, arrived in Paris in 1786 and left a convert months later after participating in a number of hands-on experiments: "I am personally acquainted with Mr. Lavoisier & have received the greatest civilities from him—I have a standing invitation to dine with him every Monday—he has one of the clearest heads I have ever met with & he writes admirably." The ever observant English agronomist Arthur Young attended in1789, and his recollection gives a sense of Marie's style as a *salonnière:*

> Madame Lavoisier, a lively, sensible, scientific lady had prepared a *déjeuné anglais* of tea and coffee, but her conversation on Mr. [James] Kirwan's *Essay on Phlogiston*, which she is translating from the English, and on other subjects which a woman of understanding, who works with her husband in the laboratory, knows how to adorn, was the best repast.

Ben Franklin, who never truly became a convert to the New Chemistry, was still charmed. In a letter to Marie, he wrote:

> I do not forget Paris and the nine years of happiness I enjoyed there in the sweet society of a people whose conversation is instructive, whose manners are highly pleasing, and who, above all nations in the world, have in the greatest perfection the art of making themselves beloved of strangers.

Thus, through her *salons* and correspondence, Marie became almost as important for the spread of the New Chemistry as her husband. She may have been considered an equal. One hint of this is the giant portrait of Antoine and Marie painted by the renowned Jacques-Louis David, who was also her teacher. This work now dominates a gallery in New York's Metropolitan Museum of Art; it is 9 feet tall, and shimmers with light and color—off glass vessels, her white cambric dress, the red velvet tablecloth before which Antoine is seated. Painted in 1788, Lavoisier is shown proofing a text, probably the soon-to-be-released *Elements of Chemistry*, and looks adoringly at Marie. He is impeccable in powdered wig, white cravat, silk coat, and black knee breeches; she shines in a silver-white taffeta gown tied with a blue ribbon. The table is littered with symbols of Lavoisier's New Chemistry: the apparatus of the twelve-day experiment; vessels and flasks used in his identification of oxygen. They are partners with national significance, for the equipment and texts symbolize Lavoisier's overthrow of phlogiston, and victory over Priestley, his *English* rival. She leans over him, like a Muse, and stares confidently at the spectator, inviting one to enter this "still, safe world of science," where reason and love are partners and chaos can never prevail.

Finally, Lavoisier showed real cunning for recruiting allies. Venel, in his entry on chemistry for the *Encyclopédie*, had said that a chemical revolution would probably spring from an Academy insider: someone of vision and influence, who understood not only the science but the fine art of persuasion. By now, Lavoisier was indispensable to the Academy, assigned to multiple investigations and commissions, and knew which strings to pull. The first important scientist to espouse the oxygen theory was Laplace, long before the guinea pig trials. A pleasant rhythm existed between Lavoisier and the younger mathematician: both were theorists, and since they worked in different disciplines, none of the jealousy existed that arose too often between Lavoisier and other French chemists. By this point, Laplace had already undertaken the task of proving that the solar system was ruled by stability, swinging back and forth in endless cycles, a cosmic perpetual motion machine. Like Lavoisier, he believed in balance; like Lavoisier, he wanted nothing less than to reinvent his entire field. Like Lavoisier, he was accused of grandiose ambition and several instances of plagiarism; and like his friend, he told his critics that he saw in the works of other men far more than the authors did themselves.

Three other alliances were especially important for Lavoisier's campaign—three fellow chemists who would help him rethink the language of chemistry; together, they formed a quartet that would be called the Four Horsemen of the New Chemistry. Claude-Louis Berthollet was actually the first French chemist to switch to Lavoisier's side. He was the best lab chemist in France, a younger, Gallic Priestley whose most celebrated discovery, the bleaching property of chlorine gas, had immense commercial potential. In 1785, Berthollet declared his acceptance of the New Chemistry and after Lavoisier's death he would take over as Commissioner of Powders. He was a mild man, less arrogant than Antoine; his demeanor and position during the Terror probably saved him from the fate of his friend.

Like Lavoisier, Guyton de Morveau was trained as a lawyer, which meant he recognized the nuance of language. Like Lavoisier, he was a chemist and economist, too. At first an ardent phlogistonist, he was appointed professor of chemistry at the University of Dijon and wrote chemical articles for the *Encyclopédie*, yet even while an opponent, he realized that chemical nomenclature was a morass of obscure, alchemical references like "liver of sulfur" and "sugar of lead." The need for linguistic reform as much as anything else swayed him to Antoine's side. He was "a man void of affectation," wrote Arthur Young, "free from those airs of superiority which are sometimes found in celebrated characters." A fine foil for the abrasive Lavoisier.

The fourth member of the quartet was Antoine-François Fourcroy. A fiery speaker and dramatist, if a lesser experimentalist than the others, his position was closer to propagandist, and so his task was to popularize the new theory. Son of a poor branch of a noble family, he was raised among the Paris *sans-culottes,* where his father held the position of pharmacist to the duc d'Orléans. Passed over by the Academy of Medicine for a degree of *docteur régent,* he attended Lavoisier's Arsenal School, where his eloquence and fiery temperament caught Antoine's eye. It was on Lavoisier's recommendation that he was appointed professor of chemistry to the Jardin du Roi upon Macquer's death in 1784. In time his textbook on the New

Chemistry would go far in establishing Lavoisier's doctrines, but this was after his death and could be seen as a sop to a guilty conscience. For Fourcroy would turn into a Judas when Lavoisier's life was in his hand.

▪ ▪ ▪

By 1780, Priestley had spent seven years with Lord Shelburne; never again would he and Mary live like the upper crust; never again would he be so productive as a scientist. During his years with Shelburne, he isolated and identified ammonia, sulfur dioxide, nitrous oxide, and nitrogen dioxide; he opened the door to understanding photosynthesis; he developed the first environmental test for air pollution; and, of course, he discovered oxygen. Yet by 1779, the tiniest cracks had begun to show in the facade. Two years earlier, in 1777, he had published *Disquisitions Relating to Matter and Spirit*: it was the second time he'd suggested that the notion of the soul was superfluous, and the book caused wide offense. "Ah, Priestley," Dr. Johnson was said to utter. "An evil man, Sir. His work unsettles everything." In February 1779, he sought an audience with George III following the success of a bill doing away with penalties against Catholics, but no other Dissenters; other scholars achieved an audience by applying to use the King's Library, but George III made clear his feelings about Priestley in a note headed "Queen's House, 33 min. Past 5 p.m., Feb. 22, 1779": "If Dr. Priestley applies to my librarian, he will have permission to see the library as other men of science have had: but I cannot think the Doctor's character as a politician or divine deserves my appearing at all." That year, Shelburne remarried, which meant new changes in the household; by 1780, he sensed an improvement in his chances to reenter the cabinet, and by 1783 he would be the minister to broker peace with the Americans. For one with political aspirations, the increasingly strident Priestley was becoming a liability.

Priestley was in an unsettled state, too. From 1776 to 1780, he hinted in his writing that his main work in science might be finished, and that he felt the call of religion more ardently than ever. As Lavoisier strengthened his position on the oxygen theory and gathered forces around him, a strange silence emanated from Priestley's quarter until at least 1783–84. The old spark was missing—if not sad, he at least seemed subdued. During these years, as the tensions heighten from the American conflict, his thoughts turned toward Millennialism, and as the social historian Samuel Eddy has pointed out, "Resort to prophecy is a universal response of beaten men." When in the summer of 1780 he heard from Richard Price that Shelburne might set him up far from London in one of his Irish estates, he asked his employer if he had somehow caused offense. Shelburne said no, but Priestley saw this as a hint that Shelburne would have no objection if he left. He handed in his resignation, perhaps for no other reason than that he thought it time to go.

Priestley parted amicably from Shelburne in September 1780 with a £150 annuity, one-half his yearly income, for the rest of his life. He settled in a house called Fairhill, on the outskirts of Birmingham, and was appointed one of two ministers in the Unitarian New Meeting House, an easy walk from his home. His household now included a daughter and three sons. Dotted about the countryside where he lived were the estates of wealthy merchants and manufacturers; he found "good workmen" to make his instruments, as well as "the society of persons eminent for their knowledge of chemistry." His brother-in-law, John Wilkinson, provided him with the house, while master potter Josiah Wedgwood supplied him with funds for experiments. The house was large: a drawing room on the ground floor for entertaining guests, front and back parlors, and a kitchen and storeroom. A wide staircase led to four bedrooms, and a back staircase led to the attic, which he converted into an extensive private library. Outside lay a brewhouse and laundry, pigsty, sheds for coal and coke, and a garden toolhouse. All that was needed was a lab, and he built this to his specifications, a two-story building near the yard wall.

Soon after arriving, Priestley was invited to join the Lunar Society—known as "the Lunatics" to locals—a shifting, informal group of a dozen or so scientists, thinkers, and local industrialists who were interested in natural science and literature. They met once a month in one another's houses on the Monday nearest the full moon, hence the name. The time was chosen so the members could return home at night by the light of the moon, both for convenience and for security from ever present highwaymen. Priestley was joining a society of philosophers, just as he had at the Royal Society, but this was no research group pursuing a fixed agenda like Lavoisier's Arsenal School. It was a society of brilliant friends and thinkers whose ideas inspired each other in more private ways: where Lavoisier's group was narrowly focused and publicly financed, the Lunar Society was, according to one member, "one of the best philosophical clubs in the Kingdom," and though rarely mentioned in print, still instrumental in merging science and industry in ways that made England the seat of Europe's Industrial Revolution. In addition to Wedgwood and Wilkinson, the members included Matthew Boulton, "the iron chieftain," whose Soho foundry employed so many local workers that he was called the "father of Birmingham"; the mercurial Scottish inventor James Watt, who with Boulton would manufacture the steam engine; Erasmus Darwin, grandfather of Charles Darwin, a scholar of immense physical proportions who was the group's de facto leader; the Quaker arms merchant Samuel Galton; Captain James Keir, a wit, finished gentleman, and man of the world; and Dr. William Withering, who studied foxglove and its use in medicine, and whose own life was in slow decline. "What inventions, what wit, what rhetoric, metaphysical, mechanical, and pyrotechnical, will be on the wing," Darwin enthused in a note to Boulton. He could barely contain his excitement in thinking of the next meeting, when ideas would be "bandied like a shuttlecock from one to another of your troop of philosophers!"

In an important way, the group replaced Shelburne as informal patrons, creating a support network that sustained Priestley's research from 1780 to 1791. Boulton, Wedgwood, Darwin, and Galton contributed an annual stipend of £200 to his work; Wedgwood also supplied a constant flow of ceramic retorts and tubes, and Boulton sent metalwork supplies. Priestley repaid the industrialists in services: he analyzed clay samples for Wedgwood for use in his pottery and iron-ore samples for Boulton. In 1781, when Watt heard of a new engine that used a recently discovered gas as fuel and was supposedly more efficient than his steam engine, he asked Priestley to analyze the gas but was worried privately whether the Doctor could keep his request in confidence. "I have raised a confidence in Dr. Priestley who had seriously promised to keep the secret and to give me all the facts," Boulton assured him. "The Dr. is under some obligation to me that you know not of and I think will not betray me, I am sure he will not."

The Lunar Society also differed from the Arsenal School in the way it was treated by officialdom. Lavoisier's fierce band of antiphlogistonists may have been loathed by rival chemists, but they were still part of scientific society and never ignored. The Lunar Society members were more isolated, pursuing their goals outside the aegis of Church and State and considered bumptious upstarts by the ruling class. As late as 1806, a reviewer of Priestley's memoirs would complain, "It has often struck us that there is universally something presumptuous in provincial genius," and bemoaned how rare it was "to meet with a man of talents outside the metropolis who does not overrate himself and his coterie prodigiously." The Lunar Men resented the power of the London elite; the effort to obtain patents and permits amounted to "licking some great man's arse and be done with you," they were told. Such patronizing attitudes "classed" provincial science as less important than the city's, and made it unfamiliar and strange. Given the right conditions, this could turn threatening—and the "right" conditions were only a few years away.

Priestley helped prepare such conditions through his combination of incendiary rhetoric and naïveté. While his scientific work did not progress in Birmingham (he published only two volumes of scientific experiments, the first taken up with work at Calne), he threw himself back into theology with a new fire. Maybe he felt freer away from Shelburne; as a Dissenting minister, he was in closer contact with radical groups than he had been in years.

In 1782, Priestley began a campaign of publication that catalyzed opinion against him. Suddenly, after seven years of quiet, he found himself attacked in print, denounced from pulpits and in the House of Commons, and ultimately considered an agent of Satan. The first work was a *History of the Corruptions of Christianity*, dedicated to "unbelievers, and especially to Gibbon," who had published the first three volumes of the *Decline and Fall of the Roman Empire* in which he blamed Rome's downfall on early Christianity. *History of the*

Corruptions was a direct attack on the central tenets of the Church of England, particularly the doctrine of the Trinity, and for this Priestley stirred up wrath and denunciation. When he challenged Gibbon to a public debate, Gibbon declined, calling Joseph an even greater unbeliever than himself.

The *Corruptions* also drew down on Priestley the organized wrath of the Anglican Church, this in the form of Samuel Horsley, Archdeacon of St. Alban's. Like Priestley, Horsley was a man of science and religion: he was secretary of the Royal Society from 1773 to 1778 and publisher of Newton's philosophical works, both of which positions gave him greater standing in the scientific world. Thus, Priestley had an enemy who could match him in credibility. The book was banned in parts of Britain, and for the very first time, the reaction to one of his theological works jumped the border when it was burned at Dort in Holland by the town hangman.

In 1786, Priestley published a *History of the Early Opinions Concerning Jesus Christ*, a continuation of the debate with Horsley in which he tried to prove through Scripture that the doctrine of the Trinity deviated from Jesus' teachings. This was his most controversial work yet, not because of anything in the text, but rather due to what he'd said at the pulpit one year earlier. The occasion for that was November 5, 1785, Guy Fawkes Day: the commemoration of the failure by Catholic conspirators in 1605 to blow up both Houses of Parliament because of harsh laws passed against practicing Catholics. The sermon would forever after be known as his "Gunpowder Sermon." While at the pulpit, he used the word "revolution" three times to describe the fall of the Roman Empire; more important was his image of Dissenters' resistance to suppression. He said he was not worried that more Unitarian churches were failing to be started because

> Unitarian principles are gaining ground every day. We are, as it were, laying gunpowder, grain by grain, under the old building of error and superstition, which a single spark may hereafter inflame, so as to produce an instantaneous explosion; in consequence of which that edifice, the erection of which has been the work of ages,

> may be overturned in a moment, and so effectually as that the same foundation can never be built upon again.

Friends had advised him not to use this phrase before the sermon; afterward, Josiah Wedgwood urged him not to publish it, as it would "cause much noise." France was still four years away from revolution, but the strident rise of democratic rhetoric in Paris and London, following the revolt in America, was seen as a threat to kings everywhere. "Be more circumspect," his Lunar friends advised, while Mary's hotheaded brother John Wilkinson warned him not to be a fool. One did not combine "revolution," "gunpowder," and "explosion" in a speech in times like these. But Joseph paid no heed. This was a long-standing debate, nothing more. "My arguments will carry the day, thanks to the soundness of my reasoning," he said.

Soon after publication, Horsley responded in an ominous way. By gathering extracts of Priestley's writings, he tried to show that the Doctor was not merely a heretic but a revolutionary, and thus a threat to the state, who would not be satisfied until his Church and King were deposed and his nation lay in ruins. Horsley could only trust in "the good town of Birmingham and the wise connivance of the magistrate . . . to nip Dr. Priestley's goodly projects in the bud."

▪ ▪ ▪

For one brief moment during the 1780s, Priestley retreated from the language of revolt and returned to science; for one brief moment, he observed a chance phenomenon that no one else had seen or taken seriously. In a return to his experimental brilliance of the 1770s, he once again unwittingly destroyed the unity of an ancient "element." But he misinterpreted the causes, leaving the field open for Lavoisier to step in to explain.

This was the Great Water Controversy, which, like so much else in Priestley's life, started innocently. After his arrival in Birmingham, in 1780–81, Joseph grew interested in hydrogen, still known as "inflammable air." He placed some red mercuric oxide in a tall cylinder containing the gas and heated it with a lens; the volume of hydrogen diminished as oxygen poured from the powder, but moisture appeared. That was strange. But at the moment he paid little attention, believing it merely a random happening.

Yet it was a troublesome detail that refused to go away. Every time he mixed hydrogen and oxygen, there was that water again. So he tried something new. On April 18, 1781, he passed an electric spark through the flask containing both gases. There was an explosion, then a fine dew clung to the glass sides. He attempted to explain the dew by means of phlogiston, but frankly, he had no idea why it was there. A fellow chemist suggested it was a constituent of one of the gases, but this didn't make sense either. As often happened when he was baffled, Priestley started talking, and the first to whom he talked was his singular friend, Henry Cavendish of London.

One could not think of hydrogen without thinking of Cavendish: the bets were out as to which was strangest, Cavendish or the gas he discovered. He isolated it in 1766, dropping zinc (and later iron and tin) into hydrochloric acid, then watching as a gas was generated that burned with a pale blue flame. Cavendish dubbed it "inflammable air" and speculated that it might be pure phlogiston. Priestley had seen the reclusive nobleman at the Royal Society and had gently entered his confidence. On May 27, 1775, Priestley and four others had been invited by Cavendish to witness experiments he'd conducted on a new instrument of war he called an "artificial torpedo." Priestley didn't like this pairing of the philosopher's art with the means of destruction, yet even he had speculated on ways of improving gunpowder's explosive yield with dephlogisticated air. Any philosopher who lived with his head above water could see the growing links between science and private industry. Every new discovery that dealt with some radical energy release—be it electricity, combustion, chemical bonding, or, in time, radiation—would be followed by an advance in ballistics, explosives, artillery, jet and rocket propulsion, or nuclear fission. If there was unity in this, it was deep-seated and primal, as if man still crouched in his cave and watched the friendly cook-fire at the entrance suddenly expand to engulf the world.

The misanthropic Cavendish shunned such ethical ruminations. It was said of him that his view of existence "consisted *solely* of a multitude of objects which could be weighed, numbered, and measured"; believed to be one of the wealthiest men in England, Cavendish never owned more than one set of clothes at a time and dressed in a periwig, long coat, and three-cornered hat, a style hailing from the previous century. "He did not love; he did not hate; he did not hope; he did not fear," wrote one biographer; if he had a passion, it was for the weight and measure, even before Lavoisier. Like Priestley, he stuttered, and perhaps it was the shared impediment that drew the men together; indeed, Joseph would be one of the few scientists with whom Cavendish corresponded. Yet even the amiable Priestley had trouble gaining his trust. Cavendish was so painfully, pathetically shy that he'd stand outside a place of meeting for minutes as if gathering the courage to enter. Once in, he'd flee with a cry of distress if addressed by a stranger.

Priestley mentioned the mist formed from oxygen and hydrogen to Cavendish, then promptly forgot it and went on. But Cavendish did not. For two years he labored over the problem, and in the spring of 1783 he announced his findings. Water was a compound of oxygen and hydrogen, mixed together in the proportion of two volumes of hydrogen to one of oxygen, today's familiar formula H_2O.

What an incredible claim! Water, so visible and wet, was born of two invisibilities? Show us the proof, demanded colleagues at the Royal Society, and so Cavendish told them quietly, without emotion. First, he'd introduced into a glass cylinder 423 measures of his "inflammable air" and 1,000 measures of common air; when they were sparked, all of the hydrogen and about one-fifth of the common air condensed into a dew. He analyzed this liquid through evaporation and condensation and proved it was pure water. Then using reverse analysis, he proved it was the oxygen in common air, not the nitrogen or carbon dioxide, that reacted with his hydrogen. Finally, he substituted Priestley's dephlogisticated air for common air and produced water in the same measures. Two parts hydrogen always united with one part oxygen to form a weight of water equal to the weight of the gases.

The Royal Society members were dumbfounded, yet Cavendish's explanation had an Achilles heel. Despite the brilliance of his tests, Cavendish explained the underlying mechanics in terms of phlogiston. Hydrogen contained water and phlogiston, he said; oxygen contained water and no phlogiston. The presence or absence of phlogiston in the two gases canceled all properties and left the water behind. Even the English found it hard to buy this explanation, and in the theoretical vacuum, Lavoisier again stepped in.

In June 1783, soon after Cavendish's announcement, the British physician Charles Blagden visited Paris, and like other foreign scientists, he visited Lavoisier. Blagden told his host about the newest developments in England, and at the moment Cavendish's findings were the talk of the town. Once again, Lavoisier repeated—with the help of Laplace—Priestley's and Cavendish's tests. He sparked a mixture of oxygen and hydrogen and found pure water lining the walls of the vessel after the explosion. Yet he saw the cause in light of his oxygen theory. He gave Cavendish's gas the name "hydrogen," from the Greek *hydra-*, or "water," and after more tests devised a working formula:

$$2H + O \rightarrow H_2O$$

Water was *not* an element, he said, just as he had proved with air years before. It was an oxide of hydrogen, the result of a chemical change.

Now Priestley's fellow "Lunatic" James Watt entered the picture, conducting the same tests as the others yet claiming the honors as his. He too produced water, and he too explained it through phlogiston. Of all the chemists eventually drawn into the long and petty controversy over bragging rights, Watt had the weakest claim. At one point, when Lavoisier decomposed steam with a hot iron, Priestley almost seemed ready to adopt his rival's views. "I was for a long time of the opinion that his conclusion was just," Priestley noted. But Watt convinced him otherwise, and Watt could be a forceful, unpleasant man. Watt talked Priestley out of apostasy; Joseph would never again waver from his belief in phlogiston.

Others were not so easily sold. After the Great Water Controversy, chemists abandoned phlogiston in great numbers: the labored attempt to fit water into the theory produced language and logic that were contradictory and surreal. To phlogistonists, oxygen was "dephlogisticated air," hydrogen was pure phlogiston (since it was so flammable), and the combination of the two put irreparable strains upon the old nomenclature. Suddenly, oxygen became "water lacking phlogiston," and "dephlogisticated air," with which Priestley had been comfortable, was in danger at one point of becoming "dephlogisticated water." It was in both the increasingly fuzzy language and damnable lack of precision that the English pneumatic chemists lost ground.

Why did Priestley refuse to see the advantages of Lavoisier's theory when he was so close to revising his own? How could someone who attacked each new experimental obstacle with such clarity fail to see the larger picture?

From a theoretical standpoint, the answer seems simple enough: for Priestley, the Chemical Revolution distilled into two fundamental opposites. Either phlogiston was real and water was a simple element, or water was composed of Lavoisier's "oxygen" and "hydrogen" and phlogiston was a lie. Yet Priestley was not a simple man, despite his public persona, and there were other reasons for resistance that included faith, friends, and national pride. In addition, Lavoisier's theory had its flaws. The Frenchman lived and died by the balance and what came to be called the law of conservation of matter, yet the balance was not yet an instrument of universal recourse in the lab. Many scientists did not place great faith in laws of conservation, particularly when dealing with subtle, immeasurable phenomena like light, heat, electricity, and even phlogiston. They reacted against it at a gut level: "You do not surely expect that chemistry should be able to present you with a handful of phlogiston, separated from an inflammable body," wrote an English supporter. "You may as reasonably demand a handful of magnetism, gravity, or electricity. . . . There are powers in nature which cannot otherwise become the objects of sense than by the effects they produce; and of this kind is phlogiston."

There were other problems, too. Caloric was too much like phlogiston: immeasurable, immutable, a convenient if invisible principle. In fact, this facet of Lavoisier's theory came under attack from the first, and would not fade from the picture until the nineteenth century when the concept of energy was more fully described. The other internal obstacle was oxygen's assigned role as a "generator of acids": there were too many anomalies, just as with phlogiston, and even Berthollet pointed out that such mild acids as prussic acid (HCN) and sulphuretted hydrogen (H_2S) contained no oxygen. Many metallic oxides showed no acidic properties and even neutralized acid, again contradicting Lavoisier's theory. By 1810, this facet of Lavoisier's reality would be discarded entirely with Humphry Davy's work on chlorine.

Yet a large measure of Priestley's blindness to *any* merits of the new system can only be explained on religious or theological grounds. Lavoisier explained chemical phenomena in terms of a multiplicity of different elements, while Priestley thought all substances were ultimately composed of the same stuff, just differently arranged. The phlogiston theory had become so entwined with his political and religious outlook that he could not detach himself: he wanted a principle of chemical action as universal as gravitation, magnetism, or electricity—as universal as liberty in the affairs of men. He associated phlogiston not so much with matter as with energy, the ultimate principle of change. And change meant everything.

▪ ▪ ▪

The war of words waged by the French and English merely reflected the tense reality of a world at war. In 1780–81, Britain was fighting wars all over the globe: against the French in India, the West Indies, North America, and Africa; against the Spanish in Gibraltar, the Balearic Islands, the Caribbean, and Florida; against the Dutch in Ceylon and the West Indies; against the Americans in the colonies. Britain might seemed unassailable, and the King was convinced that his cause was just. "I know that I am doing my duty," he told Lord North at the beginning of the American conflict, "and therefore can never wish to retreat." Thus, General Cornwallis's surrender in Yorktown in October 1781 stunned everyone. Stocks fell. "Universal dejection" ruled in London. The nation was in disgrace, and talk was rampant of political change.

France fared little better, even if it had backed the winning side. The cost of the 1778–1783 war with England totaled more than France's three previous wars together, and the nation emerged virtually bankrupt, its capacity to borrow money from other countries exhausted. From 1781 to 1789, a succession of ministers and reforms came and went; powerless liberals brimming with solutions were sabotaged by powerful reactionaries who had the support of the Queen. By 1788, the public Treasury was so fragile that Louis XVI sought help from the Estates-General in the hope they would grant him new revenues. Instead, they demanded reform.

On both sides of the Channel, liberty was the most intense political and religious sentiment of the day. Government was disliked: in England, Lord Shelburne's suggestion of an organized police force days after the anti-Catholic Gordon Riots in London in 1780 was greeted as one step from tyranny; in France, Louis was lampooned as a pudgy clown, while the hated Marie Antoinette was burned in effigy. In England, the recurring issue of a census was denounced as the end of English freedoms. In France, popular anger focused on a 30 million *livres* (about $1.2 billion) wall built around the city of Paris to curtail smuggling, and on the man who proposed it—Lavoisier.

The importance of customs duties entering Paris was no small matter to the cash-strapped kingdom; it was estimated that in 1789 the Paris customs yielded no less than 28–30 million *livres* of a national total of 70 million, or nearly 43 percent. Lavoisier, as a Farmer General, was in charge of tax collection. He watched the books as closely as he did the balance beam, and by 1779 concluded that fully a fifth of the goods brought into Paris were untaxed—thus smuggled. One of the smugglers' more elaborate techniques was to build a tall footman from wicker and stand him in the back of the thousands of coaches and carriages passing into the city. The hollow man, of course, was filled with black market goods. Lavoisier's suggestion was adopted: six years later, work was completed on his wall. The *enceinte,* a ring of fifty-four customs posts joined by a 10-foot wall, encircled Paris over a span of eighteen miles.

The dislike of Lavoisier that first found voice among rival chemists now spread into the city at large. He was accused of turning Paris into a giant prison, of placing the capital in a gourd-shaped retort through which its lifeblood dripped into the cashbox of the Farmers. The most infamous flyer suggested that

> the author of the plan should be hanged. You, vile inventor of a tyrannical project, who have no fear of sacrificing the honor and life of your fellow citizens to your insatiable greed, be assured that

> whatever the titles with which you adorn yourself, the riches with which you glorify yourself, only hate and disgrace await you. The Farm owes you gold, but the country owes you its curse.

The flyer provoked response from the police, and several libraries selling it were raided and copies confiscated. One financier offered 20,000 *francs* for the identification of the anonymous author. For his part, Lavoisier prepared a note to the newspapers declining all responsibility and blaming the wall on the government, then had second thoughts and did not send anything. He cut himself off, insulating himself from social life and taking refuge in finance and science.

But science prophesied the coming storm. Throughout the 1780s, the rate of industrial and scientific espionage intensified in a way surpassing memory. James Watt had been pirated as early as 1779 when he showed two Prussian visitors around the Soho works, then learned they had built steam engines from his design when they got home. Josiah Wedgwood heard in 1785 that "three sets of spies" were operating in the Midlands; a Danish spy was caught leaving the country in 1789 with bags filled with models, drawings of tools and machines, and even clay samples he'd bribed from Wedgwood's workers. A French marquis was discovered in 1786 sketching the rotary wheels of Boulton's Albion Mills, which he tried hiding in his handkerchief; that same year, Watt himself returned from France with Berthollet's process of bleaching textiles with chlorine. On both sides, espionage was considered patriotic if you were doing the stealing, and treason if you were stolen from.

The greatest game was gunpowder, the nuclear weaponry of the age. The first modern gun was developed in France for bird hunting, and Louis XIV reequipped his army during the 1660s from this design. For years France had been the world leader in the arms race, but after the Seven Years' War and the loss of India's saltpeter, the nation fell behind. Lavoisier was asked to correct that state when he moved to the Arsenal.

If there was one junction between the creative art of the chemist and the destructive might of the state, it was in the production of gunpowder, which had an ancient history. Gunpowder is a simple mixture of six parts saltpeter, one part charcoal, and one part sulfur:

Joseph Priestley in his thirties, while still an instructor of classics and literature at the Dissenting Warrington Academy. The sketch is believed to have been drawn after Priestley was elected a Fellow at the Royal Academy in 1766, but before he began his experiments on gases.

Antoine-Laurent Lavoisier, the French chemist and financier whose "oxygen theory" overthrew the ancient theory that all matter was composed of fire, air, water, or earth. The sketch is of Lavoisier in his twenties.

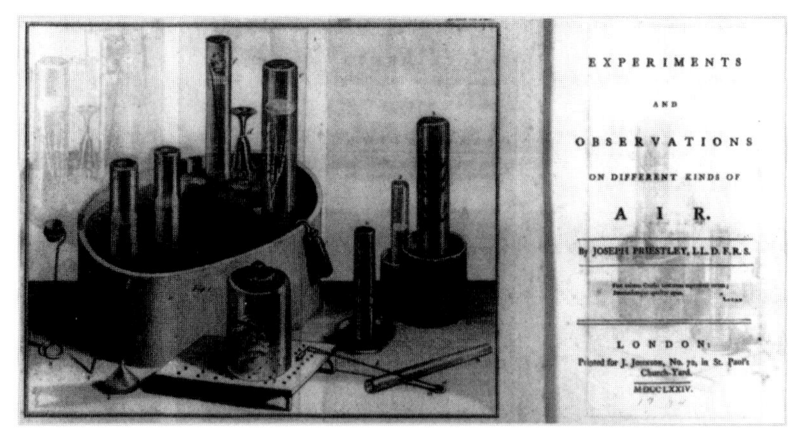

The equipment used in Priestley's gas experiments, included in his *Experiments and Observations on Different Kinds of Air.* The equipment includes the "pneumatic trough," which Priestley filled with mercury, and a beer glass in which can be glimpsed a mouse.

An etching of Lavoisier's experiments on human respiration, drawn by Marie Lavoisier. In this rendition, Lavoisier's assistant is the test subject, and Marie can be seen sitting at a table to the side as she keeps records.

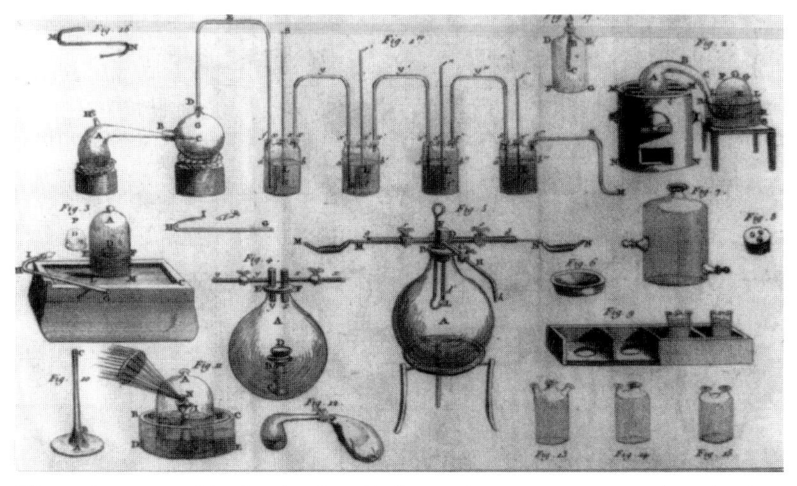

The equipment used in Antoine Lavoisier's experiments on oxygen and combustion, drawn by Marie Lavoisier.

Henry Cavendish, the reclusive British scientist who discovered hydrogen and the composition of water. A friend of Priestley's until after the Birmingham riot, Cavendish was said to be the richest man in England.

Carl Wilhelm Scheele, the obscure Swedish apothecary who apparently isolated oxygen before anyone else, but was not acknowledged as the discoverer because of his delays in publishing. Scheele may have died from the effects of breathing the fumes of chlorine gas. He is credited with discovering more new compounds than any other eighteenth-century chemist.

Antoine-François Fourcroy, one of Lavoisier's Four Horsemen of the New Chemistry, who was accused during the Revolution and Terror of failing to save Lavoisier's life. Afterward, Fourcroy's textbook on the New Chemistry helped cement the oxygen theory into modern modes of thinking.

An Experiment on a Bird in the Air Pump by Joseph Wright of Derby depicts a familiar scene during the eighteenth century—the visit of a traveling "scientific demonstrator," whose lectures included many of the scientific wonders of the day.

One of Priestley's "burning glasses" used for heating chemical substances. This one came from Priestley's lab in Northumberland, Pennsylvania; it is believed that the glass used in the famous experiments revealing the existence of oxygen was destroyed during the Birmingham riots of 1791.

One of the more famous anti-Priestley cartoons that preceded the Church and King riots of 1791. Priestley is depicted burning various documents of English freedom while treading on the Bible.

This huge commissioned portrait of Antoine and Marie Lavoisier, now hanging in the Metropolitan Museum of Art in New York, was painted by Jacques-Louis David, the French Neo-Classical painter known for his *Death of Marat*. David taught drawing and painting to Marie.

Mary Priestley was involved in her husband's work, although it is uncertain how much she participated in the lab. She handled everything in Priestley's household, freeing him for his scientific and theological investigations. Here she is depicted in her later years, cupping an ear to aid in hearing.

In this political cartoon, Prime Minister William Pitt is shown steering a small boat, *The Constitution*, which also carries Britannia, toward a castle with a flag inscribed "Haven of Public Happiness." Joseph Priestley is one of three Dissenting "sharks" following the boat.

A rendering of the sacking and burning of Joseph Priestley's house outside Birmingham during the Church and King riots of 1791.

The Treacherous Rebel and Birmingham Rioter is one of several engravings produced after the Church and King riots of 1791; it depicts Priestley being whipped and led off to Hell by demons.

Jean-Paul Marat was the French journalist and revolutionary who blamed Antoine Lavoisier for all his scientific failures. His assassination led to the Reign of Terror; historians have blamed his hatred of Lavoisier for the scientist's execution.

A sketch believed to have been drawn in prison shortly before Lavoisier's execution in 1794.

This portrait by American painter Rembrandt Peale depicts an elderly Joseph Priestley, after the scientist and his family had fled to America.

a good stirring produces a flourlike substance that burns very quickly and results in the release of products (mostly gaseous) that require about 1,000 times more space than the original gunpowder. Saltpeter is the main ingredient, but this was seldom found naturally in the West and was more common in Asia. The Arabs called saltpeter "snow from China," which suggests that its potential was first realized there; it made its way west from China via the Arabs and Romans. The West's earliest recipe for modern gunpowder was found in the writings of a certain Marcus Graecus, who was probably an alchemist; it dated from the mid-thirteenth century and was a translation of an older Greek document of unknown origin on pyrotechnics. Recipes found in the works of the alchemist Albertus Magnus, who died in 1280, and in Roger Bacon's *Opus majus* of 1267–68 were taken directly or indirectly from the Graecus recipe.

Gunpowder's use in warfare also spread west. The Chinese used rockets in the mid-thirteenth century against the Mongols, who then used rockets to capture Baghdad in 1258. By 1366, firearms were fairly widespread among the European kingdoms, and princes searched frantically for a stronger charge. But this had to wait for metallurgical advancements, since the soft metal of cannons tended to split under the increased heat and pressure of the improved powders.

By the fifteenth and sixteenth centuries, the balance was found. Firearms made obsolete the armor of knights and castles of feudal lords. Any peasant bearing a firearm could now kill a knight, lord, or king, a discovery that introduced in a violent way the modern thesis of social equality. Naturally, those in power opposed such new ideas. In 1498, for example, the Italian nobleman Paulo Vitelli ordered that all captured harquebusiers should have their hands cut off and eyes pierced because it seemed unseemly for noble knights to be killed by common foot soldiers without some form of revenge. A variation of the practice was to shoot captured cannoneers from the mouth of their own cannon. What did courage matter any longer in a world where, lamented one chronicler, "a manly, brave hero is killed by a dissolute, outlawed youngster by means of a cannon, a person who otherwise would not even be allowed to look at one or address one in a gross manner."

Lavoisier's mandate was less heroic and more technical: increase gunpowder's production while improving its explosive yield. Within three years he raised France's annual production of powder from 714 to 1,686 tons; he studied foreign methods and conducted experiments, concentrating finally on the refinement of saltpeter as the ultimate secret of yield. An account of its refinement, published by the French bureau of National Manufactories, attributed the secret to a process of "second crystallization," carried out by percolating saltpeter three times in special vats with perforated bottoms; each percolation occurred in progressively warmer water, then the saltpeter was dried in the air or warmed in a pan. One result was to create an ever finer grain of powder; this was important since more grains could be packed together with less space between the grains. More grains in the breech meant the release of greater amounts of expanding gas, and this translated into improved explosive yield.

Lavoisier's formula was so successful that by the late 1770s he produced a powder that carried a cannon ball 50 percent farther than any other powder in Europe. Its "carry" was estimated at 115–130 *toises,* compared to the 70–80 *toises* considered normal for the Seven Years' War (a *toise,* or French fathom, measured slightly over 6 feet). France's gunpowder, once the poorest in Europe, was now the best: by 1777, Ben Franklin was given secret authorization to arrange saltpeter and gunpowder shipments from France to America, and from 1776 to 1779, 1.7 million pounds were shipped to America. Historians have commented upon the importance of French gunpowder in Britain's defeat: by Yorktown, British soldiers complained that they could not get close enough to shoot colonials before they themselves were blasted from their garters.

In December 1788, the Gunpowder Administration had retrieved French self-sufficiency in munitions, its refineries producing 3.77 million pounds of saltpeter that year. It was later estimated that during the fourteen years of its existence, Lavoisier's Gunpowder Administration saved France 28 million *livres;* for his efforts, he earned a yearly salary of 17,000 *livres.*

He also nearly got killed. During the course of the crash program, his friend Berthollet discovered that when the unstable compound chlorate of potash was mixed with charcoal and sulfur, the result was highly explosive. It was so combustive, in fact, that Lavoisier wondered whether it could be used in place of saltpeter, which was still hard to procure. To test the idea, he arranged for a batch of the new powder to be made at his private factory at Essonnes, not far from Paris.

It was the morning of the experiment, October 27, 1788, a Sunday. Antoine and Marie had driven down to Essonnes to watch, along with Berthollet and a M. and Mlle Chevraud. Lavoisier had told the factory superintendent, M. Le Tort, to give the workmen added protection; they stood behind a heavy plank partition, but the mixing procedure was sloppy and some materials were thrown upon the vat walls. To make things worse, Le Tort tamped down spattered masses with his cane. The five visitors stood by the vat in the open while the workmen seemed protected. Lavoisier knew the mixture's reaction to any kind of shock, so he suggested that they and Le Tort go see the local shops while the mixing continued.

The disaster occurred as the six returned to the mill. Le Tort and Mlle Chevraud walked ahead; the others lingered behind for a moment when suddenly a violent explosion came from in front. The Lavoisiers, Berthollet, and M. Chevraud were thrown to the ground, battered and bruised, but little else. They ran up to their two companions to find that they had been hurled 30 feet through the air against a wall and were both dying. The workmen survived behind their barricade. In his report to the minister, Lavoisier wrote: "Please bring this matter to the attention of the king and let me assure his majesty that my life belongs to him and the state. I am ready to sacrifice it at any time in his service." He urged the state to continue experiments with the chlorate of potash, but the cataclysm stunned everyone and the tests were dropped soon afterward.

▪ ▪ ▪

Lavoisier's advances in chemistry proceeded more slowly and contentiously than in munitions. From 1777 to 1784, his party of "antiphlogistonists" was still a minority; but in 1785, the war against phlogiston entered a new and earnest phase. That year he was elected director of the Academy. His personal and political influence had never been greater, and he reorganized the institution under the guise of introducing natural history and experimental physics as two new disciplines. To keep the Academy's size manageable, he decreased each existing discipline by one chair, and sometimes reshuffled members to other fields; only later did others recognize his agenda. Before reorganization, Lavoisier was the lone antiphlogistonist among the chemists; afterward, the election of Antoine Fourcroy to the Academy and conversion of Berthollet to his camp saw the New Chemists controlling three of six chairs.

The new offensive began quietly as the year opened. Lavoisier's first move was subtle: he added new paragraphs to earlier papers that were to be given a second reading before spring. The theme of these insertions was the fictitious nature of phlogiston, and how the theory retarded the growth of chemistry. On January 22, 1785, he stated that phlogiston, "far from having brought light to bear upon [chemistry], seems to me to have created an obscure and unintelligible science . . . it is the *Deus ex machina* of the metaphysicians; an entity which explains everything and which explains nothing, to which one ascribes in turn opposite qualities." The initial intent of the addition was to mobilize support among his inner circle and transmit a code that war had begun. It also seemed designed to prod wavering chemists, like Fourcroy and Berthollet, to choose sides. In this it bore fruit. On April 6, Berthollet read without warning his paper "On Dephlogisticated Marine Air," in which he explained, by means of Lavoisier's theory, the generation and properties of chlorine. It was a public statement of conversion, making him the first French chemist to side with Lavoisier.

Berthollet's defection caused ripples of worry, but they were nothing compared to what followed. During the Academy meetings of June 28 and July 13, Lavoisier read out his manifesto, *Reflections on Phlogiston*. At this point, his call for revolution seemed directed at the Academy: he attacked the modified theories that had grown from Stahl's work, calling them incompatible with one another. Phlogiston was a "fatal error in chemistry," "a hypothetical being, a gratuitous supposition," which years later would be Laplace's term for God. He then presented his own theory as an alternative to a "veritable Proteus which changes its form at any instant."

He was assaulting phlogiston in the same way that Priestley assaulted Christianity, calling it a patchwork of corruptions and lies.

The reaction was explosive. Macquer complained that this young upstart would destroy all the progress of the chemical sciences; the Dutch scientist Martinus van Marum was present at the July 13 meeting when the normally sedate Academy disintegrated into shouts and chaos. Van Marum later wrote: "Then violent objections were made against this [paper], as a result of which the reading was continually interrupted. This, together with the simultaneous efforts of the reader and of his opponents to be heard, led to my understanding very little." For his part, Lavoisier was not surprised. "I do not expect that my ideas will be accepted right away," he said in his closing remarks. "The human mind inclines toward a way of seeing things, and those who have envisioned nature from a certain point of view, during a part of their career, only arrive at new ideas with difficulty."

The counterattack began. Rozier's *Journal de physique* became the rallying point for phlogistonists against Lavoisier; its new editor, Jean-Claude de la Métherie, was the grand inquisitor. In January 1786, de la Métherie introduced a feature into the journal that rebuked Lavoisier on a monthly basis. Each month, each year, the tone grew more bitter. Rephrasing the passage in which Lavoisier condemned phlogiston as a *deus ex machina,* de la Métherie wrote: "Surely anyone who has not made a specialized study of chemistry will not understand any better the combinations of the oxygen principle." He solicited the support of internationally known phlogistonists like Joseph Priestley in England, Lorenz von Crell in Germany (the editor of a sister publication, the *Chemische Annalen*), Jean Senebier in Geneva, and Felice Fontana in Italy.

Siding with the phlogiston theorists were those who despised Lavoisier on a personal level or loathed him as a symbol of academic authority. Many were scientists who'd been snubbed for admission by the Academy. Baron Etienne-Claude de Marivetz was one of this circle. When he submitted his *Physique du monde* for approval in 1781, Lavoisier wrote back: "This work . . . is a tissue of non sequiturs and absurdities; in it one finds neither cogency in its ideas, nor knowledge of the most elementary principles of physics; in a word it is the most monstrous assemblage which has ever been presented to the Academy."

Lavoisier was also beginning to hear critics call him a "damned aristocrat," but he was not blind to the suffering he'd seen as a child or during his travels with Guettard. He was in a position to urge reforms: it was only through reform that France might survive. He set up a new model farm at Fréschines on an estate near Blois, to research and teach improved cultivation, crop rotation, and soil conservation. During the famine of 1788, he dipped into his funds to buy barley for the towns of Blois and Romoratin. To avoid a recurrence of needless suffering, he proposed a system of government life insurance for the poor. Blois remembered such kindness and in December 1788 sent him to the Estates-General as its representative. He inspected Paris's jails and prisons, expressing his horror at the treatment of prisoners. The dungeons were filthy and foul; he proposed immediate fumigation with hydrogen-chlorine gas to kill the rats and fleas.

Then, in 1787, he began his greatest reform: a complete rethinking of the language used to imagine chemistry. He aimed for nothing less than a change in the discourse of reality.

▪ ▪ ▪

Again he started quietly, and for once he let his collaborators take the lead. Ever since writing the *Dictionnaire de chimie*, Guyton de Morveau had been plagued by the inadequacy of chemistry's technical language. Like many chemists, he was often at a loss when trying to identify substances. Chemicals ran through a riot of names that hearkened back to the alchemists—they were described in terms of red or green fumes, a tang on the tongue, sneezing, choking, and watering eyes. Metals were named after pagan gods: Venus for copper, Mars for iron, so that iron oxide was called "astringent Mars saffron." Others were named for properties, discoverers, or the site of origin: zinc oxide was "philosophic wool"; calcium acetate "shrimp eyes salt"; tin chloride "Libavius smoking liquor"; sodium tartrate "Seignette salt." Though poetic in its way, it was very easy to get confused. Was *natrum alkalisé,* an "efflorescence" found on rocks in Brittany, the same thing as *natrum,* a vegetable alkali? "Let us ask ourselves," de Morveau wondered, "if it is possible to find our way about in this chaos. Does not the understanding of such nomenclature require more effort than the understanding of the science itself?"

For six months, Lavoisier, de Morveau, Berthollet, and Fourcroy met daily, hashing out a new language; they reported back in May 1787 with a new book, *Méthode de nomenclature*. Perhaps more than anything else, the triumph of Lavoisier's New Chemistry was assured by this invention of a new language. His terminology, which we still use today, contained within itself the modern ideas of elements and reactions, as well as Lavoisier's own theories of combustion; it acts as an accurate boundary between the chemistry of the ancients and the chemistry of today.

Lavoisier and his collaborators understood that the lexicographer is more powerful than the experimenter; experimentation was a private endeavor, but language belongs to all. He understood the "justice" of abbé de Condillac's *Logic*:

> We think only through the medium of words—Languages are true analytic methods.—Algebra, which is adapted to its purpose in every species of expression, in the most simple, most exact, and best manner possible, is at the same time a language and an analytic method.—The art of reasoning is nothing more than a language well arranged.

Henceforth, chemistry took its form and direction because Lavoisier gave it a way of thinking and speaking. The laboratory method in chemistry was that of synthesis and analysis—of breaking down and putting back together—and a new nomenclature could express this both systematically and conceptually:

> The impossibility of separating the nomenclature of a science from the science itself is owing to this, that every branch of physical science must consist of three things: the series of facts which are the objects of the science; the ideas which represent these facts; and the words by which these ideas are expressed. Like three impressions of the same seal, the word ought to produce the idea, and the idea to be a picture of the fact. And, as ideas are preserved and

> communicated by means of words, it necessarily follows, that we cannot improve the language of any science, without at the same time improving the science itself; neither can we, on the other hand, improve a science, without improving the language or nomenclature which belongs to it.

To Lavoisier, language and action, if not identical, were at least inseparable.

The new nomenclature was based on three simple rules. First, a substance should have one fixed name. Second, the name should reflect its composition, if known (if not, the name should be noncommittal). Third, names should usually be chosen from Greek or Latin roots. If nothing else, the new system promised to demystify the old alchemical names. Common usage was sometimes substituted for the mystical, as when "*terra foliata tartari* of Muller" was renamed common potash, but the heart of the system lay in compound words and their endings first derived by de Morveau. The *-ique* (in English, *-ic*) ending on a root indicated acids saturated with oxygen; an *-ure* (*-ide*) ending meant compounds not in an acid state, often referring to the oxides of a metal. For example, an acid composed of a base and oxygen would be known by the name of the base, so that *sulfuric acid* replaced the older "vitriolic acid." If the same base formed another acid, but one combined with less oxygen, this lesser state of saturation was indicated as sulfur*ous* acid. A salt formed from sulfuric acid was *sulfate;* from sulfurous acid, *sulfite.* And sulfur-containing compounds like "liver of sulfur" (K_2S), which was not in an acid state, was now a *sulfide.* De Morveau's reforms seemed innocent enough, but adopting these suffixes meant, in effect, adopting Lavoisier's chemical view of the world.

A major part of the *Nomenclature* was a dictionary listing seven hundred chemicals, with their old names now translated anew. The adoption of Latin and Greek roots made the terms internationally recognizable, while a working advantage in the lab was the fact that the system expressed a compound's constituents: *ammonium molybdate,* for example, was the ammonium salt of the fully saturated acid of the base molybdenum. Also of importance was the recognition that the original *principe oxigène* changed to *gaz oxygène,* the word *gaz* ("gas") adopted from Macquer's *Chemical Dictionary* of 1766. All of the old physical explanations—the four elements of the ancients, the airs of the pneumatic chemists, phlogiston—were banished, damned primarily due to inefficiency. All discussion of an element's "nature," its "inner essence," said Lavoisier, were "confined to discussions of a metaphysical nature," which held little interest for him. The new definition was purely operational: an element was basically the limit of decomposition which could be reached by analysis. It was a substance broken down as far as it could go, a definition that had everything to do with man and his technology, and nothing to do with God.

The *Nomenclature* was also democratic because the rules were so simple and easy to understand; yet of all the fronts opened in the war against phlogiston, it also inspired the loudest howls of protest and provoked the greatest hostility in France and abroad. "The establishment of a new nomenclature in any science ought to be considered as high treason against our ancestors," said the English chemist Thomas Thomson, "as it is nothing else than an attempt to render their writings unintelligible, to annihilate their discoveries and to claim the whole as their own property." In reproaching the French for their presumption, Thomson accused them of something far worse: intellectual genocide. Lavoisier was not adding to the ancient store of knowledge; he was trying to supplant it, and eradicate the ancients from memory. While some critics accused Lavoisier of tinkering with the "genius" of the language, others hoped derision would scuttle the system. "So Lavoisier has substituted the word 'oxide' for the calx of metal," sneered the Irish phlogistonist James Kirwan. "I tell you it is preposterous. In pronouncing this word it cannot be distinguished from the 'hide of an ox.' How impossible! Why not use Oxat?" Anyone who adopted the new language, Kirwan said, was helping "to gratify the indolence of a beginner."

The objections to the new language were fourfold. First, what authority did *any* group of scientists have to change the language of an entire field? Second, the nomenclature was theory-laden: to speak of an "oxide" of mercury was to presuppose the theory's truth. The very word "oxygen" assumed the truth of Lavoisier's theory of acids, which *would* be proved in error. Third, there were linguistic purists who believed Lavoisier and his cronies played fast and loose with the rules of Latin and Greek. Finally, especially among the English, the nomenclature seemed a French conspiracy to replace tradition with their own ideals.

Priestley's antipathy revolved not so much around science but more around his faith in democracy. Although proficient in Greek, Latin, Hebrew, and several modern languages, he always used ordinary discourse in his scientific writing and apologized if he felt it necessary to use a new term. He saw Lavoisier as an elitist in several ways. Where no great proficiency was required to become a member of the Royal Society, the members of the French Academy drew a line between themselves and other scientists—they were the best, and the division was envied and resented. The insistence on a new language was an exercise in exclusion, and Priestley had always presented science as open to everyone of average intelligence and modest means.

By the late 1780s, Priestley had grown violently hostile to all forms of authority, and he saw the New Chemistry as an attempt to exterminate dissent rather than compromise. In 1796, he told the French chemists that "no man ought to surrender his own judgment to any mere authority, however respectable." Science represented the ultimate democracy, but Lavoisier, in his drive to suppress tradition, was reverting to a dictatorship of ideas.

In many ways, this conflict between chemists reflected a larger debate in Europe about the purpose and origin of language. Once again, the opposing sides faced each other across the Channel. Lavoisier and de Morveau had simply adopted the French dynamic theory of language, crystallized years earlier by Condillac, who held that just as a child organized its thoughts and gained reasoning powers as it learned language, so were linguistic reform and theoretical development inseparable from the progress of thought from unknown to known. The English system was more passive: language evolved with society through a system of "common usage."

Two fables prevail in the West to explain the linguistic chaos found between nations. The first is the curse of Babel, imposed on the sons of Noah as punishment for daring to imitate God. The second is that of Pandora, the first woman of Greek creation, molded by

Zeus to punish man for Prometheus' gift of fire. "Linguistic chaos" was one of the horrors unleashed from her box—whether Babel or the box, man was to be cursed forever for attempting perfection by a riot of tongues.

But this presupposes that there had been a common tongue, an idea with ancient origins that exploded on the scene as the new science of linguistics one year before the release of *Méthode de nomenclature*. In the Middle Ages a group of scholars called the Modistae believed that all languages were based on a universal grammar that reflected the structure of God's mind; they were echoed during the Renaissance by the French Port-Royal School. Then, in a speech given in February 1786 to the Royal Asiatic Society, William Jones, Chief Justice of Bengal, suggested that Sanskrit, Latin, and Greek originated in a common source, thus inspiring generations of scholars to compare sounds and meanings between languages in a search for similarities. Just as *pater* in Latin was *Vater* in German and *father* in English—as *piscis* in Latin was *Fisch* in German and *fish* in English—the similarities seemed inescapable, "so strong, indeed, that no philologer could examine them all three, without believing them to have sprung from some common source, which, perhaps, no longer exists," Jones said. Though unsupported by research, Jones's thesis was published and circulated widely, and became the beginning of a quest to find an Indo-European mother tongue.

To the English, the adoption of Lavoisier's nomenclature was part of the larger debate: whether or not to establish a national academy of languages modeled after the Académie Française, founded in 1634 by Cardinal de Richelieu. The wordsmith John Dryden complained in 1660 that English was a barbarous language, more thrown together than guided or modulated, and believed that an Academy like France's could regulate English usage, a view echoed for another two centuries by Daniel Defoe, Jonathan Swift, and even John Adams. In its youth, the Académie was an ambitious motivator of change, and in 1762, after years of work, published a dictionary that regularized some five thousand words, nearly a quarter of those in use.

Yet what is new becomes old and conservative; as academies grew older, said one critic, "they almost always exert over time a depressive effect on change." For precisely this reason, as many influential writers opposed the idea of a British academy as supported one. Samuel Johnson doubted the prospects of arresting change; Thomas Jefferson thought it undesirable, noting that if such a body had been formed during the days of the Anglo-Saxons, English would now be unable to describe the modern world. Priestley argued as much in his *Theory of Language*, published in 1762: it was one of the few times he and Samuel Johnson ever agreed. An academy, he said, was "unsuitable to the genius of a free nation. . . . We need make no doubt but that the best forms of speech will, in time, establish themselves by their own superior excellence: and in all controversies, it is better to wait the decisions of time, which are slow and sure, than to take those of synods, which are often hasty and injudicious."

Lavoisier's chemical nomenclature was one such "hasty" synod to Priestley, determined to impose an artificial order upon a naturally evolving world. The belief that "common usage" and "custom" would protect England's linguistic freedom from outside threats was inseparable from trust in England's "peculiar genius" for liberty and the evolution of a "common law." It wouldn't be changed, and especially not by some Frenchman. With the speech of William Jones, it could even be argued that the old chemical language had evolved from the mind of God.

Yet the Gallic advance in chemistry could not be stopped, and with the release in 1789 of Lavoisier's textbook, *Traité élémentaire de chimie* (*Elements of Chemistry* in English), the revolution was complete. Often called the first modern textbook of chemistry, the *Traité* offered little that was new, but rather served as a synthesis of all of Lavoisier's work and ideas. Yet it was still obviously groundbreaking. The oxygen principle, his ideas on heat and formation of acids, the new nomenclature and list of thirty-three elements . . . all were there, forged together in a theory that burned like light through the contradiction and semantic agony of phlogiston. With the *Traité*, one feels in Lavoisier a sense of peace. He had accomplished the same as Newton, harmonizing isolated facts into a world symphony. "It is *my* theory," he said simply and proudly when French chemists called it theirs. He, and he alone, had made the laws of life clear for everyone.

With the *Traité*'s publication came a new purpose, and Lavoisier made his campaign international. He sent off a number of complimentary copies to luminaries throughout Europe and America, including Joseph Black, Ben Franklin, and Alessandro Volta. Each was mailed with a personal letter describing the progress of the theory and a flattering comment that it would be a milestone if the recipient supported the New Chemistry in his own country. It would be the last nail in the coffin of phlogiston. By the early 1790s, Lavoisier's theory had swept more adherents into its embrace. Professor Thomas Hope, at the University of Edinburgh, was the first teacher to adopt the new language in his public lectures; not long afterward, his colleague Joseph Black accepted Lavoisier's explanations and taught it to his students, too. Italy and Holland became converts. To the dismay of phlogistonists in the Lunar Society, Erasmus Darwin began using Lavoisier's "oxygen" and "hydrogen" in his poems and essays. Sweden's most respected chemists wrote to offer their support. The Berlin Academy of Sciences, long a holdout, ratified Lavoisier's system in 1792. In America, scientists converted to the new faith almost to a man. Even contentious Russia endorsed the system, but claimed a forerunner: the vodka-swilling poet-scientist Mikhail Vasilyevich Lomonosov, who a generation earlier had conducted tests in "airtight vessels to ascertain whether the weight of a metal increased on account of the heat."

Undergirding the revolution was Lavoisier's faith in the conservation of matter. He took for granted throughout the *Traité* that chemical change and the processes of life were a balance sheet in which the weight of products would always equal the weight of reactants. "For nothing," he wrote, "is created in the operations either of art or nature, and it can be taken as an axiom that in every operation an equal quantity of matter exists both before and after the operation." It was this belief that became the cornerstone of modern chemistry, the principle that turned chemistry into the quantitative science it is today. All life is an equation "as simple as algebra," Lagrange would say. Nothing ever disappears in life or death, and if it seems to, the disappearance is merely a fleeting illusion, caused by our inability to measure with precision the world's greatest subtleties.

But even our limitations were subject to change. We are all children when faced with new wonders, Lavoisier said in the preface to *Traité*, and like children we can "proceed from known facts to what is unknown" and not be afraid. Like children, we learn and explore.

> In early infancy, our ideas spring from our wants . . . from a series of sensations, observations and analyses, a successive train of ideas arises, so linked together that an attentive observer may trace back to a certain point the order and connection of the whole sum of human knowledge. When we begin the study of any science, we are in a situation . . . similar to that of children, and the course by which we have to advance is precisely the same which nature follows in the formation of their ideas.

Chemistry had been a pseudoscience for too long, he was saying. Now, with the publication of the *Traité élémentaire*, a new science was born.

PART III

Reward

Thrash them, beat them, flog them for their crimes,
But most of all, because they dared outrage the gods of heaven!

—Chorus leader Koryphios, crying for vengeance against
Socrates and his disciples, in Aristophanes, *The Clouds*

CHAPTER 11

"King Mob"

THERE WERE TIMES, AS PRIESTLEY WALKED THROUGH BIRMINGHAM'S streets, that he felt his life was the happiest it had ever been—as if the years of penury and anxiety, of ill health and slander, had just been a test and this booming Midlands city was his reward. His congregation at the New Meeting House had swelled; Mary was enchanted with country life; their house at Fairhill, outside the city, was delightful.

It is surprising how tranquil life can seem when one's house runs smoothly and is well arranged. For Priestley, the things that mattered were the details. In front of the house he'd set a rain gauge, and next to that a "perambulator" of his own invention, a wheel fit with a handle and a device for reading yards and miles. Local children were allowed to play with that and an air gun, its target a block topped with one of Priestley's wigs. In the upstairs library he'd built a long room with bookshelves reaching from floor to ceiling; in this room he kept a screen for magic lantern shows and an electrical machine for giving mild shocks to the curious, though he no longer claimed to drive out demons. In the lab sat rows of stoppered bottles that looked empty but actually held specimens of air sent from around the country. The only time anyone ever remembered him growing angry was the day his daughter opened the bottles to clean them. Never again open anything in this room, he cautioned. It might be dangerous. Even when things appear safe, they may not be what they seem. Like most parents, he and Mary worried about their children as they grew before their eyes. Sally, the oldest, had always been prone to illness, but was now engaged to marry William Finch, a young man of local extraction, and anticipated setting up her own house at Heath Forge, Dudley, a good carriage ride from Birmingham. The bloom of health accompanied her love. If only the boys were so well situated. At every turn, Priestley encountered difficulty in finding positions for young Joseph and William, largely due to prejudice against his name. He'd tried many leads for Joseph but always met with an unfavorable answer; on a more positive note, the boy *was* engaged to marry the daughter of a local merchant and close friend.

Where Joseph was enterprising, if unlucky, William was another matter. A high-strung youth of "temper and high spirits," William seemed unsuited for business, and they'd even failed to get him an appointment with Mary's family, the Wilkinsons. Yet William was a brave lad, protective of his parents (when he wasn't arguing), and by 1789 Priestley thought it might be best to send him off to Paris to release that energy on the Grand Tour. That left Harry, who was most like his father—studious, a firebrand for his father's causes, an able unpaid assistant in the lab. Priestley secretly hoped that Harry would carry on his work in philosophy and theology, but for now he attended a Dissenting academy in Bristol and the parents were least worried about him.

It was hard to find one's place in the world. Priestley remembered too well his own lonely struggles. He saw those struggles echoed in his offspring. He and Mary encouraged the children to visit the house in Fairhill at any time; they all treated it like a sanctuary. A French visitor in 1784 described it as a charming place, "with a fine meadow on one side and a delightful garden on the other," where the "most perfect neatness pervades everything." In fact, Priestley's house was a waystation for much of Dissenting Birmingham: members of all sects and the Lunar circle, their friends and extended families . . . the house was in constant bustle and Joseph and Mary liked it that way. He was never as lonely as he'd been in his youth; he'd established once again the childhood home with Aunt Sarah—light and warmth, a flow of visitors. Only the grim creed of Calvin was missing, an absence he did not mind.

Priestley's day began by building the fire in the kitchen and parlor, the only chore Mary required of him, then taking the fire in an iron shovel to start the furnace in his lab. There might be tests, although by 1786 science had assumed a secondary importance for him. By midmorning or noon he'd take the one-mile walk to the New Meeting House in Birmingham. Now in middle age and still dressed in the ministerial frock and periwig that had dropped out of fashion in the 1770s, everyone recognized him. For his part, he preferred the rising middle classes, and Birmingham was a middle-class town. "There is not only most virtue, and most happiness, but even most true politeness in the middle class of life," he enthused.

Throughout England, Birmingham was considered a place for a hardworking man of talent to make his fortune. Perched high on a sandstone bluff, the Anglican Cathedral of St. Philip—its dome modeled after St. Paul's in London—looked down upon a sprawling town surrounded by fields stretching to the River Rea. The forges and workshops clustered in the low-lying areas, while wealthier citizens moved to high ground and clean air. The city directory of 1770 listed 1,252 tradesmen, a number that had grown in the years since. From 1740 to 1780, the number of houses shot from 4,210 to 8,382, the streets from 54 to 125, and the population from 25,267 to 50,295. Birmingham's property and population had more than doubled, phenomenal growth for a European city that was not a nation's capital. In 1780, there were 1,636 births recorded and 1,340 deaths, numbers that maintained the boom.

Entrepreneurs were drawn to town, imbued with the faith that "each man has his own future in his hands." Take, for instance, Priestley's good friend William Hutton, who would in later generations be called the "English Ben Franklin." Hutton was ten years older than Joseph, born in Derby in 1723 to a Dissenting woolcomber—an upbringing like the Doctor's. At age seven, he was apprenticed to the Derby silk mills; at fourteen, he apprenticed under his uncle, a stocking maker in Nottingham. He ran away from this unhappy life, but returned to finish his apprenticeship and moved to Birmingham to start a shop of his own.

What he saw of the city seemed a new way of looking at the world. "I was surprised at the place but more so at the people," he wrote in his autobiography. "They were a species I'd never seen: They possessed a vivacity I had never beheld: I had been among dreamers, but now I saw men awake. . . . Every man seemed to know and prosecute his own affairs." In a wink Hutton abandoned the idea of opening a stocking maker's shop and set up a bookshop and circulating library, an ideal choice for a town dedicated to self-improvement. From these profits, he turned to bookbinding, publishing, papermaking, and land speculation. In 1768, he was worth £2,000; by 1791, his country house alone—an estate close to Priestley's—was valued at £8,000.

Life in a mercantile society is devoted to want: its sin is to make it feel like need, so that one *must* have a thing. The flip side is its grace, where want feels like hope and a purchase is a doorway to a better dream. The logical extension is that *anything* is for sale, including spyglasses for hire in London at a halfpenny to view the Jacobite skulls spiked to Temple Bar. Primarily on instinct, Hutton hit on a drive that made men's fortunes—the desire to "improve" and "arrive." It was during this time that advertisements began to exploit envy, and Joseph's friend and patron Josiah Wedgwood was at the forefront of this wave. When Wedgwood opened showrooms in London and Bath for his dinnerware, he found that fashion outstripped merit when it came to sales. He played on aspirations to elegance by promoting neoclassical designs in his Etruscan ware, modeled on archeological finds in the newly unearthed Herculaneum.

In this sense, Birmingham had "arrived." St. Philip's Cathedral was its grandest statement, but even the New Meeting House was a foreboding edifice built in neoclassical style. The pride of the city was Thomas Dudley's Hotel, in which Edward, Duke of York, had led a dance in 1765. The hotel was now the site of most meetings, special assemblies, and dress balls. "From a handsome entrance the ladies are now led through a spacious saloon, at the extremity of which the eye is struck with a grand flight of steps, opening into an assembly room," Hutton wrote in his *History of Birmingham*. "The pile itself is large, plain, and elegant, but standing in the same line with the other buildings . . . eclipses them by its superiority."

Mary was content, and that was important to Priestley since her mood was often a source, rather than reflection, of his state of mind.

She was helped at Fairhill by two maids and a young caretaker; she spent her days weeding or planting in the garden, making clothes, and managing household finances to the point that Joseph asked for money when he strolled to town.

Priestley seemed to feel that he, too, had arrived. In 1787, he was asked by friends and benefactors to write a memoir of his life. There is no more of that smugness apparent, as when he took Lavoisier to task; no worry or anxiety, as during the dark days of childhood and Needham Market. One gets the sense of a man at peace, tranquil with the way his life has evolved. He paints a picture of himself at work, "writing on any subject by the parlour fire, with my wife and children about me, and occasionally talking to them, without experiencing any inconvenience from such interruption." His style at the pulpit also radiated peace, said Hutton's daughter Catherine.

> In the pulpit he is mild, persuasive and unaffected, as his sermons are full of good reasoning and sound sense. He is not what is called an orator; he uses no action, no declamation, but his voice and manner are those of one friend speaking to another.

Good health, abetted by sleep, were the secrets of tranquility, he said. "It has been a singular happiness to me," he wrote,

> and a proof, I believe, of a radically good constitution, that I have always slept well, and have awakened with my faculties perfectly vigorous, without any disposition to drowsiness. Also, wherever I have been fatigued with any kind of exertion, I could at any time sit down and sleep; and whatever cause of anxiety I may have had, I have almost always lost sight of it when I have got to bed, and I have generally fallen asleep as soon as I have been warm.

True, there had been "exquisite pain" from gallstones when he first moved to Birmingham; but after confining himself to a vegetable diet, he was "perfectly recovered." True, there'd been the uproar over the Gunpowder Sermon, as well as such works as *Disquisitions*, *Institutes*, *Corruptions*, and *The History of Jesus Christ*, just to name a few. But surely, he thought, such upheavals were compensated for by his discoveries in natural philosophy and the "cardinal esteem" in which he was held by his friends.

"I esteem it a singular happiness to have lived in an age and country, in which I have been at full liberty both to investigate, and by preaching and writing to propagate, religious truth," he ended up his memoir, and considered his move to Birmingham "as the happiest event in my life," conducive to all his goals both "philosophical and theological." It was for him the best of all possible worlds.

▪ ▪ ▪

How could he be so wrong? Did he sleep through those years when his reputation among the general public turned diabolic—when his orthodox critics regarded him as wicked, malignant, and destructive, and refused to attribute to him any human feelings? There'd been a lull since the stormy reception in 1785 of his Gunpowder Sermon, but the national press had a good memory and could lampoon him as "Gunpowder Priestley" and "Gunpowder Joe." In March 1787, the Dissenters tried again to repeal the Test Acts and once again failed, but the continued push for change made the Church and Crown nervous. And talk of change was everywhere. Unrest simmered in Ireland; the troubles in France ran deep; there was talk in the streets of how the necks of the last kings would be strangled by the entrails of the last priests. It was a time for tact, but tact was not always in Priestley's vocabulary.

Perhaps, as has been suggested of Ronald Reagan, another famous optimist, Priestley's intelligence was more Procrustean—one driven to produce uniformity—than Protean, or all-embracing. As the scientist Frederick Turner once described it, such an intelligence

> reduces the information it gets from the outside world to its own categories, and accepts reality's answers only if they address its own set of questions. . . . It insists on certainty and unambiguity and so is at war with the probabilistic and indeterminate nature of the most primitive and archaic components of the universe.

This could explain Priestley's dogged ignorance of certain obvious facts, such as England's hatred for him. He needed to impose order on chaos, hence his stubborn loyalty to phlogiston. He wouldn't be rushed, and didn't like surprises: everything must be punctual and contained—his measured pace, his three daily games of chess or backgammon with Mary, his regular sleep each night. In a world ruled by reason, rewards come to those who play by the rules. He'd played by God's rules, not England's, and he'd earned his just reward.

One wonders if others in the family were as blind. William certainly wasn't. His role in the records is that of one born to clash, and he clashed with most with his family. Young Joseph also seemed to smell trouble, but apparently kept his opinions inside. But Mary, Sally, Harry—these three seemed to live within Joseph's self-willed haze.

There were rumblings close to home, if he'd only notice. He founded a public library and a storm rose over controversial theological works available at the desk: the orthodox clergy condemned them as a "fountain of erroneous opinions, spreading infidelity and schism throughout the whole neighborhood." Priestley's 1788 antislavery sermon was interpreted as a broadside against suppressed freedoms at home. When he started a Sunday School for children, it became popular among the working class and interdenominational, yet another threat to the Anglicans. They feared the growing strength of dissent in Birmingham; in addition to Priestley's Unitarians, there were Baptists, Methodists, and Quakers. Each new program or reform by Priestley was seen by the clergy at St. Philip's as one more raid on their rolls.

Then, in 1788, King George "went mad." The mid-1780s had been good to him. Britain was recovering its self-confidence after the surrender to America, George was in a vigorous middle age, his figure was trim and solid. The atmosphere at court was relaxed; George studied botany and wrote pamphlets on agriculture under the name "Ralph Robinson." But then his genetic porphyria struck him down. He'd not been well in the summer of 1788, and that October went into convulsions after riding in the rain. Afterward, his behavior seemed odd and he always wanted to talk; around this time the famous Oak Tree incident took place, during which George reportedly talked to a tree as if it were the King of Prussia. The story's truth was questioned since it was first spread by a servant who had been sacked; what was unquestionable, however, was the King's obvious derangement in November when during a dinner at Windsor Castle, George attacked his son, the Prince of Wales, and tried to smash his head against a wall. He gibbered non-stop, foam flowed from his mouth, and his eyes were so red observers said they looked like currant jelly.

The treatment was as cruel as the malady. George's chief physician owned a private insane asylum, and his "cure" included lectures, threats, the straitjacket, restraint to the bed, and a special iron chair with straps that George later called his "coronation chair." There were poultices of hot mustard and Spanish fly rubbed all over his body: the painful blisters that resulted were thought to draw the "evil humours" from his pores.

The regency crisis caused worry everywhere. In November 1788, there was a panic on the stock market; performances of *King Lear* were dropped from the stage. The literary analogy between madness and social disorder was a frequent theme in Augustan satire, but now the scurrilous British press trod lightly or referred to it in code. A tone of somber horror descended on public life that was rarely witnessed. "The King is *insane,*" wrote Edmund Burke. "A certain animal is alive; the man is dead; the King is gone."

The displays of relief in February 1789 when George regained his senses reflected the depth of unease. There was an explosion of patriotism and English songs like "Rule, Britannia," "Heart of Oak," and "The Roast Beef of Old England" were struck up in an instant. That summer, during a holiday jaunt to Weymouth, George was greeted by a band that walked into the sea beside him and played "God Save the King."

In Birmingham, the celebrations were conspicuous and loud. The town elders called for fireworks; a bonfire built by the Navigation Office consumed three tons of coal. The public buildings at Boulton's Soho foundry displayed huge banners and emblems congratulating George on his good health. The crowds filled the streets despite the wind and freezing rain.

▪ ▪ ▪

Although the French Revolution is usually considered to have begun on July 14, 1789, it actually started two months earlier with the massacre of the hares. That May, the Estates-General deliberated the abolition of feudal rights, which included the exclusive hunting rights of the nobility. On June 10, Camille Desmoulins, a deputy of the body, told his father that

> the Bretons are provisionally carrying out some of the articles of their *cahiers de doléance.* They are killing pigeons and game. And here in this part some 50 young people are creating havoc among hares and rabbits. They are said to have killed between four and five hundred head of game under the eyes of the wardens in the plains of St. Germain.

But the true symbol of the Revolution was the fall of the Bastille. All that year of 1789—as Priestley slept and Lavoisier's *Traité* was published—France's financial crisis deepened, and the liberals and reactionaries schemed. On May 5, the Estates-General convened at Versailles: although anyone who was decently dressed could enter the palace, the sense of toleration did not extend to the sharing of power. For six weeks a deadlock persisted on the question of votes. The Third Estate, party of the bourgeoisie and commoners, was numerically most powerful and they wanted majority rule. When the court and clergy objected, the insurgents dubbed themselves the National Assembly and gave themselves sole power to tax. In response, Louis kicked them from the palace, and the Assembly gathered on the tennis courts, vowing never to leave until a new constitution was drafted. Louis ordered Swiss and German guards to Paris; Jacques Necker, the populist Comptroller General, was dismissed. On July 12, pricked one too many times, the mobs of Paris took matters in their own hands.

Even before the fall of the Bastille, there was evidence that horror might attend a popular uprising. Crowds swarmed the streets when Necker was sacked; the mercenaries fired and one crowd member dropped dead. So began a day-long riot: the Opéra was invaded; customshouses were torched; crowds looted a nobleman's palace. A minister rumored to have said that the people could eat grass if they were starving was hung from a lamppost outside the Hôtel de Ville. When his body was cut down, he was decapitated and his mouth stuffed with hay. The grim jest was stuck at the end of a pike and paraded through the streets for everyone to see.

Priestley's rival was in the heart of this. Unlike Priestley, Lavoisier was wide awake to danger and knew that he and Marie faced perilous days. He watched smoke rise over the city from the burning of forty of the fifty-four customshouses that lined his unpopular wall. The jobless jammed the streets, died in plain sight, and wanted someone to blame. The political factions were menacing enough, but the mob was ready to boil over, and entrusted to Lavoisier's protection were thousands of pounds of gunpowder, enough to blow up Paris. It was stored at the Arsenal, his workplace and home, both undefended.

It was this search for arms that led the crowd to the Bastille. The ancient prison was a symbol, and the crowd appropriated the King's justice as its own. It was rumored that Lavoisier had sent stores of gunpowder to the Bastille for safekeeping; the air was thick with rumors that the King's Guard was marching from Versailles, while to make matters worse the Bastille's guns were trained on the crowded slums beneath its walls. After thirty thousand antique muskets were taken from the Hôtel des Invalides across the river, the cry went up: "To the Bastille!"

Antoine and Marie could watch from their windows as the crowd surrounded the fortress: the Bastille's governor defended the fortress with 32 Swiss Guards and 82 invalided French soldiers, and before he was persuaded by his men to surrender, 150 attackers were killed or wounded. He went out to negotiate and was butchered at the gates; his head was jammed atop a pike, and his assistant was executed, too.

For all the sound and fury, few arms and little powder was taken by the crowd that day. Yet the attack had remarkable political consequences. The seven prisoners inside were freed and allowed to join the crowd; some old scores were settled when some thieves were strung from lampposts. The Bastille was razed and carried away stone by stone, each becoming a relic for sale or display. The two main aspects of justice—the right to impose death and to grant mercy—were taken over by the people, and until the Terror, the fall of the Bastille was the Revolution's most potent symbol.

It didn't take long for Lavoisier to face the judgment of the crowd. In August, a bargeload of industrial powder—not weapon's grade as used in muskets—was shipped from Metz, bound for the ports of Nantes and Rouen and from there to the coast of Guinea. Instead, it was waylaid and sent to the port of Saint-Paul near the Arsenal. Lavoisier and his assistants were bewildered when it arrived. This was not musket powder . . . there was no storage space in the Arsenal for such a huge shipment . . . where had it come from and why was it here? The Revolution was filled with comedies of error that in an instant turned deadly, and this proved to be one. Lavoisier rushed to the Hôtel de Ville (city hall) to exchange the industrial powder for a load of musket-grade from Essonnes, but when loiterers at the port spied the purloined powder being reloaded on the barge, they spread the rumor that it was being sent to France's enemies.

Again, Lavoisier appeared at city hall to explain everything. Two deputies inspected the powder and found that it was industrial grade, just as Lavoisier said. Again, the kegs were rolled onto the barge; again, the crowd jumped to conclusions; again, Lavoisier marched to city hall with the signed order of transference. Although exonerated, suspicions against him remained.

By the end of summer 1789, peace was being restored throughout France, but always at the cost of lives. Bastilles fell throughout the nation; in the provinces, peasants and townspeople were gripped by the Great Fear, the rumor that an enemy, possibly royalists, was marching and that an "aristocratic plot" was in the works. When the foe never materialized, armed peasants attacked the local lords, burning their manors and the hated manorial rolls listing feudal rents and obligations. By August, the Revolution had moved so far and fast that it was obvious to the nobility that their privileges were nearing an end. On August 4, a relation of Lafayette gave an impassioned speech before the Assembly to abolish feudal rights; soon afterward, all Frenchmen were made "equal" and Louis XVI was hailed as the "restorer of French liberty." While the Assembly began drawing up the Declaration of the Rights of Man, branches of the National

Guard taught peasants in the provinces that freedom had its limits with wholesale executions.

There was one last act of 1789—one that would have ultimate consequence for both Lavoisier and Priestley. While the Assembly deliberated and hungry crowds clamored for food, reports of a feast at the palace again precipitated mob action. On the morning of October 5, a crowd of women—and some men dressed as women—came to city hall demanding bread. There was none, and the price had been rising too long. "To Versailles!" came the cry, reminiscent of "To the Bastille!," and thus began the march of the "fishwomen," augmented by thousands who were picked up along the twelve-mile road to Versailles. Trailing behind them at a safe distance were Lafayette and his National Guard.

The Assembly appointed a delegation to go with the women, and Louis, happy to please, promised bread. During the rainy night, the women camped in the courtyards of the palace; the next morning, a group tried to force their way into the Queen's apartment. As two bodyguards blocked the way and were killed for their loyalty, the Queen escaped by another route. During the mêlée, the royal family was rescued by Lafayette and his troops, and at the crowd's insistence he escorted them back to Paris to stay at the Royal Apartments of the Tuileries, deserted for years. On the road back, the slain bodyguards' heads were displayed on the ever ready pikes. Carts full of flour from the royal granaries rolled behind.

The redoubtable Arthur Young saw the royal family soon afterward: bewildered Louis; Marie Antoinette, the Dauphin, and his older sister, Marie-Thérèse. How had it come this far? Louis had ascended the throne on a wave of goodwill; he'd seemed reform-minded and peaceful, and the liberals hoped for great things. Now he walked in these withered gardens, "plump as ease can render him," accompanied by six grenadiers. Marie Antoinette was portrayed as a sexual monster, preying on her subjects, yielding her virginity to her brother and whiling away her hours with lesbian conquests—yet on her first entry to Paris the streets were ecstatic with welcome. How had it come to this, that as she passed through the old palace, "a mob followed her talking very loud, and paying no other apparent respect than that of taking off their hats"?

Young worried most for the young heir. The Dauphin was a "pretty good-natured-looking boy of five or six years," playing alone in a "little garden railed off" for his pleasure and protection. "Here he was at work with his little hoe and rake, but not without a guard of two grenadiers," Young said, a miniature version of George III in England, happily tilling the soil as the King and kingdom went mad.

▪ ▪ ▪

Three months earlier, another boy reacted to the Revolution, and in his own way he was just as doomed. It was late July, soon after the fall of the Bastille, and across the Channel Harry Priestley burst through the drawing room door at the home of thirteen-year-old Mary Anne Galton, daughter of the Birmingham arms maker and family friend. "Hurrah!" Harry shouted, waving his hat. "Liberty, Reason, Brotherly Love forever! Down with kingcraft and priestcraft. The majesty of the people forever! France is free!"

Mary Anne was thunderstruck, while Harry barely contained his pride. His brother William "was there [in Paris], and helping"; he insisted on reading a letter from William that had just arrived. William had mailed a picture of the Bastille and two stones from its ruin, but these had yet to arrive. "But come, you must hear his letter!" the youngest Priestley declared.

This was the first flush of the news. It was the fresh beginning that Joseph had hoped for. All would be better now, the dawn of a brighter day. Corruption, tyranny, and war would be a thing of the past; the reign of universal peace and brotherhood was nigh. It was the advent of the Millennium, when the lion would lie down with the lamb and men would turn their swords to plowshares, among other now familiar clichés.

In the four months immediately following the fall of the Bastille, English public opinion about the French Revolution was much like Harry Priestley's; even today, historians believe that 1789 marks a change so momentous that many call it the true beginning of the Modern Age. "Bliss was it in that dawn to be alive," wrote William Wordsworth, "But to be young was very heaven!" During this honeymoon, the English perceived the Revolution as akin to what they had experienced one hundred years earlier: the word "revolution" did not yet imply murder, warfare, sudden death, or mass execution in the name of a "higher" good. The threads of thought that many, like Priestley, had followed were coming together. Darkness would be replaced with reason, civil liberties would blossom, a scientific Utopia would dawn.

By November, however, opinion was dividing and Priestley's firebrand friend Richard Price drew a line in the sand. The occasion was the Fifth of November, the anniversary of the landing of William of Orange, and each year the Revolution Society—composed primarily of Dissenters, but containing a few churchmen—met to celebrate the events of 1688. Price was old and sick; he had less than two years left to live. In what would be his last hurrah, he stood at a pulpit in London's Old Jewry, a small street in the City that was the site of a synagogue until the expulsion of the Jews in 1291; he preached a sermon entitled "A Discourse on the Love of our Country." To promote their nation's interests, he said, citizens should seek truth, virtue, and liberty. The last, as handed down from the Glorious Revolution, meant the right of conscience in religious matters; it also meant the right to resist abusive power and the ability to change government when power was abused. There would never be liberty in England as long as the Test Acts remained in force, Price thundered, and parliamentary representation was a fraud. He thanked God that he'd lived long enough to see 30 million Frenchmen spurn slavery, then pointed a gnarled finger straight at the King:

> Tremble, all ye oppressors of the world! Take warning, all ye supporters of slavish governments.... You cannot hold the world in darkness. Struggle no longer against increasing light and liberality. Restore to mankind their rights, and consent to the correction of abuses, before they and you are destroyed together.

That was all the spark that was needed. The next month, Dissenters, led by Priestley, held meetings all over the nation protesting the Test Acts; in response, the Anglican Church mobilized by attacking Priestley personally. The sudden change from jubilation to accusation took many by surprise. Where there'd been a dawn of hope, now there were prospects of civil strife and revolution; suddenly all the talk was of republican toasts, political clubs, "Rights of Man" radicalism, sinister muttering that priests and peers should be hung. All were forms of discontent with roots in English politics, but now they blossomed swiftly because of foreign convulsions that were still too close for comfort. For liberal Whigs like Priestley, Price's sermon tapped a libertarian rage that had flowed underground since the collapse of the Puritan revolution. For conservative Tories, the response was paranoia and fear.

Suddenly, Priestley began to see himself portrayed as a demon in the popular press, a situation resembling Lavoisier's. In February 1790, a satirical cartoon was circulated that portrayed Priestley, Price, and their Unitarian colleague, Theophilus Lindsey, being led off to Hell by demons. The next month, he was drawn at the pulpit belching columns of smoke labeled "Atheism" and other heresies. That same month, the Church passed circulars in Parliament that contained extracts of Priestley's speeches, all portraying him as an advocate of cataclysmic revolution rather than of gradual reform. He was painted as a fool, opening England up for invasion; people repeated the old saw, "When an old hound misleads the pack, the only way of treating him is to hang him at once." The sobriquet "Gunpowder Joe" was dusted off. One John Nott, a button burnisher, wrote in his *Very Familiar Letters*, "How would you like it yourself if they were to send you word they had lain trails of gunpowder under your house or meeting house?"

"My gunpowder is nothing but arguments," Priestley tried to answer back, but the metaphor was beyond his control and had assumed a life of its own.

The attacks turned a dangerous corner. As early as January 1790, Priestley was confined to the house for several weeks with extreme hoarseness; the pressure-cooker atmosphere showed up in the return of his stammer. At 2 or 3 a.m. on a Sunday, three men attempted to break into his house. Priestley's maid, who heard a noise and opened a window, was fired on with a pistol. They missed and ran away. When the story became public, John Nott again mocked Priestley: the men had been out hunting sparrows, said Nott, but Joseph was so overcome with fright that his Dissenting friends "huddled about you with their smelling-bottles and drops."

Birmingham, that happy place, was becoming distinctly unfriendly. Clergy in the town preached sermons against him; walls in town were chalked with "Damn Priestley! No Presbyterians!!" He was walking down a street when a gang of boys dogged his tracks and screamed, "Damn Priestley, damn him, damn him forever." A ballad was sung in the streets:

Gunpowder Priestley would
Deluge the throne with blood,
and lay the great and good
Low in the dust.

More harrowing was the attack of an old ally. Edmund Burke, the parliamentarian prophet of expedience, had championed the American cause in 1776, and through this Priestley and he had discovered a kinship. But now, when Burke heard of Price's sermon, his response was disbelief and anger. His *Reflections on the Revolution in France, and on the Proceedings in certain Societies in London relative to that Event*, published in November 1790, condemned the French Revolution as the beginning of mob rule by the "swinish multitude"; he argued against *all* change that did not accord with the Glorious Revolution. Whatever was not traditional was unnatural, and thus evil, unpatriotic, and un-British. In one quick stroke, opinion was polarized: liberal and conservative camps squared off, and the battle of words that Burke began created the political vocabulary we still use today—*liberal* and *conservative, socialist, anarchist, capitalist* and *capitalism, equality* and *fraternity.* All were products of the French Revolution, Price's sermon, and Burke's impassioned reply. The political clubs like the Revolution Society were "the mothers of all mischief"; for Richard Price to mention man's natural rights was foolishness—there were no "natural" rights, Burke argued, only those granted to people by the King. More than all else, Burke feared the mob's passion, and at this point he was still speaking of events between June 1789 and the summer of 1790. The confiscation of Church property, the rise of Robespierre, the Reign of Terror, and the executions of Louis XVI and Marie Antoinette—the theater of the guillotine, where the audience dipped handkerchiefs in the King's blood—all lay in the future. As they unfolded, Burke was hailed as a prophet and his voice became that of terrified Englishmen everywhere.

The response to Burke was massive, resulting in a "pamphlet war"—approximately two hundred titles pouring from the English press between 1790 and 1795 that were direct answers to his *Reflections*. Another survey estimated 1,086 titles from July 1789 to December 1793. The first attack on Burke was by Mary Wollstonecraft—feminist, novelist, and future mother of Mary Shelley. On November 20, 1790, she published *A Vindication of the Rights of Man*, in which she said to Burke that "I glow with indignation when I attempt . . . to unravel your slavish paradoxes." William Godwin followed this, then the Scottish lawyer James Mackintosh, who called Burke's *Reflections* a "hysterical reaction to events, born of fear." The most famous reply would be Thomas Paine's *Rights of Man*, published between February 1791 and February 1792 and dedicated to George Washington. Paine was already famous for his pro-American *Common Sense* of 1776; now he argued for a republican style of government as the only hope against the falsehoods and follies of monarchy. Burke's work was a hit, selling 30,000 copies in two years, but *Rights of Man* outstripped that, selling 200,000. The French liked it so much that they elected the author to the National Assembly, one of two non-Frenchmen given that honor. The other was Joseph Priestley.

Priestley surprisingly held off on his response to Burke until well into 1791: when he entered, it was on the theme of natural rights, which he considered handed down from God, not the King. He lectured Burke in the same tones as he had Samuel Horsley, the minister William Pitt, and Lavoisier. "You appear to me not to be sufficiently cool," he scolded. "Your imagination is evidently heated, and your ideas confused." He called Burke's book "a vehicle for the same poison" that princes were born to rule "and we to obey."

The smugness and anger are understandable, considering what his former friend had done. Burke had set Priestley firmly in the crosshairs; from now on, "Doctor Phlogiston" was a marked man. Burke's *Reflections* had singled out Priestley and Price as leaders of a Dissenting plot to subvert the English social order, but Price was out of reach, since he died in March 1791, leaving Priestley at the head of the traitors. Burke said of the conspiracy that "the wild gas, the fixed air is plainly broken loose," and everyone knew to whom he referred. Priestley's declarations of men's rights was nothing more than "the agitation of a troubled and frothing surface"—the bubbling of a confused and insubstantial mind.

More dangerous, Burke created an image of science that portrayed it as a threat to the established order. He argued ceaselessly that science and other forms of freethinking must be returned to the Pandora's box of evils and contained. Burke blamed the Revolution on writers, intellectuals, and men of science, all lumped together as "philosophers": it was their ambitions, claims, and ideas that formed the bloody heart of the revolt. "The age of chivalry is gone," he said while describing the arrest of Marie Antoinette; "that of sophisters, oeconomists and calculators has succeeded; and the glory of Europe is extinguished for ever." Every French revolutionary acted in defiance of Nature, like an "alchymist or empiric." British radicals were nothing but a cabal of *philosophes* who "considered man in their experiments no more than they do mice in an air pump"—and everyone knew what happened to *them.* Priestley's own mention of gunpowder proved the link between science and sedition. The brave new world of democracy and science was not merely amoral, Burke warned; by discarding all that was known and sacred, it was evil. The mad *philosophes* sought to steer the universe, and for their presumption left France littered with corpses and tombs.

These were personally dangerous words. Burke and Priestley both were classicists—they knew what happened to those who purveyed change. Socrates was executed, forced to drink hemlock, for "corrupting" the youth of Athens with his dangerous ideas. The female philosopher Hypatia was torn to pieces by an Alexandrian mob in AD 415. "Thrash them, beat them, flog them for their crimes," Aristophanes advised of Socrates and all Sophists in his play *The Clouds*. Their sins were many, but the worst was daring to question the gods.

▪ ▪ ▪

The summer of 1791 could seem safe enough, if one did not scratch the surface. This was Priestley's preference. He felt secure in Birmingham, and part of that security was couched in a vague belief that "the people" were behind him. This was not France, where the elite dined off the taxes of the peasantry, but England, and although he was part of a radical intellectual elite, the people must know he only desired their political liberation and the common good.

But there was rage in Birmingham—despair and drunkenness—and above all else, a new kind of poverty. William Hutton observed in 1781 that five-eighths of Birmingham's families lived in houses too ramshackle to be assessed for the poor rates; in human terms, this represented three-fifths of the total population of 50,000. Another "submerged tenth" of 5,000 people received weekly poor relief. On the other end of the social scale resided the bourgeois class of merchants and owners: many of these were friends of Priestley, including Matthew Boulton, James Watt, William Hutton, and Samuel Galton. Two prominent members of this set—William Russell and John Ryland—were Priestley's immediate neighbors. In 1770, 68 of these *doyens* appeared in the city directory; by 1802, the number would increase to 140. Three owners—John Taylor, William Russell, and George Humphreys—were said to employ 10,000 people between them; next to that, Boulton's 700 employees at his Soho Manufactory was small change.

Certain facts were incontrovertible. The larger the machines became in Birmingham's factories, the longer and harder worked those charged with running them. By the end of the eighteenth century, factory hands put in twelve hours a day under conditions that would seem like visions of Hell to their rural ancestors. They spent their days in the screech of gears and shafts, lathered in smoke, dust, and fumes; they went home from these labor-saving devices to hovels alive with fleas, lice, and vermin. As in the past, only the wealthy could afford meat; tuberculosis was increasing. The new disease was rickets, crippling bones for lack of sunshine and diets low on vitamin D. Josiah Wedgwood exhorted his young potters to appreciate the benefits of industry:

> Ask your parents for a description of the country we inhabited when they first knew it; and they will tell you that the inhabitants bore all the marks of poverty to a much greater degree than they

> do now. Their houses were miserable huts, the land poorly cultivated . . . with roads almost impassable.

Yet an observer with less capital at stake sketched an entirely different picture in 1795:

> The poor are crowded in offensive, dark, damp, and incommodious habitations, a too fertile source of disease. . . . In some parts of town, the cellars are so damp as to be unfit for habitations. . . . Fevers are the most usual effects. . . . I have often observed that fevers prevail most in houses exposed to the effluvia of dunghills.

Workers missed the country life of their fathers and waxed nostalgic throughout the century. The promise of industrialization was a trap: the belief that Englishmen were "free," that they did not starve or "wear wooden shoes" like French peasants, was firmly entrenched, but let them take one look at family and friends, and these workers knew they'd become a new kind of slave. Despite Enlightenment promises, science had not led them to a better life. Science and scientists built the machines that had become their new masters. Scientists like Joseph Priestley.

Maybe Priestley represented the worst of these broken promises. He was the man who said science would take them closer to the Almighty and the shining city of God. But there was no Utopia in England, and the promised Heaven was looking more and more like Hell. True, there was a growing urban middle class who would agree that life was getting better, but they accounted for 10 percent of the population while the remaining 90 percent lived from hand to mouth. Three centuries of science and technological innovation had not relieved the plight of the poor but gave birth to a new kind of misery. By 1798, the fear of revolution and plight of the "labouring class" spurred Thomas Malthus to write that poverty and misery were inevitable and would only be checked by war, abortion, infanticide, famine, plague, and a contraception that then seemed unnatural. By 1810, English factory workers would chant: "Bread or blood!"

Such were the conditions that helped spawn the so-called Church and King riots in the factory cities of Birmingham, Manchester, and Nottingham in 1791–94. These were not the radical crowds of the French Revolution that strove to replace one system of government with another. Instead, this was a deeply conservative movement that resisted the change around it and wished for a return to a golden age—one that never existed except in a warm nostalgia, where all was safely ordered and everyone was cared for by the Church and King. The mobs in France and England were opposite weights on the balance. Already by 1789, the "beggarly, brass-making, brazen-faced, brazen-hearted, blackguard, bustling, booby Birmingham mob" was notorious in England: they'd rioted in 1751 and 1759 against Wesleyans and Quakers; had gathered in response to the Gordon Riots in London to cheer "No Popery!"; had stormed grocers' shops in 1766 to protest high prices. If a conservative riot was meant to occur in England, it would be in Birmingham.

This mob was also different philosophically from the dissenting middle class: they were two tribes, with different languages, who barely understood each other. For them, work was not an idol; it did not ennoble them or get them ahead. They hardly lived outside of work—miners slept down in the pit, while apprentices worked until 2 a.m. Hours were long and wages low, so it was very hard to save. Since artisans worked by the piece, they took their own hours: Mondays were taken off ("St. Monday"), and some took Tuesday also, then worked like maniacs until Saturday, when payday brought a binge. The upper classes cursed them as feckless and irresponsible for these habits, and such regard was returned with bitterness and hatred.

The great escape was drink, especially gin. Even during work, observed Ben Franklin, they downed quarts of ale. During the period when workers were off, public drunkenness was a major problem. Spirits were cheap and amusements centered on drinking; the choir and lottery, chanting, and political clubs all met at public houses. In Birmingham, one-fifth of all businesses listed in the 1770 directory were inns and taverns; 248 of 1,252 listed establishments. This was not unusual. Midcentury Northampton, with a population of 5,000, was served by 60 inns and 100 alehouses. London boasted 207 inns,

447 taverns, 5,875 alehouses, and 8,659 brandy shops in the early 1700s. Throughout the nation, there were about 50,000 inns and taverns.

In Birmingham, the Church and King faction met in one particular public house where they ate, drank, loitered, and discussed the day's news. In 1795, an Irishman named John Binns, a sympathizer of the French Revolution, visited the pub to see for himself where the trouble began. Four years after the Birmingham riot, nothing much had changed. "Such was the furious loyalty and entire devotion to Church and King of the frequenters of that house, that they had, on many occasions, insulted and abused, cuffed and kicked, several respectable housekeepers because they declined to drink every last toast," Binns wrote in his memoirs. When he visited the place, it was still such a hotbed of feeling that his friends advised him to steer clear if he valued his skin. He went anyway, unannounced and alone, walking up three large granite steps to a plate-glass window that announced: "No Jacobins admitted here." The room inside was 60 feet long by 20 feet wide, with small tables spaced along the sides and an aisle down the middle. A Mr. Barber—a master manufacturer said to be one of the organizers of the riot—showed up and proposed two toasts: "To Church and King" and "Damn All Jacobins," since by then Binns had been recognized. When Binns refused to drink the second toast, voices swelled around him. "Turn him out," someone yelled. Binns said he left unmolested, so he was either very eloquent in his defense, a liar, or some of the earlier rancor that nearly killed Priestley had since dissipated.

The first hint of trouble came on July 11, 1791—three days before the second anniversary of Bastille Day. A notice in the *Birmingham Gazette* invited "any Friend of Freedom" to attend a July 14 dinner at Dudley's Hotel to commemorate "the auspicious day which witnessed the Emancipation of Twenty-Six Million of People from the Yoke of Despotism." Soon after that, two other notices appeared that smacked of things to come. The first was a notice in the press that "an authentic list of all those who dine at the Hotel" would be available for a halfpenny. It was signed *Vivant Rex et Regina.* The second was a little harder to classify. A few copies of a handbill found scattered under the tables of a local inn and in two factories exhorted patriots to

> Remember that on the 14th of July the Bastille, that "High Altar of Despotism," fell.
>
> Is it possible to forget that your own Parliament is venial? Your ministers hypocritical? Your clergy legal oppressors? The reigning Family extravagant? The crown of a certain great personage becoming every day too weighty for the head that wears it?
>
> But on the 14th of this month, prove to the political sycophants of the day, that You reverence the Olive Branch; that You *will* sacrifice to public Tranquility, till the Majority *shall* exclaim, *The peace of Slavery is worse than the War of Freedom.* Of that moment let Tyrants beware.

Though the handbill was incendiary, the author was never discovered. Was it an over-ardent friend of liberty, possibly from Hackney, home of Richard Price? A designing Tory, writing to incite the mob? No one ever knew, despite a reward of 100 guineas, but the handbill would be used as an excuse for the violence that followed.

Eighty people signed up for the dinner, but Priestley was advised by friends to stay away. He might be hurt, or at least subjected to insult and a good pelting of mud. Of equal or greater concern was the fear that his presence might spark a riot and place others in danger. By now, when loyal Englishmen envisioned revolution and anti-monarchist sympathy, they immediately thought of Joseph Priestley. Just three weeks earlier, Louis XVI and his family had escaped from the Tuileries in an attempt to flee from France. When their carriage was apprehended on June 22 at Varennes and the royal family placed under closer guard, Priestley was reported as saying, "Our joy at his capture cannot be described." His words were widely disseminated.

By Tuesday, July 12, trouble was already rumored. Catherine Hutton was told by her brother that "a riot was expected on Thursday, but so little was I interested by the intelligence that it left no impression on my mind. The word 'riot,' since so dreadful, contained no other idea than that of verbal abuse," she said. On July 13, William Russell—the New Meeting House's right-hand man—met one of Birmingham's magistrates, Joseph Carles. He was slightly drunk, and talkative. Carles assured his friend that "he would not have a hair of his head injured." Afterward, Russell and other Dissenters concluded that the magistrates knew mischief was up. They were willing to save the lives, if not the property, of Dissenting victims.

Thursday, July 14, dawned anxiously for much of Birmingham. Slogans proclaiming "Destruction to the Presbyterians" and "Church and King forever" were already chalked upon walls. Sometime during the morning it was decided to call the dinner off, but Thomas Dudley persuaded the organizers to carry on, just leave a little earlier than advertised. The guests arrived at 3 p.m. and were met outside by a small group of hecklers. Some shouted, "No Popery!," suggesting some confusion regarding the reason for gathering. The dinner itself was quiet enough, if a little tipsy—the diners raised their glasses eighteen times for toasts that included "the King and Constitution," the National Assembly of Paris, and "the Rights of Man." They went home separately around 5 p.m. A slightly larger crowd of sixty to seventy people had assembled outside. They pelted the celebrants with dirt and stones, but no one was injured.

What was Priestley doing during this time? For most of the day he entertained the family of Adam Walker, a lecturer in philosophy in Lancester, who stayed from 9 a.m. until 5 p.m. Of their children, only William was present, back from his tour of the Continent; he regaled the Walkers with tales of the Revolution, but on this day Joseph's mind was on the dinner in Birmingham. Mary's, too. As evening approached and the guests had departed, her curiosity apparently got the better of her and she prepared to walk the mile into town when a young girl, possibly Mary Ryland, asked if she hadn't heard rumors of a riot and broken glass. Mary Priestley's response was short but interesting, for in three short words she displayed all the stubbornness, denial, and trust in the residents of Birmingham that seemed typical of herself and her husband. "Nonsense, my dear," she said. The two walked to town together and found the gentlemen from the dinner, gathered together and drinking tea.

It was a calm, warm evening, and the stars were out when Mary got home. Her timing was fortunate, for had she visited town a little later she would have witnessed the beginning of the destruction and might have been recognized. The anti-Jacobins were holding their own meeting as the dinner let out, probably at the tavern headquarters mentioned by Binns. There was plenty of drinking, which would explain the celebrants at Dudley's giving them the slip; it would also explain their state of mind once they discovered what had happened. Finally, it suggests a lack of advance coordination, since those who witnessed the early breakup of the dinner did not rush off to inform the anti-Jacobin schemers. By 8 p.m., a large and more volatile crowd had gathered in front of Dudley's, this time "some hundreds" strong. Furious that they'd missed the celebrants, they attacked the hotel itself, breaking some windows. At about 9 p.m. they poured into Bull Street and threatened a Quaker meetinghouse, but someone pointed out that the Quakers "never trouble themselves with anything, neither on one side or the other." In a body, the crowd turned to Priestley's New Meeting House and broke down the gate and doors.

Where were the police during this early stage of affairs? If they had arrived at this point, the riot might have been quelled. In a word, there were no police at this time. Only the largest cities had anything approximating today's force, and their numbers were meager. In the late eighteenth century, as London's population neared 1 million, there were 1,000 officers and 2,000 watchmen. Paris was better staffed, with 7,000 officers and 6,000 Swiss Guards, but as the Revolution progressed, their actions in the face of chaos were guided by politics and self-preservation. The beginnings of a modern police force would have to wait one more century. At this stage, authorities relied on individual initiative to keep the peace: citizens were encouraged to become "thief-takers," the equivalent of today's bounty hunter, with a sliding scale of reward—highwaymen, for example, were worth £40 a head.

Outside of London, magistrates were charged with keeping the peace, and in Birmingham there were three: Benjamin Spencer, John Brode, and Joseph Carles, the last the tipsy friend of the New Meeting's William Russell. A later inquest suggested that these three stayed home during the early part of the evening, despite persistent rumors of trouble: you cannot prosecute what you do not see. By the time they arrived, the mob was beyond anyone's control.

In a sense, this question of accountability and coordination suffused all that followed. How much was planned and how much grew organically? Rumors were everywhere. A persistent one linked the Church of England to the destruction: an unidentified clergyman was said to have gotten the key to the building where the fire engines were kept; he gave it to his clerk, who made himself scarce, ensuring that the engines were released with much difficulty and delay. By the time the clerk was found, all of Birmingham was endangered, not just Dissenters. Outside the Meeting House, witnesses saw two men on horseback station themselves before the growing mob. One read from a document said prepared by an agent of the King. "The Presbyterians intend to rise," he roared. "They are planning to burn down the Church. They will blow up the Parliament. They are planning an insurrection like that in France. The King's head will be cut off and dangled before you. Damn it, you see they will destroy us, and we must ourselves crush them before it is too late!!" The cry of "Church and King!" went up, sprinkled with calls for Priestley's blood.

The New Meeting House was the first to go. The pulpit was torn out, the pews disassembled, the cushions, books, and wooden debris piled out front to start a bonfire. Braziers were carried inside and soon the New Meeting House was aflame. A second detachment headed for the Old Meeting House, which they attacked with bludgeons and crowbars. Here there were signs of planning: the torch was applied only after the charity children were sent home. As the fire started, surrounding homeowners told the mob that they were loyal to the King. As at the New Meeting House, pulpits, pews, and pieces of the gallery were piled outside, in the graveyard, and the crowd torched the building. When the water engines rolled up, the mob allowed the firemen to turn their hoses on the neighboring houses, but not on the church itself. The danger passed only when the Old Meeting's roof collapsed and its walls caved in.

By now, the crowd had transmuted into that thing feared by all English authorities: "King Mob," the destructive mass of Burke's nightmares, which Horace Walpole called "our supreme governors." Official England had come to realize that the poorer classes were no longer a passive force: as a mass, they were as dreaded as a Leviathan, and a lot more real. The mob itself was an agent in British politics, for only through such numbers could the disenfranchised realize the force of their power. To an individual crowd member such power could be heady stuff, and more than one observer has noted how the sudden feeling of invincibility gives rise to murderous passions. The best advice of most observers in every age has been to lay low, for when a riot starts, "an unexpected obstacle will be destroyed with frenzied rage."

It was during the Age of Revolution that the image of the crowd and fire became inseparable. The crowd was "the World Chimera, bearing fire" to Thomas Carlyle; it was the "Death-Bird of a World" that carried the "funeral flame" to consume and envelop all. Fire was the crowd and the crowd was afire: there was sameness as they spread, and nothing stood against them. Wherever there was riot or fire, spectators rushed to it, mesmerized; it was restless, contagious, joining everything in mutual destruction. "Fire is the same wherever it breaks out," notes the Nobel laureate Elias Canetti. "It spreads rapidly; it is contagious and insatiable; it can break out anywhere, and with great suddenness; it is multiple; it is destructive; it has an enemy; it dies; it acts as though it were alive, and is so treated. All this is true of the crowd."

It was certainly true of *this* mob: the apocalypse of the meetinghouses, so close in time and place, turned what had been planned as a protest into a multi-day bacchanalia that stunned all of England in one of the worst public rampages the nation had ever seen. "To Dr. Priestley's!" someone shouted, and detachments from both conflagrations formed up and proceeded to Fairhill as if they'd drilled for the maneuver on some parade ground.

The time was somewhere between 10–11 p.m. An hour or so earlier, Samuel Ryland and some compatriots got wind of events and clattered to Fairhill in Ryland's chaise. They found Joseph and Mary in the parlor, preparing for their nightly game of backgammon as if this were just another summer night in England. There was a violent rapping at the door and voices outside. Priestley admitted the young fellows, all of whom he recognized, and noticed they seemed breathless from running. He calmed them down and got the story. The mob was coming for Fairhill next, warned Ryland. "You and Mary must flee!"

The news ran so counter to Mary and Joseph's beliefs that an air of unreality ensues. "That they should think of molesting me I thought so improbable, that I could hardly give any credit to the story," Priestley later said. Bessie Rayner Belloc, who knew Priestley's descendants and their traditions, speculated that Mary declared "she would not go, abandoning her pleasant, orderly rooms, her hundred and one simple treasures, her china, her linen, her books, the house where her children had grown up." The young men who'd come running were members of Priestley's congregation and begged him to let them defend the house, but Priestley forbade any violence, saying that a minister must not condone bloodshed even in defense of his property. But that property was hard to abandon—his library of rare books, his costly equipment, his manuscripts detailing twenty-five years of research. An inventory of everything lost, later submitted to Birmingham's courts, would run to 65 portfolio pages: it listed such possessions as a large silver medallion of Isaac Newton from the Royal Society, Wedgwood dinnerware, three black inkstands, a timepiece from his friend Magellan, a large mahogany lathe, a camera obscura and other optical equipment worth £60. William promised to stay behind and protect the house within his father's parameters; with other young men and some servants, he extinguished every fire, drew the blinds, and locked the doors. "If asked about my whereabouts," Priestley told them, "say I am gone." They hoped the mob would throw some rocks, then wear itself out and leave. He and Mary grabbed some essential papers and left in Ryland's chaise for the house of William Russell, whose mansion was situated a mile up the road. They could see the glow from the burning churches as they rode.

William and the others stayed in the darkened house, awaiting the mob. While they waited, William gathered money and valuables to hide in a cache close by. The crowd arrived at midnight, crying: "Down with the dissenters!" and "God save the King!" They broke down the doors and shattered windows, piling furniture, books, papers, and debris in the yard. The shrubs and trees in the garden were pulled up or trampled, the cellar raided for the family's collections of wines and ales. With this, drunkenness escalated, and violence as well: fights sprang up between the rioters in the house and on the lawn; a town crier who'd accompanied the crowd started beating like a demon on his bell. The news that Priestley had escaped threw the crowd into an ecstasy of rage. There were hundreds in the house and on the grounds, crying that Priestley would not escape them, oh, if only he were still here! His effigy appeared—apparently made beforehand and carried from town. It was decapitated and fed to the pyre. But no flame could be found to light it, so efficient was William in dousing every fire. Rioters broke into Priestley's lab and grabbed an electrical generator; they tried to start the pyre with a spark, but to no avail. Fires were out in neighboring houses, and one rioter offered 2 guineas for a lighted candle. But finally a flame was procured and the pyre started, then torches were thrown inside Fairhill.

It was a moonless night, and Priestley walked a half-mile back from Russell's, desperate to learn what was happening at his beloved home. Those who saw him commented on his "firm yet gentle pace," his air of "serenity." Yet pacing was how he'd learned to burn off fear and worry as far back as Needham, and as he stared at Fairhill he thought he could hear every crash at this distance; he watched the intensifying glow of flames from behind a screen of trees.

It didn't take long for Fairhill to be fully engulfed once the torches appeared. A fire may smolder quickly or slowly, but once a flame leaps up—and oxygen is plentiful—destruction is rapid. The flames rise in a plume as they encounter oxygen; less dense than a room's colder air, the heat, fumes, and other combustion products form a progressively hotter layer across the ceiling, spreading like a fiery lake that radiates heat back down to the floor. In an amazingly brief period, every flammable surface in the room reaches the point of spontaneous ignition: in a room 2.5 meters high, this happens when the ceiling temperature reaches 600° C and a condition called "flashover" occurs. Since none of Fairhill's rooms was hermetically sealed, hot pyrolysis products leaked from the top of the closed doors, sucking colder air and oxygen toward them as the combustion reservoir spread. The hot gases rose wherever there was an opening—along ceilings, through cracks in walls and ceilings, up stairwells. When they ignited, Fairhill came ablaze with a roar.

The crowd seemed hypnotized by what they had done. Some rioters stretched out on the lawn to watch the fire and sleep off their drunkenness; one man was killed when a cornice from the burning house fell on his head. At about three in the morning all noise ceased. William and Russell told Priestley that the crowd had dispersed and only about twenty rioters remained, all so drunk they were no threat to anyone. Priestley returned to Russell's house: the worst seemed over, and at least they had escaped. By 4 a.m., they began to undress for bed.

But the danger was not past. No sooner did they slip between the sheets than a pounding came at the door. The mob had got wind of their location and was rushing here now. Priestley and his family packed up again and left by chaise for Heath Forge, another five miles off, where their daughter Sally lived. They'd barely arrived at Sally's when more vigilantes were said to be near. Mary declared she would not leave; Sally was pregnant and due in a month, and needed care. But all agreed that Joseph was the main target, so he should ride for London and safety. After two hours' sleep, he departed on horseback with a servant, bound for Bridgnorth and Kidderminster. Dissenting families lived all along this route, but everywhere Priestley sought shelter his presence sent his ertwhile friends into a panic, certain the Doctor would bring rioters upon them. For the next day and a half, Joseph rode back and forth along the Bridgnorth-Kidderminster road "until we could barely sit on our horses," lost in the dark as badly as those times with Shelburne on the Grand Tour. Late on July 17, he made it to Worcester in time for the last seat on the London coach. On Monday, July 18, he finally arrived to the anonymity of London, where he appeared like a lost dog at the door of his friend Theophilus Lindsey. He'd been on the run for four days with very little sleep and rumors of pursuit always at his heels. Not surprisingly, he dropped from sight for the next few days.

Priestley and his family were safe, but the violence was not ended. One wonders if the crowd's rage would have been mollified if they'd gotten their hands on Priestley; they'd missed their main target, and for the next three days the mob turned on Birmingham's Dissenters and anyone known as Priestley's friend. Early reports that the rioters were moving toward Russell's house were mistaken. Instead, immediately after leaving Fairhill, the main body swarmed John Ryland's Baskerville House, where they drank up nearly £300 worth of liquor. They brought it up in pails from the cellar as the house crashed down around them. A band of constables arrived to break up the crowd, but there was a struggle and the constables were overwhelmed and disarmed. One man was killed in the fracas and several others injured: there would be no other official effort to stop the rioters until two days later, when the military arrived. Ryland's mansion was put to the torch, but several rioters were still in the cellar among the casks of liquor and others were in the upper stories. The floor fell through and crashed into the cellar, and the house exploded in flashover. Ten bodies were pulled from the rubble when it finally cooled. Another six were badly injured, and one man, trapped in the vault until Monday, worked his way out, only to die in the grass from his injuries.

Shortly after daybreak, a landowner respected by the Church and King Party appeared on horseback and led the mob's boozy remnants back to Birmingham. But once they arrived, the magistrates lost control and the attacks began again. William Hutton's house on the High Street was attacked twice and the rioters driven off, yet that night they finally broke through. All business in Birmingham was suspended; the rioters milled and pulsed in the streets, the main body swelling like a giant single-celled organism whose pseudopods crept out to engulf the country homes of Hutton, George Humphreys, John Taylor, and other Dissenters. Some homeowners tried fending them off with muskets and pistols, others with bribes, but both expedients almost always failed.

By Friday night, the riot progressed beyond its planned intention to a madness no one thought possible, at least not in England. The original rioters were joined by thieves and drunken prostitutes with blue cockades in their hats; a band of colliers arrived, spoiling for a row. The magistrates posted large handbills that ordered everyone home, but these were torn down. The mob pulsed from house to house, levying tribute; there was scarcely a housekeeper "that dared refuse them meat, drink, money, or whatever they demanded," witnesses later said. Shops were shut, businesses closed, and everyone—orthodox and dissenter—hid their valuables. The only blessing was drink and the human limits on its intake. After awhile, rioters passed out from alcohol, fatigue, or both, and lay in the fields around burning mansions.

By Saturday, July 16, the house of William Russell, to whom Priestley had fled first, was burned to the ground. By now a ragged exodus of Dissenters could be seen streaming from town; by 2 p.m., the main body of rioters headed south where most of the mansions lay. Russell's daughter Martha was hustled out with her siblings; as she passed across a field, she could see her home burning behind her. In houses around the city there was a frenzy of packing and flight. The children and their escorts took off through fields or into woods with smoke and horsemen in the distance. The adults with them said to wrap themselves with blankets to avoid identification and walk on without saying a word. Who knew any longer who was friend and who was foe? "We now passed several houses, at the door of each the family was collected in a solemn sort of silence; they all gazed at us as we passed; not a word was spoken, except sometimes by some of them in a whisper," Martha Russell recalled. The band of children set off walking to Alcester and heard horsemen approaching behind them. They hid in the trees and as the horsemen passed one rider cried out, "I know there's a damned Presbyterian somewhere hereabouts, we'll have him before morning!" The rider meant her father, and the children pressed on.

The last attack occurred around 8 p.m. on Sunday, July 17, when a "hard core" of arsonists attacked the home of Lunar Society member Dr. William Withering. Although an Anglican, he'd fallen into disfavor because of his choice of friends. By this late date, however, he'd garrisoned his house with "a body of grinders from the blade mill" on his property and some "famous fighters" from the local taverns, hired for the purpose. The rioters were repulsed, one leader seized, and the rest driven away.

By then, after the riot had nearly played itself out, a troop of dragoons stationed in Nottingham arrived. The streets were lit up and the residents came out of their houses and cheered. The madness was finally over. A detachment of light calvary from Lichfield reinforced the dragoons on the 19th; small skirmishes popped up as the mobile troop ranged around the countryside, but by the 20th order was completely restored. Five hundred foot soldiers of the 3rd Regiment and three troops of the Oxford Blues were sent to stay as long as necessary to secure the safety of the citizens and resumption of trade, but there was no enemy to fight any more.

Horrified Birmingham counted its losses. Four Dissenting churches and twenty-seven houses had been attacked, most of them razed. Between ten to fifteen rioters were killed, depending on different accounts, plus one special constable; many more were injured. Compensation was awarded through the courts and the authorities unenthusiastically brought a few scapegoats to justice. Out of the hundreds of people involved in the riots, a mere seventeen were charged before the year ended. Of these, only four were sentenced: two were pardoned, and two hanged.

On the whole, the official response was ambivalent at best. The general feeling around Birmingham was that the Dissenters got what they deserved. Edmund Burke was elated, and George III declared to his home secretary that "I cannot but feel better pleased that Priestley is the sufferer for the doctrines he and his party have instilled, and that the people see them in their true light." The home secretary, for his part, was more ominously inclined. He told the King that the mob could just as easily turn against the court and Church next time. He ordered the Birmingham magistrates to seek out the ringleaders, but since both they and local clergy were implicated in the violence, the home secretary's edict was ignored.

And Priestley? He was exiled from the town he called the "happiest" place in the world. He stayed in London with his friend Lindsey, only appearing on the street in disguise. For the first time in his life, he carried a sword cane. By August, he preached his first sermon since the riot; by fall, he began his battle in the Birmingham courts for reparations. "He is very well," Lindsey wrote in October, "and with his wonted cheerfulness that never fails him."

But Mary would not so easily forgive. She was more shocked by the violence than her husband, and even in London did not feel safe.

"God can require it of us as a duty after they have smote one cheek, to turn another," she said to a friend, but implied that she did not do so gladly. The people of Birmingham would miss the Priestleys, she predicted. After all, "they will scarcely find so many respectable characters, a second time, to make a bonfire of."

CHAPTER 12

The World Out of Joint

THE PRIMAL MYTHS OF SCIENCE NEVER END WELL. PROMETHEUS brings light to man, and his reward is pain. Eve plucks the fruit from the Tree of Knowledge, and banishes mankind from Paradise. Pandora's name means "gift for all," and the excitement of a gift lies in its anticipation. She was not wicked, only curious—and for that curiosity unleashed evil upon the world.

The most poignant of the scientific myths is that of the master scientist Daedalus and his overexuberant son. It was Daedalus who designed the Labyrinth containing at its center the Minotaur; Daedalus who gave away its secret to free the lovers Theseus and Ariadne. For divulging that secret, Daedalus was imprisoned in his Labyrinth, but he was not worried. He designed two sets of wings and told Icarus, "Escape may be checked by water and land, but the air and sky are free." But with that came a warning: keep to the middle course, neither so high that the sun would melt the glue from the wings, nor so low that the sea became a danger. Icarus, of course, was enthralled by his newfound glory and flew too high. Despite his father's anguished cries, he dropped into the sea, never surfacing in myth again.

The French anthropologist Claude Lévi-Strauss observed that the individual tale is not what constitutes a myth, but all versions woven together. One collects as many versions as can be found and from that abstracts a general pattern, as if there is a common well of myth from which humanity dips. To Lévi-Strauss, there is no real difference between the savage and civilized; both worlds contain tribes for whom myth is an essential language. In this model of reality, the basic pattern of the scientific myth comes clear. For every advance, there is loss. The higher one flies, the further one falls. For every breakthrough, there is pain.

By September 1791, Priestley was on the run. The best of all possible worlds had come with a price and exploded in an oxygenated holocaust of fire and rage. His laboratory, rare equipment, shelf upon shelf of hard-to-find books, manuscripts and notes—the products of two decades of inquiry—gone in a blaze that could be seen for miles. Although the realization broke his heart, he dared not return to Birmingham and stayed in London; but Edmund Burke, the King and court, Church and press—all smelled blood. With Price's death in March 1791, Priestley was the scapegoat for their fears. They reveled in his ruin, launching shriller or more chilling attacks that accused him of treason, or envisioned him burnt at the stake and beheaded before a mob. Maybe, he thought, he could ride out the storm, but as he watched, the Revolution abroad in which he'd placed such hope mutated into the Terror. The proof of God's love turned into a nightmare.

Lavoisier was in the middle of this storm. When the Revolution started, he was at the summit of his career. He was surrounded by friends and admirers, honored internationally, rich, powerful, and famous; his lab at the Arsenal was the center of France's scientific life, even more than the Academy. His textbook *Traité élémentaire de chimie* was published in August 1789, one month after the fall of the Bastille, making the revolutions in politics and chemistry nearly synchronistic; his New Chemistry had burrowed so deeply into modes of thought that it was already being taught, becoming the new scientific reality. Although he accepted the Revolution as necessary, his private politics remained conservative: in a 1790 letter to Ben Franklin, he said that although the aristocratic party was broken, he feared putting too much power in the hands "of those whose function it is to obey." Events were going too far, too fast: the "physiocrats," who combined economic policy with science, knew their days were numbered.

Publicly, however, Lavoisier played an active part in the early stages of the Revolution. He was such an efficient administrator, financier, and scientist that, despite his monarchist leanings, he was seen by most as indispensable. The new Republic had too many problems, and Lavoisier proved to be an excellent handyman. Even as his position grew shaky, he continued to hope that his usefulness would save him. He became a member of the '89 Club, a group concerned less with politics and more with social reform. He established a model school for his tenants in Fréschines, and was appointed by the Bureau of Arts and Crafts to design a scheme for national education. In May 1790, the Academy was assigned the task of creating a uniform system of weights and measures, and Lavoisier was placed in charge of oversight. As funding grew erratic, he paid from his own deep pockets the expenses of the scientists, surveyors, and assistants who developed the universal system of meters, liters, and grams still used today. By 1791, he was acting as a general scientific and financial adviser for the new government. He set up a system of bookkeeping that made it possible to discover at any moment the exact state of the national exchequer; he gave advice on the paper and printing to be used for the nation's new *Assignats*, or paper bills. He compiled a careful report on *La richesse territoriale de la France*, containing statistical data on the nation's population, industry, and agriculture, the closest thing to a Census that France ever had. The new information allowed him to assess France's financial situation, of which he wrote:

> At a time when everything, good and evil alike, is liable to exaggeration, when everybody looks at things through distorting glasses which makes them seem too great or too small, too near or too distant, and no one seems capable of seeing them in accordance with their proper size and position, I thought it would be of use if someone would undertake to discuss the situation calmly, and submit the finances of the state to rigorous mathematical analysis.

He thought it best not to sign the memo, but no one was fooled: the acidic style, the insistence on unbiased reasoning freed of political emotion . . . it was obviously penned by Lavoisier. It was the kind of unpolitic statement that would come back to haunt him, and like

Priestley, he seemed incapable of realizing when he acted unwisely or why he was so disliked. Naively placing his faith in reason, he could never see why, for those who glorify power, reason is hated and feared.

More than anything else, Lavoisier lamented in his letters how the Revolution barred the progress of science. There was no free discourse or unstructured thought: everything had to fit into the Revolution's polemic, a "political correctness" more insistent than anything today. Worse, he was taken farther and farther from experimental science at a time when he felt on the verge of another breakthrough. His insights into respiration and heat during his 1783 guinea pig tests had led him to consider the clockwork of our bodies: human metabolism.

Sometime during this period—the dates are unclear—Lavoisier and his lab assistant, M. Séguin, began a study of respiration and transpiration that seemed quite surreal. He'd always used guinea pigs in his metabolic studies: "They are tame, healthy creatures, easy to feed and big enough to inspire and expire air in quantities suitable for measurement," he said. This time, however, the guinea pig was Séguin: "However laborious, disagreeable or even dangerous the experiments might prove, M. Séguin wished that they should all be performed on himself." A varnished silk bag was sown to his measurements; it was carefully weighed, and he crawled inside. An airproof brass mask fit snugly over his face; the only opening was at the mouth, and Séguin breathed oxygen from a reservoir of gas through a tube. His expired air was caught in another vessel to be analyzed for moisture, oxygen, and carbon dioxide, while Séguin himself was carefully weighed before and after the experiment. Marie sketched the test, so there must have been at least one run-through: one sketch shows Séguin at rest, while in another he performs a measured amount of work by depressing a foot pump while an assistant takes his pulse.

Lavoisier was applying the conservation of mass to the study of human metabolism. Although never completed, the research proved the relationship between oxygen consumption and the work a man or woman performed. In tests where Séguin sat still without having eaten, for instance, he breathed in 27 liters of oxygen; while pressing the foot pedal, that volume increased to 65 liters. Surprisingly, digestion also seemed a kind of work. Séguin breathed 38 liters of oxygen at rest after having eaten, compared to 27 when he had not; while pressing the foot pedal after a meal, he breathed in 91 liters, compared to 65 liters when his belly was empty. Might it not be possible to measure the effort expended by various kinds of work: a philosopher wrapped in thought, as compared to a laborer digging a ditch? Or perhaps a more subtle measurement—an author writing a book, compared to a musician composing a song?

This much was clear: when a man works, he uses more oxygen—his vital flame burns more brightly, consuming the stuff of his body. The harder he works, the more fuel he needs to replenish the flame. Fuel comes from food, food costs money, and those who labor hardest have least to spend, so that even in physiology, the inequities of class take hold:

> As long as we considered respiration simply as a matter of consumption of air, the position of the rich and the poor seemed the same; air is available to all and costs nothing; the laborer has indeed better opportunities than most of obtaining this gift of nature. But now we know that respiration is in fact a process of combustion, which consumes each instant some of the substance of an individual; that this consumption increases as the rate of the pulse and of breathing accelerates, that, in fact, it increases in accordance with the labor and activity of an individual's life; a whole host of moral questions seem to spring into being from observations which are themselves purely material in nature.
>
> By what mischance does it happen that a poor man, who lives by manual work, who is obliged, in order to live, to put forward the greatest effort of which his body is capable, is actually forced to consume more substance than the rich man, who has less need of repair? Why, in shocking contrast, does the rich man enjoy an abundance which is not physically necessary, and which seems more appropriate to the man of toil?

Where Priestley saw equality in the air, Lavoisier saw inequity in the flame.

Although the observations had the proper revolutionary slant, the work was never finished and the world would wait years before metabolism was again explored in this way. By the middle of 1791, Lavoisier, like other Frenchmen of his class, was finding his path strewn with difficulty. In March 1791, the General Farm's contract was canceled with the Assembly, and the hated institution disappeared forever. He hoped to be appointed to the post of Administrator of Customs for the Department of Paris, but the anti-monarchist Jacobins succeeded in removing his name from the list of candidates submitted to the King. Shortly afterward, he was appointed to the National Treasury; he wrote to the press to say that he accepted no salary for that appointment, since his pay as *régisseur des poudres* was more than enough for his and Marie's needs. Five months later he lost that, and was told he must find a new home. Only with difficulty was he able to secure from the King the privilege of retaining his laboratory, in which the equipment itself cost a fortune. But all of this might not have proved deadly if not for the effort of one man.

Jean-Paul Marat is not merely one of the great enigmas of the French Revolution, but of recorded history. He called himself "the friend of the people," died upon the threshold of the Terror, and would justly be called "the father of all the horrors which followed his horrible reign." From 1780 to 1794, Marat would progress from failed physician and scientist to enraged populist journalist; then, as leader of the anti-monarchists, he became one of the most powerful members of the Revolutionary government, holding the fate of thousands in his hands. He used that power to settle old scores. More than one historian has pronounced Marat responsible for Lavoisier's fate, yet he died a full year before Lavoisier rolled to the lethal theater at the Place de la Révolution. That he would have done all in his power to ensure Lavoisier's death is unquestioned, for he left no doubt of his wishes in a direct attack published on January 27, 1791:

> I denounce you, Coryphaeus of the charlatans, Sieur Lavoisier, son of a land grabber, chemical apprentice, pupil of a Genevese stockjobber, *fermier général,* commissioner of powder and saltpetre, administrator of the Discount Bank, Secretary to the King, member

> of the Academy of Science! Just to think that this contemptible little man, who enjoys a yearly income of 40,000 livres, has no other claim to fame than that of having put Paris into prison ... by a wall which cost the poor 33 million livres, than that of having conveyed powder from the Arsenal to the Bastille on the night of July 12th and 13th, a devil's intrigue! And now he seeks nothing less than to get himself elected departmental administrator of Paris. Would to heaven he had been strung up to the nearest lamppost on August 6th; then the electors of Culture would not need to blush for having elected him.

"I am the rage of the people," proclaimed Marat, and by 1791 there was a surfeit of rage. In the two years following the fall of the Bastille, the masses had realized they'd gotten nothing from the Revolution and that once again they'd been forgotten. The National Assembly provided stirring oratory; the King lived with a reduced court in the Palace of the Tuileries; the hated General Farm was gone. Yet even with such nods to reform, the rich stayed rich and the poor stayed poor. The "J-Curve of Rising and Declining Satisfactions," developed during America's own ghetto riots of the 1960s, suggests that those who have been at the bottom too long grow dangerously impatient when hope finally appears. "The J-curve is this," says its author, James Davies: "Revolution is most likely to take place when a prolonged period of rising expectations and rising gratifications is followed by a short period of sharp reversal, during which the gap between expectations and gratifications quickly widens and becomes intolerable." The Bastille's fall had not made bread cheaper; if lords and lackeys had disappeared from the streets, they had been replaced by Lafayette and his hated National Guard; the mob was still hungry, ragged, and savage. All that was needed was a voice, and it was Marat's, the incarnation of ferocity and hatred, as well educated as Lavoisier, but dressed like the mob in rags.

"I have ever sought the truth," he wrote in his autobiography, published during the Revolution, but with Marat, the truth is filtered through the ambition of one frenzied soul. Before everything else, Marat saw himself as a scientist: his first and greatest ambition was to be hailed as another Newton, and the twelve heavy volumes he published on subjects as diverse as optics, philosophy, and electricity bespoke his erudition. Yet the Academy never took his theories seriously, and for that he never forgave them; since Lavoisier was the most influential academician, he became the focus of Marat's rage. When the failed scientist came to power, it was time to take revenge.

So great was Marat's ambition that every rebuff was added to the "conspiracy" against him, and this conspiracy became the hub of his world. In Marat's mind, Lavoisier was not only the main impediment to his success but the chief conspirator in the plot against him. "This persecution began at the moment the Academy realized that my discoveries about the nature of light upset its own work," he wrote.

> Since the d'Alemberts, the Condorcets, the Moniers, Monges, Lavoisiers and all the other charlatans of that scientific body wanted to hog the limelight for themselves . . . it isn't difficult to understand why they disparaged my discoveries throughout Europe, turned every learned society against me and had all learned publications closed to me. . . .

For five years he endured such "cowardly oppression," and then the Revolution saved him. "I quickly saw how the wind was blowing, and at last I began to breathe in the hope of seeing humanity avenged and myself installed in the place where I deserved."

These are chilling words. Yet in treating Marat—his claims, envies, and unhappiness—one runs up against dichotomies that are so intense he seems more than human; an inner fire so apocalyptic he becomes more mythic than mortal. On almost no other figure in the Revolution—from Robespierre to Napoleon—is opinion as divided as it is with Marat; divided so sharply, in fact, that he serves as a touchstone for one's personal feelings about the Revolution. Was he a monster, or popular champion? Brilliant, or mad? Like Priestley, he was a human comet; unlike Priestley, anyone who came close got burned. Even his supporters described him as a monstrosity, a gargoyle leering down from the battlements of Notre Dame. He was five feet tall, with huge arms and an ugly, yellow face; he was a bundle of curious tics and tended to walk on tiptoe. "Physically," said one contemporary, "Marat had the burning, haggard eye of a hyena.... His movements were rapid and jerking, his features were marked by a convulsive contraction which affected his way of walking. He did not walk; he hopped." Said another: "His countenance, toadlike in shape, marked by bulging eyes and a flabby mouth, was of a greenish, corpselike hue. Open sores, often running, pitted this terrible countenance."

There is much to pity in such a description, as well as shun and fear. The exact nature of Marat's disease is hard to discern. The main culprit seems to have been an extreme and revolting form of eczema, with suppurations running from his scrotum to his perineum that maddened him with torment. His skin was blistered with running sores; his arms and legs were wracked with pain; his head pounded from migraines. He was often in fever, a physical burning that fed the rage in his mind. Near the end, a committee of "patriots" visited his sickbed and reported that his suffering was caused by "a suppression of too furious a patriotism."

Whatever pains beset him, he turned them into Furies to unleash upon the mob. "Rise up, you unfortunates of the city, workmen without work," he wrote in 1789. "Cut the thumbs off the aristocrats who conspire against you, split the tongues of the priests who have preached servitude." Those who read his private journals spoke of a man in delirium. "A man who is starving has the right to cut another's throat and devour his palpitating flesh," he wrote, and more prophetically, "To ensure the public tranquility two hundred thousand heads should be cut off."

One of six children, Marat was born in Switzerland in 1743, the same year as Lavoisier. His father was a chemist and language teacher; he was raised, like Priestley, in the dark glow of Calvinism, where men were of two classes—the saved and the damned. His family wanted him to study medicine; he agreed, but saw it as a doorway to fame. "From my earliest years I was consumed with a love of glory, a passion," he noted, "which has never left me for a minute." His studies took him throughout Europe; by 1772, he was in England, where he published *An Essay on the Human Soul*, a work that achieved small success in Britain but was demolished in a few sharp sentences in

France by Voltaire. In 1774, still in England, he published *The Chains of Slavery*, "an exposé of the intrigues among the princes of Europe against their peoples." This caught the attention of some English radicals, but in France was ignored. In 1775, he moved to Scotland and earned a degree in medicine at St. Andrews University; he returned to London, now "Dr. Marat," where he experimented in optics and electricity. Although touching on the same subjects as Priestley, there is no mention of a meeting in the latter's works. In 1777, he moved to Paris, where he accepted a position as physician to the guard of the comte d'Artois.

The Paris of the late 1770s and 1780s was a world apart from staid, pugnacious London. Paris was the capital of mirth, the center of the Enlightenment, a Mecca for adventurers, a haven for cults, and a magnet for innumerable get-rich-quick schemes. There is a fantastic air about it, the perfect alembic for charlatans. A review of the pulp literature of that decade reveals a public intoxicated with the power of science, but bewildered by the real and imaginary forces that seemed to people the universe. If the old gods were dead, what were the new gods? It was the decade before the Terror, when science stirred the imagination to its core. The ultimate expression of such fantastic heights arrived in 1783, when the Montgolfier brothers came to Paris at the request of the Academy to demonstrate their new invention—a huge linen bag lined with paper that was filled with hot air from a burner. The balloon inflated slowly, a breathtaking sight, then rose and floated in the air, turning man's ancient dream of flying into reality.

Somehow the Montgolfiers' flight seemed emblematic of the entire decade. Because everything seemed possible, anything could be believed. The newest *systèmes du monde* were usually described in terms of a superfine fluid that pervaded all matter and linked all existence, both physical and spiritual. It was the era of Priestley's phlogiston, Lavoisier's *calorique,* and the hotly debated oxygen that made the flame glow.

Marat's arrival coincided with that of Franz Anton Mesmer, who in 1778 proclaimed the discovery of a superfine fluid that surrounded and penetrated all bodies. Like phlogiston and caloric, Mesmer had not seen his "magnetic" fluid: it existed in the medium of gravity and bathed the universe, he said. When asked the properties of his fluid, Mesmer was vague; it was easier to feel than to describe. Newton called it the Ether; Descartes, the Universal Mover; alchemists, the Quintessence; Priestley, phlogiston; Lavoisier, *calorique,* or was it *oxygène*? All were groping for something that could not be measured in the balance—but then, what about love? Lovers felt all sorts of mysterious attractions and repulsions due to a force that could not be analyzed in the lab. Did that mean it did not exist? Love's existence was recognized by even the most empirically minded *savant.* Look at Lavoisier—even he had fallen in love.

Though the French appreciated this nod to the heart, they recognized that these were not new explanations. But Mesmer did one other thing. He put his magnetic fluid to practical use, which not even the great Lavoisier could yet claim with *oxygène.* Mesmer said his fluid could be used as an energy source, but even more amazing was its use in medicine. Sickness resulted from an "obstacle" blocking the free flow of the fluid through the body: trained practitioners could control the fluid's action by massaging the body's "poles" and inducing a "crisis," often with convulsions, thus restoring health and the harmony between Nature and man.

Mesmerism was the rage. Mesmer and his assistants put on fascinating performances, throwing his patients into fits resembling epilepsy or deep trances, claiming to cure them of everything from blindness to a lassitude caused by an overactive spleen. The mesmerist craze blazed so brightly that it became evident to even the most skeptical that this was more than just another passing fad. Mesmerism exposed the need for authority that Paris lacked, filling the gap of belief in that gray area where science faded into religion. It blurred the line between science and pseudoscience—a threat felt so sharply that in 1784 Lavoisier and a commission of academicians, including Ben Franklin, was ordered to investigate.

Lavoisier's role was significant if one considers that he was far from disinterested. Philosophically impatient with any "science" that could not be quantified, he was up to his neck in the war against the phlogistonists, both inside the Academy and in England. As the attacks of his enemies grew shrill, he became increasingly dismissive and contemptuous of all "systems" with which he disagreed. His patience grew notoriously thin for anything that showed the least taint of mysticism. He'd savaged the "hydroscopic lad," Jacques Pargne, whose eyes pierced solid rock to find buried treasure: the claims of such water diviners were nothing but "old absurdities long since proved false," he complained. "Must philosophy lose all the ground it has gained?"

Now Lavoisier's commission destroyed Mesmer. The commissioners underwent continuous mesmerizing; they observed patients falling into convulsions and trances; they tested the claims of one mesmerist outside the realm of his clinic. A false report to one woman that she was being mesmerized through a door caused her to go into a "crisis"; another "sensitive" was led to five trees in Franklin's garden, one of which had been "mesmerized," and the sensitive fainted at the foot of the wrong one. Five cups of water, one of which was mesmerized, were given to a patient: she convulsed when raising the wrong cup to her lips, but swallowed the contents of the right one without effect. The commission's conclusions were damning. Mesmer's fluid did not exist; the convulsions and trances were the effect of overheated imagination and false hope, nothing more. They ignored the use of hypnotism in the treatments, focusing, as did Mesmer, on the "physics" of his fluid. It would take later generations to investigate that side of Mesmer's findings, and when they did, it led to the blossoming of psychology.

At this unlucky moment, Marat's ambition crossed Lavoisier's anger. "Dr. Marat" conceived himself a true scientist: he neglected his patients, devoting two hours a day to sleep and one to meals to spend the rest of his energy on research in light, electricity, and fire. By the early eighties, he theorized that fire resulted from the activation of particles of an "igneous fluid" contained within a body. In his *Researches on Flame*, published in 1780, he maintained that a flame went out in an enclosed space because the air, unable to escape, compressed and smothered the flame. He claimed he could even make this phenomenon visible. The experiment took place in a darkened room traversed by a single ray of light focused in the optical lens of a microscope; by inserting the flame of a candle between the lens and a paper screen, he was able to obtain the enlarged image of the flame surrounded by moving shadows of hot air. These shadows were his fluid, he said, and produced other images of everything from incandescent cannonballs to Ben Franklin's head. Wrote one witness: "M. Franklin, having exposed his bald head to the focus of the microscope, we see it surrounded by undulating vapors that all end in a spiral."

But the Academy was not impressed, and Lavoisier, the leading theoretician in fire and combustion, least of all. True, Marat's theory diverged from his, but more than that, the entire system seemed founded on a parlor trick of light and shadows. They ignored Marat's repeated requests to observe further experiments. They dismissed him as yet another charlatan, and by the mid-eighties seemed to consider Marat a pest, or ignored him completely.

So Marat threw himself into public relations. As early as 1781, an unsigned journal article, presumably by Marat, claimed: "The revolution that M. Marat has just produced in optics has created so great a sensation among physicists who cultivate this science that they haven't yet recovered from their astonishment." Another article, discovered by Lavoisier, said Marat's theories had been approved by the Academy. Nothing of the sort had happened; it was a lie. There was a curt rebuttal in the same journal, after which Marat, once again, was ignored.

But Marat knew by now who had denounced him—who was responsible for the world not recognizing his genius. Lavoisier was indifferent to those he did not consider equals and had his own battles to fight against phlogiston. There were too many charlatans clamoring for attention. One either engaged them in endless debate or dismissed them as they justly deserved.

Only Marat refused to be dismissed, and proved impossible to ignore.

▪ ▪ ▪

By 1785, Marat was abandoned by his patients; by 1788, he published his last "scientific" work and turned to pure invective; in September 1789, he published the first issue of his weekly paper, *L'Ami du Peuple*, and tapped into a common vein. Now he had an organ to redress all the slights of his life, all the obstacles thrust up to keep him from *la gloire.* From the beginning, he denounced the Academy as a band of charlatans, reserving his choicest excoriation for Lavoisier:

> Lavoisier, putative father of all noisy discoveries; he has no idea of his own so he appropriates those of others; but since he cannot understand them he abandons them again as easily as he adopts them, changing systems as he does his shoes. In a space of six months he has picked up in turn the doctrines of fire, igneous fluid, latent heat. In shorter spaces I have seen him first infatuated with pure phlogiston then ruthlessly denouncing it. Some time ago, following the lead of Cavendish, he discovered the secret of making water from water. Then, imagining that this fluid is composed of pure air and inflammable air, he changed it into combustibles. If you ask me what he has done to warrant such praise, my reply is that he has got for himself an income of one hundred thousand *livres,* has placed Paris in prison with his great wall, has changed the term *acid* into *oxygen, phlogiston* into *azote. . . .* These are his claims to immortality. Proud of his great achievements, he rests on his laurels while his parasitic followers praise him to the skies.

In one quick paragraph, Marat painted a picture of Lavoisier that stuck for the rest of his life. Marat had finally found his gift—he had a genius for propaganda. In a one-page palimpsest of truths, half-truths, and lies, he brought back all the old accusations ever used against his enemy: the scientific opportunism, theft of others' theories, princely income, the hated Paris wall. Marat deftly slipped into the list an accusation that Lavoisier had stolen credit for his own theory of an "igneous fluid." Forgotten were Lavoisier's reforms for France, his crash program for gunpowder production, his humanitarian schemes. Instead, and most damaging, Marat painted him as part of the hated nobility—a prince among the *savants,* with the academicians his fawning courtiers.

Why did this portrait strike such a chord, serving as a warrant for all that would happen to Lavoisier? Marat's jugular instincts were the most precise of any of his day. He knew that the revolutionary crowd was a multiheaded beast, ready to strike at any time and at anyone. It has been said that the Paris mob was drawn in their overwhelming majority from the *sans-culottes:* the workshop masters, shopkeepers, craftsmen, wage earners, and petty traders of the capital. Yet within this group there was infinite variety and competing agendas.

The social psychologist Neil J. Smelser maintains in his classic *Theory of Collective Behavior* (1963) that crowd members are motivated by, and organize their actions around, "generalized hostile beliefs." If there was a common need among all these groups during the French Revolution, it was the shortage of bread and hardship due to rising prices; if there was a generalized anger, it was that someone was benefiting from this hardship and it had to be the rich—like Lavoisier. According to Smelser, individuals join a crowd with a preestablished set of villains; as more join up, a "group mind" grows. The idea of a group mind is hard to reject for anyone who has been in a riot or large political gathering: even Freud, who challenged the idea, conceded that "there is no doubt that something exists in us which, when we become aware of signs of an emotion in someone else, tends to make us fall into the same emotion."

Marat's skill was to eloquently provide a villain—the royalist, the rich—and Lavoisier was not the sole target of his pen. Lafayette, in charge of the National Guard, felt his lash as well. "I'll tear the heart out of that infernal Lafayette right in front of his battalion of lackeys," Marat threatened, but this time he picked on the wrong man. As early as September 1789, the Assembly issued an order for the arrest of the rabble-rousing journalist, and thus began for Marat a series of midnight flights, concealment in cellars and attics, always hidden from the light of day. For awhile he sought refuge in the Cordeliers District of Paris, a warren of cul-de-sacs and crooked streets that proclaimed itself a separate republic, a hotbed of revolutionary fervor where Marat was treated like a king. He renewed his attack against Lafayette from this bastion, and in January 1790, the government dispatched three thousand men to flush Marat from his hideout and throw him in chains.

He escaped in disguise and slipped into England, then by the summer of that year had returned. The hunt for Marat had a romantic urgency that kept Parisians glued to their papers: in September, the police were again on his trail, finding his shop and smashing his presses, an action that drove his hatred beyond all bounds. His presses were his voice; without that voice, he was no one again. His health became seriously affected by the chase, his nervous ailments exploding into the hideous torment that contemporaries would remember long after he was gone.

By 1792, a new madness had insinuated itself into France's social mix, a panic that was impossible to ignore. That year, Lavoisier was appointed treasurer of the Academy, but it was a hollow honor. Already there were rumblings of discontent that the Academy, a symbol of royal patronage, should be dissolved. There seemed no middle ground anywhere. The Assembly was split between the Girondins, who advocated transforming the monarchy into a federal republic like that in America, and the Jacobins, who wanted a highly centralized regime. Enemies seemed everywhere, inside the borders and without, and on April 20, 1792, France declared war on Austria, beginning a series of conflicts known as the French Revolutionary War that would not see an ending until twenty years later with the Battle of Waterloo.

That same month, the chaos spread as feared into Lavoisier's beloved Academy, but the blow was from within. Antoine Fourcroy—Lavoisier's disciple, one of the Four Horsemen of the New Chemistry—forwarded a motion that certain members should be removed from the register. There was general horror at this introduction of politics, as well as nervous recollection of several uncomplimentary comments about the new regime. Several members objected strongly to Fourcroy's proposal on the grounds that the Academy had no concern with members' politics, but then *everything* was political these days. As a compromise, it was proposed that the list of academicians be sent to the minister of the interior, who could revise it as he saw fit. The only change made would be that some noble émigrés were eventually stricken from the rolls, yet the Academy was a monarchical institution and for that reason its days seemed numbered.

Why Fourcroy? His former student's betrayal seemed to hit Lavoisier harder than any other single event of the Revolution. Even

Marat's attacks had been weathered with grace; even the struggle to keep his lab at the Arsenal. Later events would show that Fourcroy had more highly developed political instincts than his fellow academicians; he also had a more finely tuned sense of survival. In putting forward his motion, he clearly saw which way the winds were blowing. Fourcroy was a chemist of some distinction, the first in France to teach the New Chemistry, and so he realized before all the others that such distinction made him a target, too.

If Fourcroy didn't agree with Marat, he was the most well disposed of all the academicians to see the direction of the man's mind. Like Marat, he was a propagandist: it had been Fourcroy's job to advance the New Chemistry in journals at home and abroad. Like Marat, insults from the aloof and aristocratic Lavoisier wormed into his soul. Just like Marat, he turned viciously against the early leaders of the Revolution, which included the Enlightenment *philosophes*. As other scientists did what they could to save themselves, or stared stunned like deer in the headlights, Fourcroy would turn rapacious, seeking and gaining a post on the important Committee of Public Instruction, then later becoming an agent of the powerful and deadly Committee of Public Safety.

That autumn of 1792, Antoine and Marie retreated to their estate at Fréschines. He was safe there. Around Blois, he'd set up the experimental farm, helped increase the farmers' yields, loaned the peasants money without interest in times of crop failure, acted as magistrate during the disputes that arose in village life. He'd founded a school for the peasants where none had existed, fought for their rights in the provincial assembly at Orléans. None of the intrigue and envy, the misery and ambition that so poisoned life in Paris had followed him here. He had already politely turned down salaried positions offered to him by Louis XVI, and had made up his mind to avoid all future posts that would put him in someone's crosshairs. He needed rest; he thought it wise to retire entirely from the public arena. The murderous events of that autumn made him hesitate ever to return.

A letter to Robert Kerr, once an opponent but now planning to translate the English edition of *Traité élémentaire de chimie*, illustrates his frame of mind. The letter asks Kerr to delay publication and apologizes for not responding sooner. "But today, with France given over to the strife of factions, it is becoming extremely difficult to get anything done in this field, and the man who aspires to a great position must either be very ambitious, or very crazy."

Crazy, indeed. That summer, before they left the city, Marie and Antoine had watched as Paris crumbled around them, the thin tissue that held the city together dissolving like the acid paper alchemists used to record secrets. On June 20, 1792, there was a demonstration at the Tuileries and the royal family was humiliated. On July 11, Sardinia and Russia joined the war against France and the Assembly declared a state of national emergency. On August 10, Charles William Ferdinand, Duke of Brunswick, declared that he would destroy Paris if the royal family was mistreated and put all rebels to the sword. In response, insurgents stormed the Tuileries, massacred the King's Swiss Guard, and imprisoned the royal family in the Tower of the Temple. Louis was ordered to stand trial for conspiracy.

Rumor piled upon rumor—rumors of royalist forces coming to save the royal family, of a religious revolt in the Vendée district spreading through the nation. With the imprisonment of the King, a new and violently republican organization came into power: the self-elected Paris Commune. Marat was one of the leaders. His time had finally come.

He immediately struck at his old enemies under the guise of national security, and there is evidence that one of the first strikes was intended for Lavoisier. On August 15, Antoine and Marie left the Petit Arsenal, in transit as their new apartment was being readied at 243 Boulevard de la Madeleine. While waiting, they stayed with a former colleague of the General Farm. Three days after their departure, government agents raided the Arsenal and arrested the officials who'd taken over from Lavoisier. One of the detainees committed suicide in prison, while the other was released five days later by order of the Assembly.

Throughout that month, the Commune lengthened its reach under the guise of patriotism. A general search for arms and suspects was ordered: on August 30, the city gates were closed, every street lit, and the National Guard went from house to house, searching for enemies. It was the first time during the Revolution that the government invaded people's homes; the search, controlled by Marat, threw

2,000–3,000 suspects into the Paris jails. Most of those arrested were royalists and anti-revolutionaries, but a number were taken to settle old scores.

The problem of dealing with so many new prisoners was a grave one; simultaneously, rumors flowed of advancing allied armies and plots to overthrow the Commune. And so, the new government showed how it dealt with such problems. A proclamation dated September 3, 1792, signed at the bottom by Marat, described a solution:

> The Commune of Paris hastens to inform its brothers in the Departments that many ferocious conspirators detained in its prisons have been put to death by the people—acts of justice which seemed to be indispensable in order to terrorize the traitors concealed within its walls. . . . The whole nation will without doubt hasten to adopt this measure so necessary to public safety. . . .

From September 2 to 7, between 1,100 and 1,400 prisoners were slaughtered in what became known as the "September Massacres." Small groups of armed men broke into the prisons and worked their justice: surprisingly, only one-quarter of those executed were priests, nobles, or "politicals," while the rest were thieves, vagrants, and prostitutes. At Bicêtre, a prison-hospital for the poor and insane, 170 derelicts and 33 teenage vagrants were killed; at Salpêtrière, the "ferocious conspirators" were women and girls. The massacres were blamed on a "popular effervescence," but for those who watched closely, they were a harbinger of worse terrors to come.

Lavoisier was one who saw it coming. Throughout early September he strove to escape Paris; by September 10, he and Marie obtained passports, and they arrived at Fréschines on September 15. He presided at a ceremonial planting of a "Tree of Liberty" in the village square, and paid a cost overrun out of his pocket. He read letters from friends in Paris: the educated and wealthy society he'd known and loved was disintegrating, and with it the Academy. Each morning, as a golden light spread across the Cisse Valley and the *allée* of lindens that led from the château to the stables lengthened in shadow, he wondered whether it was wiser to stay or flee. Pierre du Pont had stood with a contingent of the National Guard to protect the King in the attack in which the Swiss Guard were massacred; he'd barely escaped with his life and remained in hiding, planning to emigrate to America with his son. Du Pont urged Antoine and Marie to go with him; land had been purchased northwest of Philadelphia for royalists to spirit away Marie Antoinette and the King. Even Lafayette had escaped across the border into Belgium to save his skin. The monarchy was doomed, that much was certain—but how imperiled were those like him who'd lived in Louis's favor? How far could a madman like Marat actually go?

Only one thing could take Lavoisier back, and that was the threat to French science. The Academy was scheduled to meet again in late November, and its very existence was under threat from the hatred of Marat and betrayal of Fourcroy.

Antoine and Marie returned to Paris in mid-November 1792, just in time to learn that London's Royal Society had awarded him the Copley Medal for his work in oxygen, the highest award it had to offer—the same that had been awarded to Priestley nearly twenty years earlier. The last bastion of phlogiston was conceding the soundness of his vision, and in the process rejecting Priestley. But now was not the time to be honored by France's once and future foe.

Luckily for him, the papers elected not to print the honor. More momentous news filled the page. The trial of the King had started: Louis XVI, now simple Citizen Louis Capet, was being tried for treason. Some argued that such a trial was too dangerous; the foundation of the new Republic could be weakened if the King were found innocent, they said. Maximilien de Robespierre, once a monarchist who opposed execution, wrote on December 3, 1792:

> Louis cannot be judged, he has already been judged. He has been condemned, or else the Republic is not blameless. To suggest putting Louis XVI on trial . . . puts the Revolution itself in the dock. After all, if Louis can still be put on trial, Louis can be acquitted; he might be innocent. . . . But if Louis is acquitted . . . what becomes of the Revolution?

Yet there was a more forceful argument in favor of a trial and execution. The destruction of the monarchy takes physical form in the destruction of the body royal. The death of one gives birth to the "body of the People." Lavoisier would recognize this as a redox reaction, the ultimate give-and-take, balanced out in the pan. According to revolutionary rhetoric, the King's blood would "seal the decree which declares France a republic." More rhetoric: "The blood of Capet . . . washes from us a stain thirteen hundred years old." The tally of the final vote was 361 for execution, 288 against, and 72 for some sort of delay. On Monday, January 21, 1793, Louis rode in a carriage to the scaffold, built next to the empty pedestal of his grandfather's triumphal statue.

▪ ▪ ▪

The death of Louis changed everything. The republican Revolution had reached its logical climax; the destruction of the *ancien régime* was complete. With regicide, there could be no compromise and all the powers of horrified Europe joined the coalition against France. Revolt flared anew in the Vendée; there were bread riots in Paris; the power struggle between the Jacobins and Girondins entered a homicidal phase. With Louis, all of France and Europe lost its head.

It was as if Louis had always awaited execution, as if he were crowned knowing this would come. Time stopped as he waited at the Tuileries, and on this January day, it inches forward. Sleet covers the ground and makes the steps up the scaffold slippery. Every vantage point on the two-mile ride from the Temple is filled with *le Peuple.* All accounts agree on one detail: the silence of the crowd watching the King's last journey.

Louis XVI was not the first king to be killed by his subjects during the 1790s. That honor went to Gustavus III of Sweden in March 1792. His killers were nobles, outraged by his democratic leanings, and they assassinated him at a masked ball in Stockholm. But Louis is the first king to die by the guillotine. He stops at its foot, looking with interest at this tower and blade that he'd signed into existence less than a year earlier.

Like so much in the law, the guillotine appealed to precedent. One was used in China centuries earlier, and in the early seventeenth century there was an attempt to introduce the machine in Halifax, England. But it never caught on. Perhaps it lacked finesse: a contemporary account claimed that "the head blocke wherein the ax is fastened douth fall downe with such a violence, that if the necke of the transgressor were so big as that of a bull, it should be cut in sunder at a stroke, and roll from the bodie by an huge distance." Revolutionary authorities do not want this; the physics of death are refined. A basket is placed beneath the head, which is then lifted and displayed, not kicked around like a soccer ball. This is what the crowd awaits: the moment of discharge, like a spark from Priestley's demon-banishing machine.

The French version is first mentioned after the fall of the Bastille, on October 10, 1789. Dr. Joseph-Ignace Guillotin, a foe of the death penalty and deputy for Paris, proposed six articles regarding executions, including a measure that the same penalty be inflicted on all offenders regardless of rank. No penalty should be more severe than decapitation, traditionally reserved for the nobility. His motion was tabled until May 3, 1791, when Dr. Guillotin's humanitarian machine was officially approved.

But what would be its specifications? Guillotin asked France's royal executioner, Charles-Henri Sanson, for advice, and after much thought, Sanson concluded that the machine must fix the victim in a horizontal position and work more precisely than any man's hand. Chroniclers introduce a German musician to the saga, a violist named Schmidt who played duets with his friend the executioner. Between tunes Schmidt told Sanson that he worried too much and sketched out a machine that met all the conditions. The prototype was tested on corpses and living sheep, and on March 25, 1792, the new method of execution was passed into law. Three weeks later, on April 11, the machine made its successful debut, dispatching highwayman Nicolas-Jacques Pelletier before a crowd in a way that did not "stain the hand of a man with the slaughter of his fellow-creature."

Now Louis stares up at the painless machine. Royalists and republicans give differing accounts of his walk up the steps; there was a two-minute delay at the foot of the scaffold that spawned varying tales, but the most accurate was the executioner's. Louis stepped from his four-wheeled carriage at 10:20 a.m. and mounted the stairs 120 seconds later. "As he got out of the carriage," wrote Sanson,

> he was told to remove his robe. He objected to this and said he could be executed as he was. It being represented to him that this was impossible, he himself helped to remove his robe. He made the same objection when it came to tying his hands, but offered them himself when the person who was with him said that this was the last sacrifice. He then inquired whether the drums would continue to beat. He was told that no one had any idea. And this was true. He mounted the scaffold and wanted to go right to the front as if to speak. But it was made clear to him that this was impossible too. He allowed himself to be led to the place where he was bound, and here he cried very loudly: "People, I die innocent!" Then he turned to us and said: "Sirs, I am innocent of everything I am charged with. May my blood cement the happiness of the French people. . . ." [I]f the truth be told, he bore all this with a calm and perseverance that astonished us all.

The blade falls; the head is held up. In death, it is a head like any other, and yet it is not. There are reports of crowd members soaking white handkerchiefs in the King's blood, of trying to dip knives in the stream. The hair of the King, cropped at the foot of the scaffold, is put on sale, as are slivers of his clothing. It is a collective behavior that none had predicted and defies understanding; it is hard to find in these accounts any proof of dignity or rationality in the crowd.

Other royals would not die as serenely and would not wish the crowd well. Nine months later, in October 1793, Marie Antoinette bared her throat and spit in their faces. Louis-Charles, the eight-year-old heir, was shut in a darkened *oubliette,* verminous with lice and fleas, his body covered with ulcers. When he finally died in June 1795, his heart was cut out and bottled in alcohol. In time, the liquid evaporated. The heart dried out and became hard as stone.

As the winter of 1793 turned into spring, Lavoisier felt his heart harden, too. On February 1, Great Britain and the Netherlands declared war on France; on March 7, Spain joined the allies, and the French government voted to draft 300,000 men into the army. Mounting civil war, the enemy's advance, a dearth of bread—all added to the nation's desperation. During this time, Lavoisier labored to keep the Academy solvent; if he could hold on a little longer, maybe the Academy would outlast the madness and French science would survive. As treasurer of the Commission on Weights and Measures, he must transmit salaries and expenses to the chief surveyors, Jean-Baptiste-Joseph Delambre and Pierre-François-André Méchain, entrusted with the task of re-measuring the meridian between Dunkirk and Barcelona while trying to survive the war and paranoia around them. As treasurer of the Academy, he must pay the salaries of the academicians. The Commune was of two minds about the Academy: the government finds it convenient to have a tame body of scientists at hand, yet Marat would love nothing more than to see the downfall of the institution on which he blamed his failures.

Such strangulation dragged on, week after week, through spring 1793 and into summer. The fate of his scientific colleagues, many of whom fought him so virulently over phlogiston, was all that mattered now: these men were impractical, retiring, often poor and old, and they depended upon Lavoisier for their livelihoods. The Commune was trying to save the nation from bankruptcy, and despite promises, there was little love for an institution they considered monarchist and reactionary. When June arrived and no pay was forthcoming, Lavoisier paid his colleagues' salaries from his own savings.

In July, there were strong rumors that the Academy would be abolished, and Lavoisier threw himself into the offensive. The country was at war, he stated in a long memorandum, and science is power. He argued for science on strategic grounds: science allowed France to regain world prominence in gunpowder production; all mass industry, which gave England its great wealth, was due to science. Strip France of that, and the nation had no industrial future. In mathematics, France had stolen the lead from England; French chemistry had "given law to all nations." The Academy was responsible for this; without the Academy, France would fall behind.

Next, he appealed to the Commune on humanitarian grounds. The continuance of salaries was "demanded by justice":

> There is not an academician who, if he had applied his intelligence and means to other objects, would not have been able to secure a

> livelihood and a position in society. It is on the public faith that they have followed a career, honorable . . . but hardly lucrative. Many of them are octogenarians and infirm; several of them have spent their powers and their health in travel and investigations undertaken gratuitously for the Government; the sense of rectitude of Frenchmen will not allow the nation to disappoint their hopes; they have at least an absolute right to the pensions decreed in favor of all public functionaries.

Then he made a unique argument: "Even under the old régime, science was . . . organized on republican lines." The search for truth was in itself revolutionary, predating and inspiring the Revolution; the respect with which science was held "protected it against despotic interference." Research was a republican effort because the day of the amateur scientist was over; equipment cost too much, but more important, there was no longer "a despotism" by any one field. The physicist, chemist, and engineer all required one another's help, while the mathematician was needed to check and recheck their findings. The world was a complicated place, and our understanding of it was no longer a solitary effort. Men needed each other to make sense of the world.

It was a grand and humbling performance, for Lavoisier was saying that the days of men like Priestley and himself—of Galileo, Newton, Copernicus, and the others whom history considered giants—were over. The man with an ego as big as the Academy had in the end prostrated himself to prevent what he called the "annihilation of the sciences." It could be argued that his motives were selfish: that he was merely trying to save himself. If so, he could not conceive of life apart from the sciences. Without them, there was no meaning for him. The image of the lone researcher would continue long after this memo, and stays with us still, but the work of one *savant* was built on the lives of too many others. There would be other geniuses, other revolutions—but henceforth, no investigator could ever say that he truly worked alone.

But in this new world, science is also political. The popular image of the Academy was expressed by Marat: "A collection of vain men . . . they are like automatons accustomed to following certain formulas and following them blindly, just like a horse who, in turning a mill, makes a certain number of circles before stopping." On August 8, 1793, the painter David, who'd painted Marie and Antoine together so affectionately, told the Convention that "these dismal academies . . . are incompatible with the reign of liberty." Carried away by the words of the artist, the Commune confiscated the Academy's property and ended its charter. The congregation of *savants,* established in 1666, was dead.

▪ ▪ ▪

Marat, who'd lived for that moment, would not savor it. A young Norman woman thrust a knife into his heart; in so doing, she unleashed the Terror and rang up the last act for Lavoisier.

In many ways, Marat had reached his apotheosis. On June 2, 1793, Marat's party had ended the two-year struggle with the Girondins: of the 136 representatives of the ousted party, 29 were arrested and the rest fled. France was now led by a triumvirate of Marat, Georges Danton, and Robespierre. On July 13, Marat sat in his shoe-shaped bathtub, the only place he could find relief from his physical torment, a bandana soaked in vinegar wrapped around his forehead and a bathrobe slung over his shoulders. A long board, supporting an ink bottle, quill, and some paper, lay across the tub as he corrected proofs and composed new copy.

"I killed him in cold blood," Charlotte Corday later said. She walked through the assassination like a somnambulist. Manipulated by the Norman Girondins as their instrument of revenge, she slipped into his bathroom and sat on a stool beside the tub. She was beautiful, tall, with cold blue eyes, and Marat was mesmerized. He asked what he could do for her; she replied that she had come from Caen with the names of Girondins involved in an uprising. "Excellent!" he cried. "In a few days' time I shall have them all guillotined in Paris."

His words condemned him. She reached inside her bodice, unsheathed the knife, and drove it into his breast, puncturing a lung and penetrating the aorta. "Help, dear friend; help!" he cried hoarsely. Corday left the room as others rushed in to help, but she was apprehended as she crossed an antechamber to the front door.

An enormous crowd, undeterred by a storm, turned out for her execution on July 17. The rain soaked Corday's dress, outlining her body; the crowd jeered, and she turned away her eyes. "It seems like a long trip, doesn't it?" Henri Sanson asked, perhaps feeling a touch of sadness at the way a beautiful young woman was being treated. "We're certain to get there eventually," she replied. Sanson tried to block her view of the guillotine when the tumbril stopped. "Please step aside," she asked. "I've never seen one of them before. In my situation, I am naturally curious."

She was brave, no doubt, but her gesture breathed air into the already glowing fire. Lavoisier watched firsthand as a strange hysteria rose. At first, hearing of Marat's death, he must have sighed in relief, thinking that his foe could not persecute him any longer. But then, as a member of the National Guard, he was obliged to stand at attention during Marat's funeral, held the day before Corday's execution. He had a glimpse of what was in the offing. He had stood, musket at shoulder rest, before Marat's body; the painter David, the ceremony's organizer, clad Marat in a toga and crowned him with laurels. His body rested on a raised couch, drawn by twelve men; young girls surrounded the couch, all dressed in white and carrying wands and branches of cypress. The grief for the "martyr Marat" was as violent as the jeers for his murderess. "Oh heart of Jesus!" people sobbed along the route, some falling to their knees. "Oh sacred heart of Marat!" As Lavoisier marched, he knew that Marat's followers would finish what their martyr had begun.

Thus began the Reign of Terror. On September 17, 1793, the Convention passed the Law of Suspects, empowering the arrest of anyone who "by their conduct, their contacts, their words or their writings, showed themselves to be supporters of tyranny, of federalism, or to be enemies of liberty." The people would be governed by reason and the enemies of the people by terror, Robespierre proclaimed. "Terror is nothing else than justice, prompt, secure, and inflexible! It is, therefore, an emanation of virtue . . . applied to the most urgent wants of the country." In fact, practically anyone could fall foul of the sweeping law. Those who refused to call each other "Citizen" instead of the traditional "Monsieur" fell under suspicion; politician guillotined politician as an act of survival. On October 17, Marie

Antoinette was the first famous victim, followed two weeks later by the Girondists arrested in the June purge.

For the next nine months, Europe watched in fascinated horror as Revolutionary France disposed of its enemies with cold, mechanical efficiency. In all of France, between 16,000 and 17,000 people are estimated to have died by the guillotine from October 1793 to July 1794. If one includes those who died in overcrowded prisons or were executed for treason on the field of battle, the count leaps to 30,000–40,000. It was not the decade's bloodiest episode—the same number died in Ireland in 1798, but in a country with one-sixth of France's population; up to 20,000 people were slaughtered in Warsaw on November 4, 1794. But the juxtaposition of "liberty" and the guillotine lifted the Terror to another plane.

The irony was that the greatest number of victims came from those meant to be protected. Of those on whom death sentences were passed, 8 percent were nobles, 6 percent clergy, and 14 percent bourgeoisie. The greatest group, 70–72 percent, were workers and peasants—people whom the Revolution was meant to save. They were charged with desertion, draft dodging, hoarding, and rebellion; they were guilty of wrong choices, when indifference itself was a crime. The roll of the condemned gives a snapshot: Martin Alleaume, seventeen, a hairdresser's apprentice; Marie Bouelard, eighteen, a domestic servant; Jacques Bardy, eighty-five. In Paris alone there were 2,639 executions, over half in June and July 1794. A strange numbness fell over the city, a trance divorced from reality. Even Robespierre and the Public Prosecutor, Antoine-Quentin Fouquier-Tinville, seemed shaken by the torrent of blood. "I am not well," said the prosecutor one evening as he momentarily staggered. "Sometimes I imagine that I see the shadows of the dead following me."

The long-expected blow fell on Lavoisier in November, the second month of the Terror. The order went out to arrest all former members of the General Farm; the warrant for Antoine's arrest was dated November 24, 1793. But there was a window of confusion, which gave him some time. The original warrant was signed for his lab and residence at the Arsenal, from which he and Marie had moved more than a year earlier. The arresting officer was told that the Lavoisiers had moved to 243 Boulevard de la Madeleine de la

Ville l'Evêque in the Section des Picques but instead of visiting this address, officers knocked on the door of a M. Lavosière at 1295 rue Ville l'Evêque on November 29. It is not known whether the wrong man was arrested, but police confiscated a rifle, bayonet, and pistol from the unlucky fellow. During the same period, a Godefroy Elizabeth Lavoïsier (no relation to Antoine) was arrested in the Section des Picques, and at least two more Lavoisiers showed up in the records.

Such mixups could be deadly. Common surnames were enough to get one executed. A woman named Mayet was confused with a Mme de Maillé and taken to the Revolutionary Tribunal; the mistake was pointed out, but the prosecutor said, "Since she's here, we might as well take her." She was executed that day. Two women named Biron were imprisoned in the Conciergerie; the prosecutor's clerks called up "the Biron woman," so both were tried and executed. The poet André Chenier was confused with his brother Sauveur. He, too, died.

In this case, however, the confusion gave Lavoisier some time. On the day that police found the right house, he was on duty with the National Guard. Alerted to the pursuit, he did not return home. He wandered Paris's streets, first taking refuge with the Academy's former usher, then hiding in the deserted Academy itself. For four days he crept through the echoing storerooms and lecture halls, and no one seemed to think to look for him there. He wrote a memo to the Committee of Public Instruction, on which Fourcroy sat, mentioning in defense his work for the Treasury and Commission on Weights and Measures. He wrote to the Committee on General Security, circumspectly hostile to Robespierre. But neither note was effective.

He had two choices left: turn himself in, or flee. Other friends would run. Condorcet, in safe hiding as he wrote his last book, *Progress of the Human Mind*; Rouchefoucauld, though he would be killed by a stone hurled through his coach window as he fled. From all sides came news of friends fleeing, hiding, dying. Antoine equivocated, unsure of his route, but when he heard that his father-in-law, Jacques Paulze, had been taken in the roundup, he turned himself in.

Perhaps it wouldn't be so bad, he told himself. There were twenty-seven Farmers in jail, and all were being promised they would only stay in prison until they settled their accounts with the government.

After that, they would be released. Antoine expected to be stripped of his fortune, but he could always work as a pharmacist. He'd still be a chemist, and he'd still be alive.

▪ ▪ ▪

Not every Farmer was as confident as Lavoisier. Etienne-Marie Delahante, the youngest of the prisoners, had joined the Farm after many of the alleged abuses but had no confidence in his captors' assurances. "I foresaw that the commissioners would accuse us of fictional abuses," he wrote, "that we would not be allowed to defend ourselves against the charges, and would be judged guilty of these alleged corrupt practices: thus, we would be doomed." The Farmers were accused of stealing 107,819,033 *livres* from the state; another charge against Lavoisier was that of watering down tobacco, the crime he'd struggled against so many years ago.

His home for the next six months was the Port-Libre prison, a former convent. The first story was reserved for the rich, and it was here that Lavoisier spent his stay. The Farmers roomed in the cells of the nuns; Antoine and his father-in-law were put together. Paulze, seventy-five, suffered greatly in the cold. The prisoners had money for meals but lived frugally, and all seemed to gather in Lavoisier's room.

Marie did what she could. She brought food and drink, communicated with friends, wrote letters of support to committees. As she did so, the state confiscated their property in Paris and the livestock at Fréschines. Some of Lavoisier's old colleagues—Fourcroy and Guyton de Morveau especially—occupied positions of influence in the new government, but de Morveau stayed silent out of fear, and Fourcroy out of fear and hostility. During these months, the number of executions steadily increased: 69 in December, 71 in January, 73 in February, 127 in March, and 257 in April. Marie herself was in danger, since the wives of other Farmers had fled; by remaining, she presented a tempting target. She had little money now for bribes, her property confiscated and assets frozen. As early as December, Antoine grew worried and wrote:

> My dear one, you are giving yourself a lot of trouble, exhausting yourself both physically and emotionally, and, alas, I cannot share

> your burden. Do be careful that your health is not affected. . . . I have had a long and successful career, and have enjoyed a happy existence ever since I can remember. You have contributed and continue to contribute to that happiness every day by the signs of affection you show me. I shall leave behind me memories of esteem and consideration. Thus, my task is accomplished. But you, on the other hand, still have a long life ahead of you. Do not jeopardize it. I thought I noticed yesterday that you were sad. Why be so, since I am resigned to everything and I consider as won all that I shall not lose. Besides, we can still hope to be together again. . . .

Yet he seemed to know hope was fleeting, too, and so came to peace with his remaining time. As winter turned to spring, he wrote nearly two volumes of his projected *Mémoires de chimie*, somehow managed to hand the manuscript to a printer, and corrected the proofs. He did what he could for his father-in-law, who grew more feeble. According to an anonymous account that was probably written by the young *fermier* Delahante, Lavoisier reconciled with the Church and took communion at least five times.

By April 1794, it was certain that their fates would soon be decided. The chief investigator for the state, Antoine Dupin—a deputy from Aisne and former comptroller for the Farm, best known in Paris for luxurious parties—was finishing his report, but before doing so, he asked to see Marie. According to all accounts, he was more well disposed toward Lavoisier than to the other Farmers, and in exchange for leniency wanted tears and gratitude. The meeting showed promise since it was arranged by Dupin's sister-in-law.

But the meeting did not go as planned. Marie refused to play the game; she refused to bow to anyone. She stood in Dupin's office and snarled, "I have not come to humble myself by soliciting the pity of a Jacobin. My husband is innocent and only a villain would accuse him." Her back straightened and she looked him coldly in the eye. "Lavoisier would be dishonored were he to allow his case to be separated from that of his companions. You want the lives of the *fermiers généraux* because you want their money. If they die, they die innocent."

Outraged, Dupin sent her away. Afterward, he was deaf to all entreaties for the famous scientist. That had been Lavoisier's last chance, and it was gone.

On May 5, 1794, the ax fell. Dupin submitted his 187-page report, accusing the Farmers of graft, cheating the state of millions, doctoring tobacco for personal profit, delaying payment to the National Treasury, and using the nation's taxes as a private bank. Without discussion, the Convention decreed that every Farmer be tried by the Revolutionary Tribunal.

It was a sentence of death. A spectator who'd been listening hurried to the prison and told the first person he saw—Lavoisier. Antoine told his fellow prisoners. The financiers began burning their personal papers or writing farewell letters to family and friends. Some argued for suicide as a way to cheat the executioner; opium had been smuggled inside for that purpose, but when someone offered the option to Lavoisier, he turned away in scorn. That night they were taken to the Conciergerie, a hive of dark passages and cells in the midst of the Palais de Justice. On May 7, the indictments were read in court to each man; at 10 a.m. on May 8, the Farmers were taken as a group to the "Salle de la Liberté" to face the Revolutionary Tribunal.

If one slim hope remained for Lavoisier, it was his old disciple, Antoine Fourcroy. But how much did Fourcroy try to save his mentor? Borda and Haüy, two members with Lavoisier on the Commission on Weights and Measures, risked their lives to plea for Antoine's life, but the Academy had treated the scientific work of the martyr Marat with contempt, and no one forgot that Lavoisier was the most contemptuous. Now Marat's seat in the Assembly was filled by Fourcroy, and he was in a position of power.

A great debate rages between historians as to whether he tried to intervene. Most believe he did not, based on earlier actions. "I made efforts for the unfortunate Lavoisier, but I failed," Fourcroy wrote, but he'd made self-serving statements before. Yet according to Fourcroy's student, André Laugier, the former disciple did take steps that placed his life in danger:

> A day or so before Lavoisier was slaughtered, M. de Fourcroy, although he had no right to do so, dared to enter the Assembly hall

> where the Committee of Public Safety was meeting. He spoke in favor of Lavoisier and explained with the passion so natural to him what a dreadful loss for the sciences the death of this great chemist would be. Since Robespierre, then the president of the committee, did not answer, no one else ventured to speak, and M. de Fourcroy was obliged to leave without anyone seeming to pay the slightest attention to what he had said. Hardly was he outside the door when the president complained of his gall, menacing Fourcroy, and so terrified Prieur de La Côte d'Or that he ran after Fourcroy and urged him not to do anything more if he wanted to save his own head.

There would be no further intervention. On a raised platform sat three judges: the president, Jean-Baptiste Coffinhal, a huge, black-eyed man of thirty-one, his chest crossed by a tricolor sash to which was pinned a medal with the words *"la roi";* and two assessors, dressed in black robes, white ties, and black caps surmounted by three black plumes. Before them stood a bottle of wine and a glass; behind them were busts, including one of Marat, watching this last revenge. The proceedings took three hours; three were released, including Delahante. The prosecutor gave an eloquent account of the Farmers' frauds and thefts, painting a picture of treachery and greed. "The crimes of these vampires cry aloud for justice," he declared. The rest were found guilty, including a charge of counterrevolutionary conspiracy to supply the Republic's enemies with vast sums withheld from the Treasury. The execution was to be carried out immediately.

Now begins a famous scene. One of Antoine's defenders managed to bring to the Tribunal a testimonial of Lavoisier's services to France in one last plea for his life; he cited his administrative past, his establishment of the New Chemistry, and his importance to the state as a scientist. To this was added a plea that was unique to the Salle de la Liberté: Citizen Lavoisier was engaged in experiments on respiration and metabolism which would be of great benefit to humanity. Would the Tribunal defer sentence for two weeks to allow him to reach an end?

There is no documentary evidence of the request or reply, but the story arose a few months after trial. Coffinhal grew impatient with the request and blasted back: *"La République n'a pas besoin de savants; il faut que la justice suive son cours"*—"The Republic has no need of scientists; justice must follow its course."

The request is refused. Although the comment was in character for Coffinhal, the famous line is believed apocryphal. At the time, however, it was thought to be true. It doesn't matter. The guillotine nears. We enter the realm of the "terrible instant," and the idea of the "thinking head."

▪ ▪ ▪

Part of the guillotine's horror and power lay in the fact that so little could be seen by the crowd. What is not witnessed must be conjured by the mind. By May 1794, Dr. Guillotin's "humanitarian machine" had become "appalling"; the prolonged ordeal of death by torture or hanging was replaced by instantaneous decapitation, yet it was this rapidity and unerring outcome that people found so chilling. The machine towered over the crowd, a singular, stark presence in the Place de la Révolution: it weighed 1,278 pounds, while the weight of the blade alone was 88 pounds. The height of its twin posts was 14 feet; within the cradle formed by those posts the blade dropped 7 feet 4 inches. It fell in 1/70 second at 21 feet per second. The beheading took 2/100 second: the blade fell so swiftly as to seem invisible.

Such a blade towering over crowd and victim would be rife with symbolism. It came to represent the terrible *instant* of death, when every hundredth of a second has meaning. Even the word is potent: it hails from the Latin *instans,* that which hangs over, or threatens. Yet the instantaneousness of the fall and beheading begged a more terrible question: Was death simultaneous with decapitation? It was widely believed that the soul resided in the pineal gland, the precise point encountered by the blade. Did the severed head immediately lose consciousness and the soul depart, or was it possible for one to know what had happened afterward?

The question had enormous consequence for the Revolution, and was the cause of medical upheaval in the years after 1794. The "humanitarian machine" was supposed to be painless; it ended the torture of death imposed by the nobles and lifted the means of execution to that reserved for nobility. Yet was a severed head aware of its fate? What could be more terrible than an awareness of death *after* it happened?

There were already uneasy whispers by May 8, when Lavoisier climbed in the tumbril with his colleagues. The most famous story was of Charlotte Corday. The executioner's assistant, a man named Legros, fished into the wicker basket where her head had fallen, picked it up by the hair, then dealt it a savage slap across the cheek. Horrified spectators claimed to see the face blush, while others claimed she was indignant. A scandal erupted, all the way to the Assembly. Another tale concerned two rivals in the National Assembly: when their heads were placed in a sack after the execution, one bit the other so badly they could not be separated. A brochure entitled *Anecdotes sur les décapités* collected rumors of decapitated heads that spoke. Though surely "inspired inventions of poetry," the author dwelt at length on an experiment during that century in which a head was "brought back to life," as well as on the anecdote in which the severed head of Mary Stuart was said to have groaned.

With this, the Revolution entered the realm of the sacred and taboo. For the "enlightened" mind, all reasons for hope and progress, even during the current bloodbath, were set on end. The Enlightenment and Revolution led to the guillotine and Terror, and these produced a monster: a head without a body, aware of its own death—a "thinking head." By 1795, a German surgeon, Soemmering, raised the likelihood of survival after beheading, seconded in 1796 by a French surgeon, Sue. According to both, although the unity of mind and body was severed, consciousness, seated in the cranium, remained. The sudden change of state was pregnant with horror: the victim knew what *until that point* in human history had been unknowable. He was aware of his own death *after it occurred.* The motto of the Enlightenment, "I think, therefore I am," was suddenly turned into "I think, but I am not," or "I think, but I am dead." Added Sue, "I am convinced . . . the heads would speak" if the vocal cords were not destroyed. The violent refutations by the French medical establishment suggest the two had touched a deep-seated fear.

The question of lingering consciousness has never been answered, and still lingers today. In 1836, a murderer named Lacenaire agreed to wink after decapitation, but failed to keep his promise. In 1879, attempts to elicit a reaction from a guillotined murderer named Prunier also proved fruitless. In 1880, a doctor pumped blood from a dog into a murderer's head three hours after execution—the lips trembled and eyelids twitched, but little more. The question seemed settled; but then in 1905, a doctor claimed that when he called the name of the murderer Languille, the severed head opened its eyes and focused on him. Two French doctors in the 1960s wrote that the brain was capable of breaking complex sugars into oxygen for as long as six minutes: "Death is not instantaneous," wrote Drs. Piedlieure and Fannier. "Every vital element survives decapitation . . . it is a violent vivisection followed by premature burial."

In the end, all is speculation. This much is known. The real cause of death in decapitation is *exsanguination,* Latin for "bleeding out"; a rapid loss of approximately one-half to two-thirds of the body's blood volume is sufficient to still the heart. The brain is particularly sensitive to deficiencies in oxygen and glucose and fails quickly when these are shut down. When the pressure and volume of the blood reaching the brain becomes too low to sustain consciousness, one falls into a coma. The cerebral cortex is the first to fail, but the "lower brain"—the brain stem and medulla—hold on longer, and with them certain reflexes. Appearance changes, and within a minute the face takes on the gray-white pallor of death as if one's essence has fled. The victim is toneless, no longer inflated with *pneuma,* the Greek term for the vital flame.

Lavoisier was facing a death from lack of oxygen. Like the mouse in the jar, long ago.

▪ ▪ ▪

This much is also known. By May 8, 1794, when Lavoisier was loaded on the tumbril, the executioner was tired. Charles-Henri Sanson had been involved in executions at least since 1757, when he was assistant to his father, Charles Jean-Baptiste Sanson. That year the execution of Dimiens, the attempted regicide of Louis XV, was a horrible failure: he was to be hung, drawn, and quartered, but the horses would not dismember him, he screamed horrifically throughout, and eventually he had to be cut apart with a knife. In 1793, Charles-Henri finally inherited the chief executioner's post from his father, and he vowed to bring professionalism to the job. By May 1794, although the number of executions demanded by the Tribunal had not yet reached the 1,500 mark of June and July, the number of condemned in each "batch" had grown daily. By the height of the Terror, Sanson would boast that he and his assistants needed only "a minute per person" to get the job done. Yet by May 8, he'd not yet reached that level of efficiency. It would take thirty-five minutes to execute the twenty-eight victims in this batch, or one and a quarter minutes per person.

Sanson was a celebrity in his own way. It was he who executed Marie Antoinette, her sister-in-law Elizabeth, and the duc d'Orléans. It was he who executed Charlotte Corday, warning her of the bumps in the road so she would not pitch forward in the cart, shielding her view of the guillotine until she asked him to step away. Life cannot help but be seen as a tragedy in such an occupation, but he always tried to maintain a shred of dignity for those about to die. In August 1792, death hit close to home when his younger son, serving as assistant (as he had for his father), tripped and fell to his death from the scaffold while displaying a condemned man's head. Sanson was always first on the stage and last to leave; he opened the spectacle and gave the crowd what they desired, but such duty took its toll. "The law has been passed," he wrote to the minister of justice when the Terror was at its height.

> The executioner asks nothing and will make no demands for himself in his capacity of executioner. This post is supposed to be worth 17,000 *livres*—but when the cost of his assistants has been deducted, as well as all the different and numerous expenses that he pays out of his own pocket, it will be seen that he is very unlucky to have such a post. And indeed the executioner cares little for the post. He has fulfilled his duties for forty-three years. The overwhelming work that it entails makes him wish to bring his services to an end.

Since Sanson rode with each batch on its way to the scaffold, Lavoisier saw a little man, delicate and dapper, with carefully combed hair, an elegant coat, and a hat in the current English fashion. The executioner was vain, but he was not cruel or rude. The condemned had their hands tied behind their backs: Sanson helped seat them in the cart and advised them how to weather jolts along the way. The iron gate separating the large cobbled courtyard of the Palais de Justice from the street swung open. The tumbril lurched forward at 4 p.m.

As the Farmers were not personally famous, the crowd at the beginning of the route was not as large as it had been for Marat's assassin or the King. Still, they were known as a group, and spectators trickled out to mock them. It was a warm, still afternoon; the spring of 1794 had been unusually humid, and from the blood-soaked stones at their destination it was said that a terrible vapor was rising, sickening all who lived near. The cart was preceded by horsemen and musketeers, clearing a way through the crowd. Old, tall houses look down on them along the route; they emerged from the rue du Roule and turned left into the rue Saint-Honoré. At this point, the flood of onlookers grew enormous due to the proximity of the markets. The carts were held up in the narrow street, and old Papillon d'Auteroche, one of the Farmers, remarked, "What grieves me most is to have such unpleasant heirs."

Lavoisier rode silently to the Place de la Révolution. Looking across the Seine, he could see the Collège Mazarin where he'd spent his early years. When the tumbril rolled to the foot of the scaffold, it was 5 p.m. and still light. A crowd stood in a circle around the guillotine: a police report described the Place de la Révolution "filled with people running with all their might, lest they should miss the sight." Almost all had opera glasses. Three assistant executioners stood on the platform, putting final touches to the machine.

The Farmers stepped from the tumbril, said their good-byes, and were lined at the foot of the steps. Sanson turned their backs to the machine so they would not have to see "what it was like." When all were in order, he drew a bloodstained leather apron over his clothes and gave a signal to his assistants, who grabbed the first victim in line and helped him up the stairs.

The Farmers were executed in the order in which their names appeared on the indictment: Lavoisier's father-in-law, Jacques Paulze, was third. Antoine was fourth on the list.

Since he could not see the proceedings, this is what he would hear. The executioner held Paulze by the right arm, an assistant by the left. A third would take his feet. Paulze was thrown against the plank onto his stomach. There were three dull thuds from above: the plank sliding forward. The neck clamp falling into place. The blade coming down. Then it was Antoine's turn.

There is a story that persists today, two hundred years later, of these last few seconds of Antoine Lavoisier. In one version, he tells a friend or assistant to watch when the blade descends and he will test the mystery of the thinking head. He'll start blinking before the ax is released and see how long he *keeps* blinking when his head is severed. In the second version, Sanson himself recognized the great scientist and said he had always wondered how long life and consciousness remained. Would M. Lavoisier help him with this truly metaphysical problem? Sanson is said to have requested. In both versions, Lavoisier takes the challenge: he blinks rapidly before the latch is sprung; he hears its release and concentrates on research as he never has before. The blade falls. He blinks ten times after his head falls in the basket. In other versions fifteen, or twenty. At least one story swears to thirty blinks before his mind dims.

But this, like Coffinhal's statement, is also apocryphal. Sanson did not recognize the great scientist; he would have recorded it otherwise, as he did Papillon d'Auteroche's comment about his heirs. The bodies were stacked in the wagons, the heads collected in a large wicker basket. All were taken to a large common grave that would contain the 943 victims guillotined at the Place de la Révolution between March 25 and June 9, 1794. The graves were dug in a wasteland called *Errancis,* which meant "maimed man."

But the myth outweighed the truth. Death itself couldn't keep Lavoisier from research, people found comfort in thinking, and this was the greatest experiment.

Could even death be quantified? How did it measure up in the scales?

CHAPTER 13

The New World

LONDON LOOKED AS DANGEROUS TO PRIESTLEY AS PARIS DID TO Lavoisier. Like William Godwin's title character in *Caleb Williams*, Priestley felt as if his life had gone from happiness to "a theatre of calamity. I have been the mark for the vigilance of tyranny." The only difference was that the English preferred the gibbet to the guillotine.

After his escape from Birmingham, it seemed he might be able to start again. Soon after his arrival in London, he thought he might inherit the Unitarian congregation of his deceased friend Richard Price; he also lectured in science and history at the Dissenting college in Hackney, in north London, and by September 1791, he and Mary set up house nearby.

But there were signs of trouble. Shopkeepers, workers, and servants had nothing to do with them. Young Joseph was forced to give up a business partnership in Manchester. William was urged to tarry longer in France. Priestley's friends in the Royal Society shunned him, especially Cavendish, with whom he'd enjoyed an informal scientific partnership for nearly twenty-five years. He'd thought scientists were above the hatreds of common men. He withdrew, understanding that "the chased deer is avoided by all the herd." But that didn't make exile any easier.

He felt a loneliness like that in Needham Market and as a child. The exile swiftly extended to every part of his life. He was exiled from London science, from his old Midlands haunts, and it was soon evident that he could never go back to Birmingham. "In this business, we are the sheep and you are the wolves," he wrote in a public letter to Birmingham on July 19, 1791, five days after being forced to flee. "We will preserve our character and hope you will change yours." But it was a vain hope. Letters of sympathy from friends in the Midlands all warned him to stay away. The Philosophical Society of Derby wrote that "almost all great minds in all ages of the world, who have endeavoured to benefit mankind, have been persecuted by them," yet urged him not to risk his life by returning. James Watt, on hearing that Joseph might preach in Birmingham's temporary Union Meeting House, reminded him that he had duties to family, friends, and humanity "that should direct you not to risk a life so valuable to them all."

Yet he was not completely alone. Supporters in England and Europe were vociferous in their praise. Old friends among the Dissenters stood by him, as would congregations and constitutional societies across the land. French patriots offered him a furnished house two miles from Paris; the French Academy sent its sympathy, and the National Assembly offered him a seat as representative from Orne, an honor he declined. John Wilkinson made investments for him in France and America in case he had to emigrate. His old friends in the Lunar Society sent money and equipment to reestablish his lab.

But the fact remained that he'd become the man that England loved to hate, and the realization hit him hard. There was always something likably naive about Priestley, a childlike need to be loved, and now that was stripped away. He was repeatedly burned in effigy with Thomas Paine; the Reverend Dr. Tatham of Oxford wrote, "Long have you been the Danger of this Country, the Bane of its Polity and Canker-Worm of its Happiness." Almost every issue of *The Gentleman's Magazine* contained some screed or parody. *The Times* ran an inaccurate account of the riot, claiming that the mob had been outraged by Priestley's toast to "the King's Head upon a Charger," and included with the story a political cartoon. Although the London paper was forced to retract its account, enemies continued to circulate the cartoon for the rest of the year. Thousands gazed upon the satirical prints in shop windows, clubs, workhouses, and taverns. Priestley could not walk outside his house without seeing himself strung up from the gallows, portrayed as a wolf or a shark, or counseled by Satan to pen tracts entitled "Assassination" and "Treasons." His ultimate reward was a popular theme of the illustrators; they liked to draw him being led off to Hell by triumphant demons, or dragged behind a wagon festooned with the words: "The man who denies the Messiah, may this be his fate noble or great."

Matters didn't get better in 1792. The fact that he tried to reclaim £4,083 10*s* 3*d* in damages for the riot unleashed a flood of contempt, yet Priestley wasn't alone in excoriation. The enormous success of Paine's *Rights of Man* created fear of home-grown revolution, and a government-supported propaganda campaign began. Its theme was simple, crude, and effective: There was no difference between reform and revolution, and any challenge to authority would lead to chaos and mob rule. Radicals like Priestley, Thomas Paine, and William Godwin were foreign puppets, manipulated by France. As everyone knew, France was a place of madness, mayhem, and horror, an alien and violent world where reason had long since disappeared.

The campaign worked, especially as events in France continued to unfold. The arrest of Louis XVI and the September Massacres created real fear in England; there were shortages and riots in the provinces, and the crown responded by issuing fresh proclamations against seditious writing. It mobilized the militia and fortified the Tower. Robespierre's announcement on August 26, 1792, that Priestley would be made an honorary citizen of France did not play well at home. Around that time, Priestley himself wrote to Lavoisier that in case of more riots, "I shall be glad to take refuge in your country, the liberties of which I hope will be established notwithstanding the present combination against you. I do hope the issue will be as favorable to science as to liberty." Misery made strange bedfellows. The bitter rivals had become mirror exiles, uncertain if there was any place on earth where they could be safe and free.

That August, Priestley wrote to William in Paris. His second son had applied for French citizenship on June 7, a fact the English papers lampooned viciously. "Remember you are to be a *man of business,*" Joseph advised his son. "I hope you will not let the attention that has been paid to you by the National Assembly hurt your mind."

Louis XVI's execution put an end to all hope of staying in England. Panic struck the English: rumors sprang up of secret plots to overthrow the monarchy and destroy the constitution. Citizens thronged the courts to profess their loyalty and swear oaths of their patriotism. No abuse was thought unjust against Dissenters, and Priestley most of all. For the first time since the Birmingham riots he began to feel unsafe, especially after war was declared in February 1793. Burke continued his attack in the House of Commons; public tracts detailed the Doctor's death or exile. "Fly from the country you hate, and which, with greater reason, hates you," cried a letter from Dr. Tatham in March, the learned don who'd earlier dubbed him a "Canker-Worm." A *Second Heroic Epistle to Joseph Priestley L.L.D.* imagined his death atop a burning pile of his books:

And the bright flame ascends the kindling skies
Yet while thy face now this, now that way turns,
With black'ning features writhing as it burns,
Death grins in mockery, nor relieves thy pains.

In May, he was accused of poisoning the waters of the New River, and it was only when the company supplying the water actually denied the accusation that the public believed the river had not been the target of mad Priestley's terrorism.

It was obvious now that he would be the scapegoat of every wild panic and fear. In June, he told a friend that the outlook was "peculiarly dark and discouraging" and there seemed no fresh air anywhere. Mary was sick and spitting blood, and Joseph feared she had tuberculosis. William wrote that he was leaving for America, but Priestley had no way to confirm this. There was talk of a colony for English fugitives starting on the Susquehanna River in Pennsylvania, but the prospect of being separated from Sally—who refused to go—threw him into the doldrums. His old battle with boils and gallstones flared up; the court settlement over damages still hadn't come through. Maybe the New World *was* the only answer, as his sons insisted, but how could he emigrate without funds?

It was at this point that English politics took a particularly sinister turn. In Nottingham, the houses of suspected French sympathizers were attacked; in London, harsher measures were adopted against those who spoke out for reform. The Edinburgh lawyer Thomas Muir was sentenced to fourteen years deportation to Botany Bay for advocating reform of Parliament; the Unitarian clergyman Thomas Palmer, a friend and convert of Priestley's, was sentenced to seven years for his association with a publication that called the war tax extravagant. And then there was the case of Richard Brothers, which scared everyone.

Brothers's sin lay not in his friendship with France but in his fervid hope for the end of the world. He was a Millennialist, like Priestley, but where Priestley used reason to interpret the Bible, Brothers wrote letters to the government in which he claimed that God revealed to him directly that the French Revolution was the fulfillment of ancient prophecy and should not be challenged by British arms. In 1794, he published the first of his two-volume *Revealed Knowledge of the Prophecies and Times*. The government responded by charging him with treason and throwing him in a madhouse, where he stayed until his death in 1806.

Maybe Brothers was mad, but what broke loose was the greatest outpouring of apocalyptic writing that England had seen in decades. Anthologies were choked with prophetic visions. Many described the fall of King Nebuchadnezzar, apocalyptic shorthand for George III. Burke claimed such rants were dangerous and used the prophecies to paint all Dissenters as unstable and mad. The poet William Blake rose from this ferment, part of the radical circle that included Priestley, Thomas Paine, Mary Wollstonecraft, and William Godwin, and introduced by the publisher Joseph Johnson, for whom Blake worked as an engraver. By 1794, Blake's publications included *Songs of Innocence*, *Songs of Experience*, *The French Revolution*, and *Marriage of Heaven and Hell*, in which themes of revolt were mixed with visionary ecstasy. In "London," he described the embattled city as a place where every face he meets shows "marks of weakness, marks of woe."

There was little doubt to Priestley that he would be tarred with the same brush and share the same fate as Muir, Palmer, and Brothers if he stayed in England. But this was just the beginning. In London, the office of the London Corresponding Society, which published the works of the radicals, was sacked and its membership rolls seized for later persecutions. Leaders of the society were tried for treason but acquitted thanks to sympathetic judges and hailed by the people as heroes. The government's response was to pass the Gagging Acts, which increased censorship, and the Seditious Meeting Acts, barring meetings of more than fifty people not approved by a local magistrate; and to suspend *habeas corpus.* With these measures set, the government detained radicals of all stripes. Blake quit writing about politics and turned to his vision of the Bible. Priestley's friends fled, laid low, or were charged and jailed.

Joseph knew it was only a matter of time before he too would be jailed. It was easy enough to make an unwise and unwary utterance at the pulpit, and that would be his undoing. By early 1794, the compensation from Birmingham finally came through—it was less than he'd claimed, but all that he would get, and he forwarded £2,000 to Philadelphia for settlement on his sons. He booked passage on the *Samson*, "Capt. Smith for New York": Mary and he packed their belongings and said their good-byes. "I do not pretend to leave this country, where I have lived so long, and so happily, without regret," he sadly wrote, "but I consider it as necessary." Maybe he would one day receive justice from his countrymen. But not now.

On April 7, 1794, Priestley left Britain forever. His ship of exile had the English equivalent of the name of Lavoisier's executioner. As Priestley sailed to the New World, Lavoisier rode the tumbril with the state's axman at the helm.

▪ ▪ ▪

There are brief moments when nations come to their senses and realize the evil they have done; when individuals grow ashamed of their homeland and mourn the madness committed in their name. No nation is immune to insanity: it is most disturbing when the injustice springs from an excess of idealism. In France, Pierre Laplace bemoaned Lavoisier's death when he heard of his execution: he remembered the guinea pig experiments he and Antoine had labored over, and how they had given some of the creatures names like little humans. They were all guinea pigs now. He turned away at the news and spat how it only took his nation a second to chop off his good friend's head—but to find another containing such brilliance would take another hundred years. As for his murderers, they too would be hunted down: Robespierre, Coffinhal, Fouquier-Tinville, all buried in the same mass grave as Lavoisier. The public funeral for Lavoisier would be as lachrymose as Marat's. His giant bust would be lowered as if by angels from the ceiling, while friends who aided the executioner now gave eulogies. The folly stretches through time in every direction, from the death of Socrates to America's invasion of Iraq—each carried out in the name of national security. The madness cannot be blamed solely on leaders and their henchmen; it is embraced by entire populations, and the next decades are spent trying to figure out what went wrong, and how to undo the ruin.

Englishmen wondered what kind of nation drove its best minds into exile. As the *Samson* and its one hundred passengers made ready to leave the harbor at Gravesend, a few friends and admirers sadly said good-bye. Joseph was given a silver inkstand with the inscription: "To Joseph Priestley, LL.D., etc., on his departure into Exile, from a few members of the University of Cambridge, who regret that expression of their Esteem should be occasioned by the ingratitude of their country." The Society of United Irishmen of Dublin, no lovers of Mother England, envisioned a scientific future of war and devastation, and bemoaned a world where

> all the arts and sciences are put under a state of requisition, when the attention of a whole scientific people is bent to multiplying the means and instruments of destruction and when philosophy rises in a mass to drive on the wedge of war. A black powder has changed the military art, and in a great degree the manners of mankind. Why may not the same science which produced it, produce another powder which, inflamed under a certain compression, might impel the air, so as to shake down the strongest towers and scatter destruction.

But he was going to a better place, "a country where science is turned to better uses" and where thousands of Irish had fled. They hoped that all would find peace on that far shore.

Priestley was hoping for more than just peace: he hoped to create with his sons an ideological Utopia. Many exiles from the ruins of the Enlightenment had set their eyes on the New World as "the only great nursery of free men left upon the face of the earth," an "asylum for the friends of liberty." When Priestley sailed, the hope of founding a Utopia had progressed farther than just a pipe dream. Soon after the *Samson* departed, a correspondent for *The Gentleman's Magazine* reported that Priestley, his sons, and a mutual friend "have contracted for a large quantity of land, estimated at 300,000 acres in Northumberland and Luzerne counties in the state of Pennsylvania, about 120 miles from Philadelphia, situated on the West Branch, north of the Susquehanna River."

By May, as the Priestleys were halfway across the Atlantic, Joseph Jr. and his partner, Manchester-based radical Thomas Cooper, were sending copies of the land proposal to friends in Europe to pass to whoever might be interested. The 300,000 acres would sell at a dollar an acre, and the capital of the new company would consist of 100 shares at $3,000 each. Another $20,000 would be raised for town improvements, houses for mechanics and artisans, and other amenities needed to create a home away from home. Although the land selected was rough and isolated, it was also rich in timber and close to water; the future Utopia was located in a state that frowned on slavery and situated in an area at peace with the native tribes. Sympathetic investors—including William Russell, who'd hidden Joseph and Mary on the night of the riots, and John Vaughn, another Birmingham friend—bought vast tracts which they offered for sale in 150-, 300-, and 400-acre lots. The place names echoed towns in Britain and Ireland (Cambridge, Dublin), names of patriots (Franklin, Washington), and ideals (Hope, Liberty, and Equality).

There had been other European Utopias. To French philosophers, especially, America and its revolution seemed to embody Enlightenment hopes, and its politics were entirely secular. Denis Diderot saw America as an asylum from tyranny; Turgot saw Americans as "the hope of the human race." In 1788, the Genevan banker M. Clavière and two partners sent the famous French journalist J.-P. Brissot de Warville to America to investigate such a scheme. Brissot's report, *New Travels in the United States of America*, called the project a "Eutopia," where "men of wisdom and information should organize the plan of a society before it existed, and extend their foresight to every circumstance." The project never came to be: Brissot returned to France on the eve of the Revolution; a Girondin, he was guillotined by Marat's party on October 31, 1793. Afterward, there were aborted plans to set up an asylum for Marie Antoinette and fellow royalists far to the north of the Priestley lands, and in the following century myriad Utopias would be proposed for this and nearby areas. Civil and religious liberty were the planned hallmarks of most; but Priestley's "pantisocracy" included the very British tradition of land ownership as a means for "improving" one's station in the world.

The most vocal enthusiasts were the young Romantic poets Robert Southey and Samuel Taylor Coleridge, who dreamed of a society of poets far from the cares of the world. "To live in a beautiful country, and to inure myself as much as possible to the labours of the field have been . . . my dream of the day, my sigh at midnight," Coleridge declared. Southey, not to be outdone, imagined himself as a philosopher-farmer "wielding the axe, now to cut down the tree, and now the snakes that nestled in it." Echoing Priestley, whom they both admired, the Romantics proclaimed, "Let us be free ourselves, and leave the blessings of freedom to our posterity." They figured that twelve pantisocrats could clear 300 acres of land in four or five months; for $600, a 1,000-acre plot might be cleared and houses started. They raised money by conducting a series of popular lectures to packed houses in Bristol. Coleridge seemed especially adept at such missionary work: at one point, two giant butchers danced around the room while drinking from a large glass of brandy and shouting, "God Save the King!, and may he be the last."

America in 1794 had room for all types of Utopian schemes. From a distance—far from the harsh realities of climate, terrain, isolation, insects, and angry natives—the new nation seemed an incredibly vast space in which to carve out any community. Millions of acres were claimed by the original colonies that stretched to the banks of the Mississippi; although these claims were disputed by the British, French, and Spanish (as well as native tribes), new settlers paid such niceties little heed. According to the 1790 Census, there were 3.9 million people in America: 1.6 million men, 1.5 million women, and

757,000 slaves (their gender unrecorded). Indians were not counted. Only 5 percent of this population lived in cities, while the remaining 95 percent lived off the land. True, there were "Indian problems," but most of these were located to the west, in the Ohio Valley beyond present-day Pittsburgh. True, politics could be savage—debates between Alexander Hamilton's Federalists, who favored a strong central government, and Thomas Jefferson's anti-Federalist Republicans, often dissolved into brawls. Still, brawls were better than prison or the guillotine, and given enough distance between a man and his neighbor, it seemed everyone could be happy.

Joseph and Mary were at sea on the *Samson* for eight weeks, and seasick the first five. By any standard, theirs was a rough voyage. There were gales and sudden squalls; the captain cursed and the food was bad. The passengers were crowded in the dark hold. Since much of the voyage was stormy, long hours were spent below decks, where the smell of bilge and sickness was magnified. "I would recommend every one to have motives strong enough to overbalance every inconvenience they meet with on the voyage," Mary later wrote. "I would also advise them to lay in a great stock of patience; and where so many are to be so long together in so small a compass, they should make up their minds to bear and forbear."

Up on deck was scarcely less worrisome. "We passed mountains of ice, larger than the captain had ever seen before," Mary recalled. In fact, the captain was so shaken that he kept watch for two nights straight for fear they might collide with one. One day was filled with sightings of waterspouts, the worst of which could tear a wooden ship apart during the age of sail: "I saw four at one time," said Mary, "but, happy for us, they kept a proper distance." They saw mountainous thunderclouds, which at night seemed "all on fire." All in all, said Mary, "we had all that could well be experienced on board, except shipwreck and famine."

Once, they had an extremely close call. For three days they plowed forward against a head sea that beat upon the bow like a sledgehammer. Boarding seas flew in green sheets over the deck and wind whistled through the rigging. A tremendous gale shook the *Samson* while the ship was under full sail; the great mainsails billowed up and out, threshing and roaring against the mast, while the sailors in the yardarms beat and pulled on the canvas till the ends of their fingers bled. In such havoc the wind made a snarl of the rigging and carried away the topsails. This was the kind of situation where cargo suddenly shifted and a ship, unable to right herself, quickly went down. For about thirty minutes their lives hung in the balance, but then the winds abated and the crew regained control. "I found myself more vexed than frightened," said Mary, "as I fancied it might have been lessened by care."

Through it all, Joseph maintained his strange distance from worry, as if in a huge game. He wrote two sermons on infidelity; conducted services for the passengers; read the Hebrew Bible, Ovid's *Metamorphosis*, the *Dialogues* of Erasmus, and the second volume of David Hartley's psychology. He seemed on a different voyage than Mary, and wrote:

> We had many things to amuse us in the passage; as the sight of some fine mountains of ice; water-spouts, which are very uncommon in these seas; flying fishes, porpoises, whales, and sharks, of which we caught one; luminous sea-water, &c. I also amused myself with trying the heat of the water at different depths, and made other observations, which suggest various experiments, which I shall prosecute whenever I get my apparatus at liberty.

They arrived in New York on June 4 amidst "thick fog and rain, which they say are very unusual here at this time of the year." It was an inauspicious if prophetic beginning. They were met at the dock by Joseph Jr. and his wife, and spent the night at Mrs. Loring's Boarding House on the Battery. As they dined and rested, they could watch ships sail from the harbor to sea on one side and up the Hudson on the other. The next day, Priestley was given a hero's welcome: there were speeches by Governor George Clinton, the Bishop of New York, several societies, and swarms of dignitaries. The American Republic would be a "safe and honourable retreat" for the Doctor in his declining years: one day England would regret "her ungrateful treatment to this venerable and illustrious man." Not everyone was impressed. The Federalists, who desired closer ties to England, did not appreciate his arrival. Priestley's presence could ruffle diplomatic feathers, and his ties to France were worrisome as the young Republic tried to steer a neutral course in the war between the European superpowers. His religion was suspect, too: the day after Priestley's arrival was Trinity Sunday, yet none of the city's many churches invited him to speak. On the whole, however, Mary reported that "Dr. P." was "wonderfully pleased with everything," and for her part, "I never felt myself more at home in my life than since my arrival here."

They left after two weeks and headed for Philadelphia, but found the capital city excruciating. It was expensive by English standards, the summer heat was oppressive, and Priestley thought the wealthy Quakers smug and pretentious. He wrote:

> Probably in no other place on the Continent was the love of bright colours and extravagance in dress carried to such an extreme. Large numbers of Quakers yielded to it, and even the very strict ones carried gold-headed canes, gold snuff-boxes, and wore great silver buttons on their drab coats and handsome buckles on their shoes.

The women, too, were vain. In no other place but Paris had he seen the female of the species "so resplendent in silks, satins, velvets and brocades, and they piled up their hair mountains high." Yet even here there were amenities. He met George Washington and admired his unpretentious manner, renewed his acquaintance with John Adams (from when Adams was ambassador to England), and started a friendship with Benjamin Rush and other members of the American Philosophical Society, to which he'd been elected in 1785. Nevertheless, Mary did not like the city, and neither did he. By October, they started out for Northumberland, the closest settlement to their planned Utopia, five days journey up the Susquehanna.

Traveling the 160 miles from Philadelphia to Northumberland was quite a different experience than taking the civilized coach road from London to Leeds, and soon Priestley began to realize they were "almost out of the world." The inns were poor and bug-infested, the streams so swollen from the heavy rains that several times they had to be ferried across in canoes. In Harrisburg he hired a "common wagon" in which he and Mary slept for two of the five nights. As they passed through Sunbury, a traveler noted that Priestley had already started to go native: "Dr. Priestley . . . has left off his periwig, and combs his short grey locks in the true style of the simplicity of the country."

Northumberland was beautiful, set between two branches of the Susquehanna and circled at the edges by rock bluffs and hanging woods, but life here was so very different from what he'd known at home. He wavered between homesickness and fascination with the back-to-basics lifestyle he encountered. He wrote a friend in Britain:

> Here every housekeeper has a garden, out of which he raises almost all he wants for his family. They all have cows, and many have horses, the keeping of which costs them little or nothing in the summer, for they ramble with bells on their necks in the woods, and come home at night. Almost all the flesh meat they have is salted in the autumn, and a fish called *shads* in the spring. This salt shad they eat at breakfast, with their tea and coffee, and also at night. We, however, have not yet laid aside our English customs, and having made great exertion to get fresh meat, it will soon come into general use.

Sometimes he wrote his widowed sister Martha that he preferred life in England; sometimes he wrote to Birmingham requesting more scientific equipment, which hinted at his plans to stay. At one point he was tempted to accept a chair in chemistry at the University of Pennsylvania in Philadelphia, but Mary was charmed with Northumberland, and that turned the tide.

One fact was certain: his Utopia fifty miles to the north was not to be. The late eighteenth and early nineteenth centuries were replete with failed American Utopias, and by late 1794 it was obvious that Priestley's pantisocracy would be joining them. The land itself was more hard and craggy than advertised: buyers were not coming, and money, the final arbiter in a Utopia's success, was hard to find. Thomas Cooper, who acted as the main promoter back in England, persuaded only a handful of investors to take the American gamble. Few details of these pioneers survive, although of those whose stories were recorded, many did well. There was James Wheatley, a schoolteacher, who became Northumberland's justice of the peace and a state senator; James P. DeGruchy, a French distiller living in Birmingham at the time of the riots, who became a large landholder; and John Cowden, Northumberland's first storekeeper and post-master, of whom it was said that when he traded with the Indians, he never took along a scale. Instead, he swore that his hand weighed 2 pounds and his foot 4, and he used these as measures.

In England, the Romantic poets and potential émigrés were running into problems, too. Coleridge, Southey, and fellow poet Robert Lowell, the newest convert, raised money throughout 1794 and continued to plan. All three decided to marry before coming to America since they knew eligible young women would be scarce, so they proposed to three willing sisters, the "Miss Frickers." But by 1795, Lowell would die of fever, leaving his widow to care for their infant, and Coleridge eloped with another woman, leaving poor Southey with all three Frickers on his hands. The hopes of the Romantics were scuttled by a different kind of romance and its inevitable consequences.

By then, Priestley had reconciled himself to ending his days in Northumberland. There was a point when the town was favored as state capital; there was talk of starting a college, an idea Priestley promoted. If anyone remained true to the pantisocratic ideal, it was his youngest son, Harry, who began clearing 300 acres of cheap land outside town. Joseph took pleasure in his growth. "He works as hard as any farmer in the country and is attentive to his farm, though he is only eighteen," he wrote. "Two or three hours I always work in the fields along with my son."

Then there was Mary. "I am happy and thankful to meet with so sweet a situation and so peaceful a retreat as this place I now write from," she told her English friends. Gone were the attacks on her family's home and beliefs, gone the debilitating fear she'd felt in London. The woods of Northumberland were the longed-for Paradise she'd kept secret to herself when Joseph went seeking a better world; the people here were "plain and decent in their manners," and that was always important to her. They bought for £100 eleven acres at the edge of town with a view of the river and township. Joseph began to buy cattle and became a wealthy farmer. She designed a house, to be set a quarter of a mile from the water. It was to be three stories high, of modified Georgian design, loaded with closets and pantries, decorated simply with whitewashed wainscoting and plaster walls. A fireplace heated every room on the first and second floors. The central mansion was balanced at both ends by two wings. The eastern wing would be Joseph's lab: it would be the best on the continent, equipped with a furnace on each end of a long work table, a fume hood running the length of the room, and a loosely constructed exterior wall that would bear the brunt of any explosion. The western wing was Mary's kitchen, a room as large as his lab. With this house and this design, she was making a statement. All was in balance, and such symmetry proclaimed the equal importance of each partner's labor.

▪ ▪ ▪

But as so often happened in this New World, Nature could be harsh, disrupting hope and dreams. The winter of 1795–96 was cruel, and when disaster visited Priestley, it came in threes.

Harry died first, at the end of 1795. "We have lost poor Harry!" Joseph wrote to a friend. The boy had grown accustomed to working hard in all weather, finding joy in a labor he'd never known in England, and that winter he'd contracted a chill. At first, Priestley and a local doctor had not thought it serious, and perhaps the first bout was not. But Harry was living in a one-room log cabin of his own construction like an early Henry David Thoreau, and he was struck down a second time. Once again he recovered, and started building a two-room stone house for which he had to mix the lime and build the kiln. During this effort, he was felled the third, fatal time. He died on December 11, 1795, and a friend, a Mr. Bakewell, followed the funeral train to a lonely spot,

> and there I saw the good old father perform the service over the grave of his son. It was an affecting sight, but he went through it with fortitude, and after praying, addressed the attendants in a few words, assuring them that though death had separated them here, they should meet again in another and a better life.

Nine months later Mary died, on September 19, 1796. She was fifty-two, and Joseph's companion for thirty-four years. After Harry's death, her sadness and the cold sent her into decline; her tuberculosis returned and she spit up blood once more. At times, she described the pain as being stuck with knives. During the last two weeks she suffered an intestinal blockage with internal bleeding and endured "great and excruciating pain." But after a while the pain abated and she "died apparently without any pain and without any struggle," her son Joseph later said. Priestley looked at the plans of the house she'd dreamed of finishing and knew that his course was set: he could not go back home. *This* was his home. He thought of how she'd always calmly handled life and realized that "I was only a lodger in her house." Everywhere he looked, he saw signs of Mary. "I feel quite unhinged," he told John Wilkinson in a letter dated January 25, 1797, "and incapable of the exertion I used to make. Having always been very domestic, reading and writing with my wife sitting near me, and often reading to her, I miss her everywhere." As he told his friend Theophilus Lindsey, "I never stood in more need of friendship than I do now."

He needed a friend all right, for the New World was proving to be a lonely place and the domestic temple he'd carefully built around him felt as if it were being pulled down. It is hard to determine today which came first, the trouble with William or with finances, but both occurred almost simultaneously, curling about one another in a tightening stranglehold.

William had always been the most unsettled of the brood. He faced the mob in Birmingham, witnessed the Revolution in Paris, and became the whipping boy of the British press for being there too noticeably. Everything he did seemed to end in failure. Everywhere he went, he was not his own man but Joseph Priestley's failed son. One senses an anger building over the years. He'd followed his brothers to America in time to greet his father in New York, but while the rest of the family headed to Philadelphia, he stayed in the Northeast and worked for a year outside Boston for a farmer. A year—his usual tenure before moving on.

By 1795, he'd returned to Pennsylvania and was working part of Harry's land. Sometime before Harry's death, he married the daughter of a local farmer, "the agreeable Miss Peggy Foulke, a young lady possessed with every quality to render the marriage state happy." Yet their existence in the forest worried Priestley. "The life they lead, quite solitary in the woods, is such as you cannot easily form an idea of," he said.

Shortly after Harry's death, Priestley was called to Philadelphia for a series of lectures that helped establish Unitarianism in America. He liked these visits to the capital, and tried to make them an annual event, just as he had with his trips to London. William's new wife accompanied him on the journey, and at this point they all seemed to get on well together. Priestley was invited to tea with George Washington; gave his Unitarian lectures at the new chapel on Lombard Street, where he was pleased to see members of Congress and even the conservative John Adams in the audience; renewed his scientific membership at the American Philosophical Association, where he still defended phlogiston. Preaching on religious subjects was especially gratifying since he'd only been invited to one pulpit in Northumberland; afterward, the minister stated that if he ever allowed Priestley again to his pulpit, the congregation would probably relieve him of his job. Even in Philadelphia there'd been "much jealousy and dread of me" during his first visit, but his mild manner and humor won him sympathy and friendship if not converts to his Unitarian ways.

Yet during this visit, relations with William suddenly soured. One wonders whether it was sparked by the reapportionment of family land after Harry's death, or whether older resentments were involved. In March, while still in Philadelphia, Priestley was stricken with pleurisy and his friend Dr. Benjamin Rush attended him several times, bleeding him as was common. During one visit, Priestley "in his sickness spoke of his second son, William, and wept very much," Rush later said. It was the only public mention of trouble at this point; when they returned to Northumberland, Mary's health had deteriorated considerably, and she was attended by the wives of both brothers until she died.

Mary's death led to a realignment of both family living arrangements and finances. As her house neared completion, Joseph Jr. dissolved the Utopian partnership, which had ultimately acquired some

700,000 acres, and moved back to Northumberland to be with his father. He set up a brewery and, later, a nursery for English garden plants and fruit trees. When Mary's mansion was complete, Joseph Jr. and his brood, which included two children, moved in with Priestley.

This was disruptive enough, but John Wilkinson, who'd carried on a long legal battle with his brother and was short on funds, presented his own surprise. Without warning, he mailed Priestley an expense account—the Doctor owed Wilkinson $56,219 for funds he'd drawn since coming to the New World. Joseph was stunned, and suddenly threatened with bankruptcy. He couldn't possibly meet Wilkinson's demands. At the same time, in England, his daughter Sally's husband had failed in business, too. The only way out of this mess seemed a trip to France to retrieve the funds invested there in his name. Only by this expedient, he thought, could he repay his debts and help poor Sally.

Where William figured into all the family negotiations and finance is hard to determine; after Mary's death, his influence diminished as Joseph Jr's became more steady. There are no records of relations between the brothers, yet of all the siblings, only Joseph Jr. seemed to inherit his mother's business sense while William inherited his father's luckless inability to rub two cents together. Presumably, for awhile, William lived in his house in the woods and kept to himself. Joseph Jr. dissuaded his father from rushing to France in the midst of a world war, but not before rumor of the journey reached Paris and bounced back to England and America.

Once again Priestley found himself in a political hailstorm, but this time his only sin was that of trying to iron out a financial emergency. In fact, ever since coming to America, he'd tried to steer clear of political entanglements and hoped to spend his last years immersed in science and theology. Left to himself, he might have succeeded. But strange satellites orbited the old scientist, and now they were set to collide.

The catalyst was John Adams's assumption of the presidency on the Federalist ticket in March 1797. Apart from the Civil War, the clash of the Federalists and Republicans in the late eighteenth century was one of the most contentious political periods the nation has known. Priestley, a supposed friend of France, became a scapegoat for the American press as much as he had for the British. The loudest voice against him was that of former soldier–turned–political pamphleteer William Cobbett, who wrote under the pen name "Peter Porcupine." Cobbett began attacking Priestley in his Philadelphia-based *Porcupine's Gazette* from the moment he arrived aboard the *Samson*. He accused Priestley of being a French sympathizer and French citizen—and when word drifted back of Priestley's intended visit, of conspiring to commit treason as well.

The other satellite spinning around Priestley was Thomas Cooper, Joseph Jr.'s partner in the failed land company. Now he was editor of the *Sunbury and Northumberland Gazette* and a frequent houseguest at Priestley's mansion when his funds ran dry. Cooper's *Gazette* had become one of the nation's strongest organs for Republican Party propaganda, continuously accusing the Federalists, Adams, and his administration of mischief, misconduct, and secrecy. "Is it a crime to doubt the capacity of the President?" he wrote. "Have we advanced so far on the road to despotism in this republican country that we dare not say our President is mistaken?"

As a matter of fact, conditions were that bad. In July 1798, the U.S. Congress passed the Alien and Sedition Acts, and though less draconian than those Priestley had fled in England, the new measures were still a powerful threat to freedom of speech and other civil liberties. The ultimate sanction for "seditious" acts was deportation, and as in England, sedition could mean nothing more than a writer's criticism of policy. The Alien and Sedition Acts were a political weapon and little else, and though no one was ever deported, several Republican writers—including Thomas Cooper—were charged with treason for criticizing the Federalist administration. Cooper himself was fined $400 and sentenced to six months in jail, the only person to be imprisoned under the law. When Adams's secretary of state, Thomas Pickering, suggested prosecuting Priestley too, Adams was more compassionate: "He is as weak as water, as unstable as . . . the wind," he answered. "His influence is not an atom in the world."

Adams was correct, if uncharitable. Priestley's was a voice crying alone in the wilderness, and no one took him seriously. He seemed undone by his multiple tragedies, and out of touch with modern thought because of his isolation. His theological writings turned increasingly apocalyptic; he stayed away from the learned societies in Philadelphia in 1798–99 as yellow fever raged through the city and the streets became "lifeless and dead." The only place he was able to hold a Unitarian meeting was in his own drawing room. He wanted to be part of the international scientific scene that he'd inhabited so long, but advanced research passed him by and he insisted on studying problems that other chemists considered solved—the composition of water, the nature of combustion, and the doctrine of phlogiston. He wrote a letter defending phlogiston to the French chemists in 1796, but the challenge was ignored. It seemed that all the world had been converted to Lavoisier's system, and Priestley, whose discovery of oxygen initiated the Chemical Revolution, kept insisting that the new science was wrong. This might explain in part his refusal to acknowledge Lavoisier's death: he was bitter, and felt betrayed. He was the last phlogistonist, banished to the very edge of civilized life, and no one had time for him now.

Into this turmoil, William made one final appearance in a family scandal that would prove the bitterest of Priestley's long life. Early in April 1800, the Priestley household fell ill after eating a pudding, and the Federalist press accused William of trying to poison his father and the rest of the household by adding arsenic to the meal. Only Joseph Jr. was absent, away on business in England. The episode, made public by an anonymous correspondent in an April 17, 1800, letter to the pro-Federalist *Reading Advertiser*, is unclear even today. That the family fell ill is unquestioned; that William left the state immediately afterward is certain; that Priestley never publicly blamed his son is also true. According to the account:

> On Monday last, Doctor Priestley, Mrs. Priestley (wife of Mr. Joseph Priestley jun.), her two children, a hired girl, and a little bound girl, all of them were poisoned; they are however so far recovered, with their own exertions (by drinking warm water) and the assistance of the medical gentleman of this place, that they were supposed to have overcome the most imminent danger. The hired girl made a pudding for dinner, took the flower [flour] as usual out of the meal chest, but discovered some shining particles of some substance intermixed with the flower—she acquainted

> Mrs. Priestley thereof, who thought little or nothing of it—the girl however, and a hired man, went to the chest, and took off the top which appeared to have most, and threw it away; otherwise they all must have inevitably fallen an instantaneous sacrifice. The poison intermixed with the flower is said to be arsenic, and was so strong, that after the Doctor and family had discharged a quantity from their stomach, by vomiting, the poultry eating thereof almost instantly died.

These facts seemed clear enough, yet the innuendo began immediately. Who was to blame? Was there a motive? The maid who cooked the pudding seemed innocent since she also fell ill, and "therefore it can't be her." The finger pointed next at William, "the ordinary drunken wretch," who although unnamed in the report was still clearly identified. Priestley and his son were full of "French principles," claimed the writer, principles which led many sons to massacre parents during the Revolution. Joseph Jr.'s wife Elizabeth "said it was him," the writer continued. "The hired girl, for several days before [had] seen him about the meal chests, opening them, asking her who eats Indian meal, and who eats wheat meal, etc?"

It was a curious mix of fact and possible slander. The allegations were hotly denied by father and son, but not for almost two months after the affair. By then, William and his family were nearly thirty miles downriver in the township of Middle Paxton, where he said he'd been since early April, having left Northumberland five days before the alleged poisoning. Priestley said he tested the flour for arsenic and found no traces and a local doctor seemed confident that the foreign substance was "Tartar Emetic." Yet questions remained. Elizabeth Priestley always blamed William. And William left the state, resettling eventually in St. James Parish, Louisiana, just north of New Orleans on the Mississippi. He became a sugar planter worth several hundred thousand dollars; he was also a slaveowner with twenty-six slave dwellings on his plantation and fifty-eight slaves. His fugitive life was a direct assault on his father's values—but then, he'd never succeeded by taking his father's road.

By then, however, it little mattered. "I feel more compassion than resentment on [William's] account," Priestley wrote John Wilkinson in June 1800, the same month he defended William in the newspaper. "I do not expect, or wish, to see him any more, but I shall continue to write to him, and give him my best advice." Father and son never met again.

▪ ▪ ▪

With such misfortune, one wonders how Priestley continued. His was an optimistic faith, grounded in a belief in progress and reason, the very optimism satirized by Voltaire in *Candide*. Young Candide's former tutor, the philosopher Pangloss, reasoned that all misfortune was for the best since this was the best of all possible worlds; he refused to change his opinion through shipwreck and the Lisbon earthquake of 1755. Nothing hurt Joseph more than a comparison to Pangloss or to be called a hypocritical "Tartuffe," as he was by William Cobbett, and he admitted in a letter to Lindsey that every year he spent in the New World was attended by a trial—death, betrayal, continued controversy, threats of financial ruin. "I hope, however, I feel solid consolation for the full persuasion that every thing is ordered in the best manner; and though not now, I shall, one day, see how it is so."

If there was anger at his fate, it came veiled in his theology. After 1801 and William's departure, he immersed himself in the study of the prophetic books of the Bible. His letters to others began to abound in references to the Woman in the Wilderness, the Restoration of the Jews, the Personal Reappearance of Jesus, the Millennium. He read the newspapers with the Book of Daniel by his side. When news arrived of Napoleon's expedition to Egypt, he wrote, "Something favourable is promised in Egypt in the *latter days,* which I think are at hand." When his English friends complained of his elaborate conclusions, he apologized: "I find a great disadvantage in being alone, having no person whatever to confer with on any subject."

There was one person who still took him seriously. On February 17, 1801, Thomas Jefferson was declared the winner of the presidential election, held the previous December, after the thirty-sixth ballot of the Electoral College. The bitterness of the campaign still enraged John Adams: after attempting to pack the judiciary, he rode out of town before Jefferson's inauguration on March 4, 1801. With Adams out, Priestley was suddenly in. The wheel had turned. For the first time, he found himself living in a country of whose leaders and government he fully approved. One of the first letters Jefferson wrote from the Presidential Mansion was to the ailing Priestley:

> As the storm is now subsiding, the horizon becoming serene, it is pleasant to consider the phenomenon with attention. We can no longer say there is nothing new under the sun. For this whole chapter in the history of man is new. The great experiment of our Republic is new. Its sparse habitation is new. The mighty wave of public opinion which has rolled over it is new. But the most pleasing novelty is its so quickly subsiding over such an extent of surface to its true level again.

All things find their balance. The philosophy, so central to Antoine Lavoisier's life and work, brought peace to the old man. In the last years of his life, Joseph Priestley corresponded frequently with the president. Jefferson asked the Doctor to draw up a course in the sciences for educational reform so that the nation could attract professors from Europe; he conversed with Priestley about the tenets of Unitarianism. He'd read the old man's *Corruptions of Christianity* and said he used the book as the basis of his own personal faith. Priestley was so touched by the attention that he dedicated his last book, *Jesus and Socrates Compared*, to Jefferson. When the president heard in 1801 that the old man had been struck down with a "bilious fever of pleurisy," he wrote a long and consoling letter. "Yours is one of the few lives precious to mankind," he said.

Such kindness bucked Joseph up, at least spiritually. Perhaps that was what he sought most, to be appreciated. In 1801–02, there were a few final stabs at science. He took up the study of botany, and conducted studies on the growth of algae; he carried out tests on the Voltaic pile. Most important, he discovered carbon monoxide by passing steam over a hot bed of charcoal. In 1802, he told a friend that "I have lived a little beyond the usual term of human life, and am content and thankful. Few persons, I believe, have enjoyed life more than I have done." Yet by 1803, he knew his time was limited: his feet were always swollen, his teeth were falling out, the stomach ailments he'd attributed to indigestion forced him to subsist on soft food. Weakness so often overcame him that he could no longer hold his Sunday services in the drawing room. That year, his daughter Sally died of tuberculosis in England, and with this last tragedy he seemed ready to go. In January 1804, the weakness increased and he fought against it, rising daily and shaving, walking through the side door of the house to visit his laboratory. But by the end of the month it was a struggle even to light the fire and start the experiments, so he closed the door behind him and left the lab for good.

On January 31, 1804, one of his grandchildren asked him for a 5 cent piece he'd promised. Priestley smiled and handed over the money, but when he tried to make a joke, found he could not speak. The end, he knew, was near, and he began winding up his affairs. He recovered his speech, but on Thursday, February 2, made a final entry in his diary: "Much worse: incapable of business: Mr. Kennedy came to receive instructions about printing, in case of my death." That Sunday night, February 5, he felt himself sinking. When the children came into his room, he told them: "I am going to sleep as well as you, for death is only a good long sleep in the grave, and we shall meet again."

As he declined, he spent time with friends. One was John Binns, the Irish radical who would eventually return to Birmingham to visit the site of Priestley's destruction. They sat in the front room of the mansion, playing board games: it was a large room, with a high ceiling and a broad fireplace piled with burning and sparking logs. They played chess and backgammon by the three windows overlooking the lawn that sloped to the river. Priestley was Binns's superior in games, but he always let the younger radical win the last they played.

Sometimes the old man would look out the window and talk about this new country of his. In 1784, long before he considered coming to the New World, a friend had written that "the people of America have a saying—that God is Himself a mechanic, the greatest in the universe; and He is respected and admired more for the variety, ingenuity and utility of His handiwork than for the antiquity of His family." It was true. If there was one common faith he'd observed in the people of this nation, it was a faith in efficiency—in handiwork and invention. "The American people have a wonderful talent at making small tools," he marveled. They lived to invent, even more so than his friends in Birmingham, which everyone knew was one of the "great towns to manufacture iron, tin, and papier-mache in the world."

Yet if there was a flip side, it was a tendency for waste and recklessness. When he first came to Northumberland, there'd been thousands of shad that were caught a little above town and sold for $6 per the hundred, but as dams and canals were built to improve navigation on the Susquehanna, their numbers dwindled. By now, Northumberland was the greatest trading town on the river, with several stores, three hatters, five tanneries, five brewers, a hotel, a nail factory, and 161 households. But all this growth and comfort came at a price. This was the hardest lesson he'd learned in a life spent espousing science: it was another redox reaction, on a human scale. Man never realized what he had until he lost it. By then it was too late and could never be regained.

Now, on this last day, February 6, he asked to be moved again to this favorite front room. Joseph Jr., Elizabeth, and Thomas Cooper were with him. At 4 a.m., he called to Joseph Jr. in a weak voice and asked for wine with a tincture of bark to soothe his throat.

"How are you feeling, Father?" his son asked.

"I have no pain," he replied.

He lay quiet until 10 a.m., when he asked to see some final revisions on *Jesus and Socrates Compared*. Cooper brought down the alterations from his second-floor bedroom, and Joseph dictated in a clear, distinct voice the changes he wished made. Joseph Jr. read it back, and Priestley objected quietly that Cooper had put the changes in his own words. "I wish to have it in mine," he said.

So Joseph Jr. took up pen and ink, and Priestley repeated his revisions word for word. When his son read it back, Joseph smiled. "That is right," he said. "I have done now."

It was cold in the large room with the high ceilings, and either Joseph Jr. or Cooper started the fire. Priestley watched the sparks rise in the chimney, and no doubt remembered his own fires—the distant glow from the burning house, the sudden flare of the candle that led to his greatest discovery. Would anyone remember these things, overshadowed as they were by Lavoisier? Would anyone remember the 150 pamphlets, books, and manuscripts that had flowed from his pen? Near the head of his daybed was set a small table; on it rested the manuscript and inkstand. Within reach were some bottles with powders of different color. He smiled as his eyes rested on them. One thing about chemistry . . . it had always been colorful.

About thirty minutes after finishing his manuscript he called to his son Joseph and Elizabeth one last time. He would like to be moved from the daybed to a cot a little closer to the window so he could gaze outside; he would also like it if his legs could lay flat and his head be upright, so he could better see. He looked down the lawn to the river, past his son's newly planted apple orchard to the rushes by the shore. People who've reported back from near-death experiences have said that in these final moments objects seem to brighten in color, a beautiful, even peaceful parting vision, as if all things burned from within. That in itself is ironic, for the underlying and ultimate cause of all death is a breakdown in the body's oxygen cycle. We recognize the spark most clearly at the moment that we lose it. The flame burns brightest in the darkening room.

"He died in about ten minutes after we had moved him," his son later wrote, "but breathed his last so easy, that neither myself or my wife, who were both sitting close to him, perceived it at the time." He'd put his hand to his face, as Socrates had done as he lay dying, preventing those he loved from observing that final diminishment, shielding his eyes from a world that glowed from within.

EPILOGUE

The Burning World

LONG BEFORE THE TERROR, MICHEL DE MONTAIGNE ASKED whether a surfeit of death could make its approach easier. Witnesses observed a quiet fatalism in many of the guillotine's victims, as if, like Lavoisier, they regarded the last spectacle as a fitting finale for their lives. "I fancy there is a certain way of making [death] familiar to us," Montaigne ruminated, "and that may render us more confident and more assured. If we cannot overtake it, we may approach it and view it, and if we do not advance as far as the fort, we may at least discover and make ourselves acquainted with the avenues."

Eleven weeks after Priestley's flight and Lavoisier's execution, the reaction to the bloodshed set in. The Terror ended with the *coup d'état* of 9 Thermidor and the execution of Robespierre. Coffinhal, the judge credited with the line "The Republic has no need of scientists," was found disguised as a boatman: despite the pouring rain, spectators crowded his route to the scaffold and poked him in the ribs with umbrellas, daring him to "parry that thrust if you can." Fouquier-Tinville, the prosecutor who called Lavoisier a monster, was called the Monster himself: the executioner held his severed head aloft for several long minutes, turning it slowly so that all could see. All three were tossed in a mass grave like their victims.

As Priestley struggled in exile, the intellectual reaction to the Terror set in. In England, Burke remained the main spokesman, and he saw the Revolution as the work of a conspiracy of thinkers. Too much free thought would destroy the world. The power of reason was too small in men, said Burke, and they were best guided by custom and tradition. They should revere and bow to old institutions and time-tested formulae if they hoped to survive in this harsh new world.

The Continental reaction was just as unforgiving. To Frenchman Joseph de Maistre, a leading theorist of the Counter-Enlightenment, the *philosophes* were guilty of pride, the worst of the Seven Deadly Sins. All attempts to place man on a level with God could only lead to destruction. "The mad *philosophes,*" those apostles of tolerance, humanity, and wisdom, had sought to steer the universe and instead left France littered with corpses.

The reaction continues today. The historian Carl Becker, who once claimed that "Man is but a foundling in the cosmos, abandoned by the forces that created him," criticized the Enlightenment as an unrealistic quest for an earthly paradise. To the sociologist Herbert Marcuse, the Enlightenment dream to establish a Utopia only led to a more tightly controlled "administrative society." The French phenomenologist Michel Foucault observed that the era gave rise to the modern prison, asylum, and school—all improved and efficient institutions of state discipline and control. The age that resisted domination discovered new ways to dominate; the era that celebrated life ended in the first modern assembly-line method of mass execution. "The fully enlightened earth," write Theodor Adorno and Max Horkheimer, "radiates disaster triumphant."

Fourteen years after Priestley's death, Mary Shelley published *Frankenstein, or, The Modern Prometheus*, the tale that became the archetype for how the products of discovery return in unexpected, terrible form. One of the early reviewers commented on the tale's "air of reality": on its famous night of conception, Shelley and her fellow Romantics ranged in their talk over recent experiments, from Volta's electric batteries that seemed to vivify corpses to the dried worms of Erasmus Darwin that wriggled violently after soaking. Above all else, they spoke of chemistry—the new and modern chemistry that Priestley hoped would reveal the design of God and Lavoisier believed might reshuffle the building blocks of Nature. In Shelley's tale, it was that same chemistry that solved the mystery of life and death, inspiring Victor Frankenstein to give birth to his tortured creature.

There is a turning point in her book, when Victor listens to Professor Waldman's "panegyric upon modern chemistry," words that were "enounced to destroy me," he later said.

> The ancient teachers of this science [says Waldman] promised impossibilities and performed nothing. The modern masters promise very little; they know that metals cannot be transmuted and that the elixir of life is a chimera. But . . . they penetrate into the recesses of nature and show how she works in her hiding places. They ascend into the heavens; they have discovered how the blood circulates, and the nature of the air we breathe.

"Knowledge is power," said Francis Bacon, and scientists had listened for centuries. Victor Frankenstein listened, too. That night his mind grappled with these "words of fate" and he thought that "so much has been done . . . more, far more, will I achieve." At no other moment does the fictional scientist sound so like the living. Victor's ambition echoes a young Lavoisier's.

What are the moral costs of ambition like this, "to tear aside the veil of nature," as Priestley's disciple Humphry Davy often said? In *Frankenstein*, innocents paid the price of progress with their lives. Victor's younger brother William is murdered by the Monster; Justine, a maidservant, is wrongly executed for his death; Victor's bride and cousin Elizabeth is killed by the Monster when she enters the bridal chamber alone.

Mary Shelley was the daughter of William Godwin and Mary Wollstonecraft, two of Priestley's most able defenders when Burke attacked him in his *Reflections on the Revolution in France*, so she was not raised as a reactionary, afraid of change. Yet as the literary critic Harold Bloom has noted, the "shadow"—that deadly double of the self, the antithesis of pure intention—is the dark thread running through *Frankenstein*, as well as in the work of those Romantics who greeted the Revolution as the hope of the world. For every Prometheus there was a Pandora, and the Hope left waiting at the bottom of her box was the greatest evil of all. The discovery of oxygen, the beginning of modern chemistry, did not immediately turn to horror, but two centuries later the atomic theory that sprang from it would spawn the bomb, the "Apocalyptic Beast" that Priestley foretold. And for Priestley and Lavoisier, the fame that came with discovery exacted a terrible toll. The age that nurtured ideals of discovery and freedom destroyed the very ones that it inspired.

Priestley and Lavoisier were not the only idealists whose lives ended in murder or exile. "The time will therefore come when the sun will shine only on free men who know no other master but their reason; when tyrants and slaves . . . will exist only in works of history and on the stage," wrote their mutual friend Condorcet in his last book, *The Progress of the Human Mind*. Soon afterward he was arrested disguised as a peasant and slated for the guillotine. Before that appointment, he died in his cell, most probably of the poison he carried in his ring. Pierre du Pont fled to America, followed soon by his family. Bayen, Cadet, Lalande, Laplace—chemists and physicists who'd supported or opposed Lavoisier in the far less lethal battle over phlogiston—all fled or disappeared until the Terror was over. Tom Paine was imprisoned for challenging Robespierre over the execution of Louis XVI; he was saved from the guillotine by a jailer's mistake and died in exile in America. Lavoisier's assistant Séguin laid low like his colleagues, but abandoned chemistry forever. After the Terror, he made a fortune from producing waterproofed shoes for the army—a success that resulted in arrest under Napoleon.

Yet given the murder and exile of its crafters, the Enlightenment still realized its goals. Crowned heads fell or were diminished; the Church no longer ruled. In America, Church and State were separate; in France, the state took over Church property and made civil servants of the clergy. From the ashes of the *philosophes* rose the things they dreamed: the establishment of parliamentary institutions; the Declaration of the Rights of Man. The Enlightenment saw the end of all the hated ancient institutions—the feudal restrictions on individual activity, primogeniture, obligatory service to the lord of the manor—which were replaced by the two new secular faiths, Science and Liberty. Both challenged the ancient authority of King and Church, pledging allegiance instead to the individual and his "natural right" to unfettered thought. But all new faiths need martyrs, and Priestley and Lavoisier were ready-made.

▪ ▪ ▪

What is the role of a martyr?

Not in human terms. In that respect, martyrdom means grief and waste of human potential. Bitterness and anger remain for the survivors: those who feel within themselves a sense of destiny illustrate with their deaths the odd fact that martyrs cannot live easily in the society they hope to save. Pride is at work, for as Ernest Becker observes, "both sin and neurosis represent the individual blowing himself up to larger than his true size," and martyrs historically suffer from conflated opinions of themselves. There is also anger at the martyr for making those choices that lead to a defining suicide. Lacey Baldwin Smith has noted in her hagiography of Western martyrs that suicide is "an act of supreme egotism whether it be viewed psychologically—the most 'brutal way of making sure you will not readily be forgotten'—or theologically—the usurpation of divine authority." What is more single-minded than a planned and public suicide?

No, what is the role of the martyr in establishing a religion or cause? Would Christianity be remembered without an angry rabbi's high-profile act of defiance? Would modern India have fused together at a critical moment without the assassination of Mahatma Gandhi, or the American civil rights movement without the death of his philosophical offspring, Martin Luther King Jr.? And what about the Arab terrorists, often no more than boys or girls, in America, Palestine, Israel, Iraq, and Lebanon, certain that the bomb strapped to their chest or in the vehicle with them guarantees a place in Paradise? Taken from the viewpoint of the group, the "pale martyr in his shirt of fire" is an effective tool. As a rallying point for the uncertain, a larger-than-life symbol for the yet-to-be-convinced, the martyr becomes the most credible possible exponent of a religion's importance with his or her very sacrifice. Life's value inflates when juxtaposed to death; the martyr loses his individuality to become a process that justifies the group or creed. It has been said that martyrs create faith, not the opposite. In that sense, a martyr is the oxygen needed to feed the dwindling flame.

Martyrs are also a phenomenon that rarely appears singly. They tend to be the offspring of a society in conflict with itself; their defiance to the status quo broadcasts serious rifts in that society's definitions of justice, allegiance, duty, and reality. War and discord are the breeding ground for martyrs, and the Age of Revolution was filled with both. As a result, new martyrs were frequently proclaimed. Marat was a martyr to the Jacobins; Charlotte Corday to the Girondins. The poor Dauphin was a martyr to monarchists and, some said, to innocence. Soon after their deaths, Priestley and Lavoisier received similar labels: Lavoisier in the name of Science, and Priestley in the name of Liberty.

Neither sought martyrdom. Priestley fled to escape imprisonment, while Lavoisier would have scoffed at such romanticism. But both were aware of the theater of their lives. Priestley seemed most aware of the "martyr-potential" of his struggles because of his religious training, and seemed at times to model himself after two predecessors. The first was Michael Servetus, a Spanish physician who first described the blood's circulation through the lungs. Servetus was burnt at the stake in 1553 for his criticism of the Holy Trinity, a fate that cast him as an early Unitarian. Prior to the Birmingham riots, Priestley was advised in print that his story might echo Servetus's if he didn't choose his words more wisely. Priestley responded that he would welcome such an end, since the Spaniard's death signaled in Europe the birth of religious toleration.

Priestley's more intriguing model was Socrates. The Greek Sophist cemented for the West a martyr's proper performance, and Priestley's last work as he was dying was a comparison of Socrates and Christ—their lives, works, and martyrdom. Priestley's last words concerned the accuracy of this text; his final act was to cover his face, as Socrates had done. The English blamed Priestley for introducing "dangerous" foreign influences into mainstream thought, as the Athenians did with Socrates. The reasons given for Socrates' execution—"Refusing to recognize the gods recognized by the state," "Introducing other new deities," and "Corrupting the youth"—could be lifted from those blamed on Priestley. Both men dared inquire into all things "under the earth and in heaven," an act considered treasonous since it weakened faith in the Church and State during times of war. And both had a gift for attracting attention, a strength and fatal flaw.

Soon after Priestley's death, and in the years following, there were tributes to his life and death, pilgrimages to significant sites, and memorials thrown up in Leeds, Northumberland, and Birmingham. In 1813, nine years after his death, John Adams and Thomas Jefferson still argued about him in their final correspondence. Who could understand the old scientist's religious beliefs, complained Adams: his references to the "Apocalyptic Beast," his increasing mysticism? There was an *otherness* to Priestley that Adams found deeply troubling . . . somehow not made for this world. And yet, Adams said, he missed the old man, and the world would have been poorer without him: "This great, excellent, and extraordinary Man, whom I sincerely loved, esteemed, and respected, was really a Phenomenon: a Comet in the System, like Voltaire."

Of the two rivals, Lavoisier's treatment after death was the more ironic and absurd. In August 1796, Lavoisier became doubly memorialized as a martyr to science and Jacobin brutality. That month, although Lavoisier's remains were still hopelessly mixed with those of his accusers and fellow victims, the Lycée des Arts arranged a huge funeral with three thousand in attendance. Above the Lycée's entrance hung the words *"A l'immortel Lavoisier"*; in the great hall, the windows were darkened with black tapestries hung with wreaths and ermine. Twenty funeral lamps and a chandelier decorated with cypress branches lit the scene. Every pillar in the hall held a shield with an inscription to some of Lavoisier's discoveries. As a curtain was drawn to reveal a tomb surmounted by a statue of Liberty, a choir one hundred strong sang a doleful cantata. When the last verse ended, a large bust of Lavoisier crowned with a wreath was lowered from above.

Lavoisier's former colleagues, now accused of abandoning France's greatest scientist, tripped over each other to turn him into a hero. Modern chemistry was a French creation, and *its* creator was Lavoisier. The loudest in praise was Antoine Fourcroy, who gave the funeral oration, hounded even then for failing to protect his mentor. "Carry yourself back to that time," he pleaded, "when the least word, the slightest mark of solicitude for the unfortunate beings who were preceding you along the road to death, were crimes and conspiracies." Just as there were no Nazis after the fall of the Reich, there were no Jacobins after 9 Thermidor.

> Think back to those dreadful times when Lavoisier perished along with so many other illustrious martyrs to liberty, knowledge, talents, and virtue . . . when terror created a vast distance between people, even friends; when it isolated families from each other, even within their own homes. . . . Reread those fatal pages of our history and reply to those who dredge up from those horrible sacrifices, perfidious doubts, or still more criminal slanders against men who supposedly had some power or influence to stop these executions. From the tyrant's viewpoint did not these men, by their works and lives completely dedicated to public service, merit the same fate as Lavoisier? Were they not already under the shadow of arrest? . . . Had not the judge-executioner proclaimed that the republic had no need of scientists, and that a single intelligent man sufficed to run its affairs?

Marie did not attend. She never forgave her husband's former colleagues for deserting him; she would not honor them in their charade. She'd retired to her home on the day of execution, having lost her husband and father simultaneously. A month later she was jailed and her furniture confiscated; the only thing that saved her from execution was timing, for the Jacobin tyrants were judged and sent to the guillotine while she still lingered in jail. When released on August 17, 1794, she was penniless, homeless, and saved from total destitution only by the kindness of an old servant, who gave her a roof and food.

In December 1795, the wheel turned again. That month, the widows and children of the executed Farmers received full restitution for their confiscated properties. Now she was a wealthy widow, and she dedicated herself to canonizing Antoine's memory. She resurrected his unfinished *Memoirs of Chemistry*, which he'd worked on to the last, and published them finally in 1805. She had many suitors, including a spurned Pierre du Pont, but in the same year as the *Mem-*

oirs' publication, she married a second chemist, the American Benjamin Thompson—an expert in the physics of heat and light better known to the world as Count Rumford.

Maybe she thought chemistry held the secret to happiness; maybe she thought she could once again be a research partner, as she had with Antoine. Perhaps the count tired of comparisons, for he was neither a modest nor a sociable man. Whatever the reasons, the marriage turned into a famous disaster. A few months after the wedding, Rumford wrote to his daughter that "peace no longer dwells in my habitation." He called Marie "imperious," "tyrannical," and a "female dragon"; her opinions of him are not preserved, but can be guessed from her actions. Once, when discovering that Marie had invited yet another large party of guests "for the sole purpose of vexing me," Rumford stormed to the porter's lodge and ordered the man to allow no one on the grounds, then locked the gates and took the keys. Marie came down next and yelled to her arriving guests across the brick wall. Enraged, she poured boiling water on Rumford's flowers. The count liked his flowers, and the marriage ended after four years.

She lived another twenty-seven years, presiding over a thrice-weekly gathering of artists, scientists, and poets that was described as France's last surviving example of an eighteenth-century *salon.* Almost up to the day of her sudden death in 1832, at age seventy-eight, she greeted her guests while curled up on a loveseat, dominating the conversation with a mixture of charm and rudeness that captivated everyone. It was like stepping back in time, one visitor remembered—David's huge double portrait of Antoine and Marie hung in the drawing room above the *salon,* dwarfing the attendees as if time had not passed and the Revolution never occurred.

One hundred years after his death, Lavoisier, like Priestley, got a statue. And just as art imitates life, Lavoisier's effigy did not survive as long as his rival's. Funded by an international subscription begun in the 1890s to commemorate Lavoisier's work and death, it was discovered after the unveiling that the face was really that of the marquis de Condorcet. It didn't matter, joked the French: all men in wigs look the same. It was eventually melted down for bullets during World War II.

Lavoisier and Priestley would appreciate the irony. Modern chemistry was the most fitting tribute to their work, not monuments, spectacle, or tributes to martyrdom. "It's true that by blundering about we stumbled on gold," wrote Francis Crick, co-discoverer of DNA's structure, "but the fact remains that we were looking for gold." Priestley and Lavoisier blundered in the same way, and they, too, delved for the secret of life. Their lucky strikes made possible all the wonders that followed: from the molecular biology that allowed Watson and Crick's vision of the double helix to the strontium carbonate, magnesium, barium chlorate, and sodium oxalate mixed and lit in the red, white, and blue fireworks for Bastille Day and the Fourth of July.

Few other relics remain of that rivalry: Lavoisier's Arsenal lab in Paris, Priestley's quiet manor on the Susquehanna, some scientific instruments in archives. What matters in the end are ideas. The relics barely speak to the new world the two brought into existence, a modern reality whose chemical manipulation began with the discovery of oxygen and brought with it comfort and carnage. New modes of seeing and thinking begin in such simple ways: one has to wonder what future revolutions stare us in the face, but which we refuse to see. This one began with a candle and a mouse, and led in ways that are beautiful and deadly to a burning world.

Marie could see it very clearly when she died. "The Old Turk," young men called her, with her aged, masculine face, her enormous calves and extra pounds. The years were not kind to her, as they had not been kind to Priestley or Lavoisier. But she was determined to enjoy herself to the end. "She often gave elegant balls that we enjoyed more than the [social] visits," said Adrien Delahante, grandson of the surviving Farmer General, "in spite of her active surveillance and the severity with which she chased us away from the buffet to have us dance the quadrille."

One does not linger in the dance, or sulk on the sidelines. As she told a guest: "You have to have lived under the vacuum pump to appreciate the luxury of breathing."

DRAMATIS PERSONAE

Jean Le Rond d'Alembert (1717–1783)—French mathematician, Encyclopédiste, and *philosophe,* whose work in mechanics is still considered groundbreaking.

Francis Bacon (1561–1626)—English natural philosopher and author, who developed the inductive method and whose work *Novum Organum* influenced the use of accurate observation and experimentation in science.

Anna Laetitia Aiken Barbauld (1743–1825)—One of the most popular British poets of her day, she was the friend of Mary Priestley when both were young girls at Warrington Academy. One famous poem imagined what it must be like to be a mouse in a Joseph Priestley experiment. Largely forgotten today, she was admired by Wordsworth and Coleridge.

Pierre Bayen (1724–1798)—French chemist and apothecary for the French army, Bayen was Lavoisier's rival in the race to discover oxygen. He may have isolated the gas before Priestley or Lavoisier, but did not realize that he had done so.

Claude-Louis Berthollet (1748–1822)—One of Lavoisier's Four Horsemen of the New Chemistry, he was the first French chemist to accept the oxygen theory. His method of bleaching textiles with chlorine became a standard of the industry.

Joseph Black (1728–1799)—Scottish professor of natural philosophy at the University of Edinburgh, who discovered carbon dioxide, or what he called "fixed air." One of the founders of the British branch of chemistry known as pneumatic chemistry.

Matthew Boulton (1728–1809)—British industrialist whose foundry was one of the largest in England and who developed, with James Watt, the steam engine. A member of the Lunar Society, he was called "the iron chieftain" by James Boswell.

Robert Boyle (1627–1691)—British physicist and chemist; an early proponent of the scientific method. Boyle was the first scientist to isolate a gas, and his improvements on von Guericke's air pump led to the British field of pneumatic chemistry. Boyle's *The Sceptical Chemist* (1661) was one of the first works to attack Aristotle's theory that all matter is composed of fire, air, water, or earth, and propose instead an early form of atomic theory.

Edmund Burke (1729–1797)—British statesman and orator. His *Reflections on the Revolution in France* (1790) encouraged European rulers in their hostility to the French Revolution. It attacked Priestley and other English radicals as traitors to English freedoms, and initiated a pamphlet war that spawned much of our political vocabulary today.

Henry Cavendish (1731–1810)—A reclusive British scientist who discovered hydrogen and the composition of water. A friend of Priestley's until after the Birmingham riots, he was said to be the richest man in England.

Jean-Baptiste Coffinhal (1762–1794)—Chief judge of the Revolutionary Tribunal, who reportedly said during the defense of Lavoisier, "The Republic has no need of scientists."

abbé Etienne Bonnot de Condillac (1715–1780)—French philosopher whose theory of "sensationalism," in which he argued that all human knowledge and experience are derived from sense perception alone, influenced the later science of psychology.

marquis Marie-Jean de Condorcet (1743–1794)—French mathematician, philosopher, politician, and mutual friend of Priestley and Lavoisier, who outlined the progress of the human race, beginning with the primitive, through nine stages. Through education, a tenth stage—perfection—could be attained. Condorcet was arrested while trying the flee the Terror. He was thrown in jail, and the next day found dead, presumably from poison he carried in his ring.

Thomas Cooper (1759–1839)—English radical and friend of the Priestley family, Cooper was a co-partner with young Joseph Priestley in the failed land scheme to establish a "pantisocracy" in northern Pennsylvania. He settled in Northumberland, near the Priestleys, and became such an effective editor of a pro-Jefferson newspaper that he was the only person jailed during the Adams administration under the Alien and Sedition Acts. Cooper ended his days as a professor of chemistry at Dickinson College in Carlisle, PA, and at the University of Pennsylvania in Philadelphia.

Charlotte Corday (1768–1793)—French patriot who assassinated Marat.

Georges-Jacques Danton (1759–1794)—French minister of justice during the Revolution and Terror, who ruled the Commune with Marat and Robespierre. The last hope of the moderates, his execution in 1794 led directly to the overthrow and death of Robespierre and the end of the Terror.

Jacques-Louis David (1748–1825)—French neoclassical painter, known for his *Death of Marat* and his huge portrait of Antoine and Marie Lavoisier; taught drawing and painting to Marie.

Humphry Davy (1778–1829)—Renowned English chemist and disciple of Priestley, best known for his experiments in electrochemistry and his invention of the miner's lamp. He discovered boron, proved that diamonds are composed of carbon, showed that "rare earths" were oxides of metals rather than elements, and proved that hydrogen—not oxygen—caused the characteristics of acids.

René Descartes (1596–1650)—French mathematician and *philosophe,* known as the father of modern philosophy, whose statement "I think, therefore I am" became the motto of the Enlightenment.

Denis Diderot (1713–1784)—French *philosophe,* novelist, dramatist, and critic, whose vast, 35-volume *Encyclopédie*, based on reason and rationalism, became a powerful propaganda weapon against ecclesiastical power and semifeudal privilege.

Pierre-Simon du Pont de Nemours (1739–1817)—French economist and statesman, friend of Antoine Lavoisier, and rumored lover of Marie. He fled to America during the Revolution, urging the Lavoisiers to flee with him. His son, Eleuthère Irénée (1771–1834), stayed with the Lavoisiers as a kind of unofficial nephew and worked with Antoine in the Gunpowder Administration. When Pierre-Simon fled to America, he began the successful Du Pont gunpowder mills in the Delaware Valley.

Antoine-Quentin Fouquier-Tinville (1746–1795)—French politician and Public Prosecutor at Lavoisier's trial.

Antoine-François Fourcroy (1755–1809)—One of Lavoisier's Four Horsemen of the New Chemistry, he was the publicist of the New Chemistry. During the Revolution and Terror, he rose to a position of power and was later accused of failing to save Lavoisier's life. Afterward, his textbook on the New Chemistry helped cement the oxygen theory into modern modes of thinking.

Benjamin Franklin (1706–1790)—American statesman, journalist, and scientist. He was Priestley's first patron and his sponsor for election into the Royal Society of London. Franklin became the apotheosis of the "enlightened" man during his later years, and a cult of Franklin included children named after him.

George III (1738–1820; reigned 1760–1820)—The British monarch during Priestley's adulthood, who saw Priestley as an enemy of the state because of his Dissenting faith and civic protest. Stricken with the genetic disease porphyria, his first extended bout of "madness" in 1788 led to the regency crisis as, across the English Channel, France was on the brink of Revolution.

Otto von Guericke (1602–1686)—German physicist who invented the vacuum pump and electrical spark generator.

Jean-Etienne Guettard (1715–1786)—French geologist and mineralogist who encouraged the young Lavoisier in the sciences and took him to the Vosges Mountains to survey mines and natural resources for a nationwide atlas.

Joseph-Ignace Guillotin (1738–1814)—French physician and opponent of capital punishment, who convinced the National Assembly to approve a "humanitarian machine" for painless executions. The resulting *guillotine* was named after him, to Guillotin's lifelong chagrin.

abbé Nicolas-Louis de Lacaille (1713–1762)—French astronomer and Lavoisier's teacher at the Collège Mazarin. He was so efficient as an astronomer that he named 15 of the 88 constellations, and from 1750 to 1754, while studying the stars of the southern hemisphere from the Cape of Good Hope, observed and catalogued over 10,000 stars using a ½-inch refractor.

comte Joseph-Louis Lagrange (1736–1813)—French mathematician and astronomer, in charge with Lavoisier of the commission to establish a new system of weights and measures.

Joseph-Jérôme Lalande (1732–1807)—French astronomer who predicted that the world would come to a cataclysmic end in 1789.

Pierre-Simon de Laplace (1749–1827)—French mathematician, astronomer, and physicist who was an early proponent of Lavoisier's New Chemistry; worked with him on the guinea pig trials on respiration; told Napoleon that God was a superfluous hypothesis.

Antoine-Laurent Lavoisier (1743–1794)—French chemist and financier whose "oxygen theory" overthrew the ancient theory that all matter was composed of fire, air, water, or earth. His oxygen theory initiated the Chemical Revolution and led eventually to the atomic theory of matter.

Marie-Anne Paulze Lavoisier (1758–1832)—Lavoisier's wife and lab partner, whose scientific *salon* was one of the chief instruments of propaganda in the war to establish the New Chemistry. She went on to publish Lavoisier's works and correspondence after his execution. A brief marriage to Count Rumford was notable for its bitterness; she never married again, but was known for holding one of the last eighteenth-century *salons* in post-Revolutionary France.

Theophilus Lindsey (1723–1808)—A convert of Priestley's to Unitarianism, Lindsey was the founder in 1774 of England's first Unitarian church at the Essex Street Chapel in London.

Louis XVI (1754–1793)—Grandson of Louis XV, ascended the throne in 1774. Promised to liberalize France, but was instead guillotined during the Revolution.

Pierre-Joseph Macquer (1718–1784)—Foe of the New Chemistry, with Guillaume-François Rouelle, one of the most influential French chemists and phlogistonists in midcentury. He demonstrated phosphorus and wrote a treatise on dying wool that became a standard in the industry.

John Hyacinth Magellan (João Jacinto de Magalhães) (1722–1790)—The former Portuguese monk, said to be descended from the Great Navigator, who befriended Priestley and became one of France's most effective scientific spies.

Jean-Paul Marat (1743–1793)—French journalist and revolutionary who blamed Lavoisier for all his scientific failures. His assassination led to the Reign of Terror; historians have blamed his hatred of Lavoisier for the scientist's execution.

Franz Anton Mesmer (1734–1815)—Austrian physician whose demonstrations of "Mesmerism," or manipulations of a magnetic fluid said to pervade the universe, took Paris by storm during the 1780s. A commission of French academicians led by Lavoisier and Benjamin Franklin demolished his claims during their investigation. Both Mesmer and the commission overlooked Mesmer's inadvertent discovery of hypnotism, however; its reinvestigation during the nineteenth century would lead to the science of psychology.

Jean-Charles-Philibert Trudaine de Montigny (1733–1777)—A noted French chemist and friend of Lavoisier, Trudaine de Montigny was director of the Bureau of Commerce and spy handler of John Hyacinth Magellan. From Magellan's letters, he was the first to realize the strategic importance of Priestley's discoveries and directed the French chemists, and especially his friend Lavoisier, to investigate the Englishman's findings. In effect, he began an arms race between France and England, and set Lavoisier on his course for the study of gases that resulted in the oxygen theory. When Priestley visited Paris in 1774, Trudaine de Montigny was a constant host, assuring that the talkative Priestley gave away his secrets of "pure air."

Guyton de Morveau (1737–1816)—One of the Four Horsemen of the New Chemistry, a trained French lawyer who was the driving force behind Lavoisier's 1787 *Nomenclature*.

Marie-Joseph du Motier, marquis de Lafayette (1757–1834)—French general and statesman who served in America; he was in charge of the National Guard during the early stages of the French Revolution before fleeing the country.

Isaac Newton (1642–1727)—English mathematician and natural philosopher, considered one of the greatest scientists in history: he was the co-inventor of calculus, solved the mysteries of light and optics, formulated the three laws of motion, and derived from them the law of universal gravitation.

Jacques Paulze de Chasteignalles (d. 1794)—Farmer General, director of the French East India Company, father of Marie-Anne Paulze, and father-in-law of Antoine Lavoisier. Paulze was executed immediately before Lavoisier, on May 8, 1794.

William FitzMaurice Petty, 2nd Earl of Shelburne (1737–1805)—British prime minister who was sympathetic to the American colonies and one of the most liberal politicians of his day. For this reason, he was hated by George III and loved by Dissenters. Shelburne was Priestley's patron from 1773 to 1780, during which time Priestley did his most important scientific work, including the discovery of oxygen.

Richard Price (1723–1791)—Dissenting minister and friend of Priestley, whose sermon "Tremble, all ye oppressors of the world!" led Edmund Burke to publish his *Reflections on the Revolution in France*. Price was a member of the Royal Society and an early proponent of insurance, writing pamphlets on *Life Insurance*, *Contingent Reversions*, and the national debt.

Harry Priestley (1775?–1795)—Priestley's youngest son, but the first to die, of fever, in America.

Joseph Priestley (1733–1804)—English chemist, natural philosopher, and theologian, credited with the discovery of oxygen as well as at least eight other gases, and called one of the modern founders of the Unitarian faith. Fled England in 1794 due to increasing persecution, and resettled in Northumberland, PA, where he planned to start a Utopian "pantisocracy" with his sons.

Joseph Priestley Jr. (1768–1833)—Priestley's oldest son, who managed the failed land company with Thomas Cooper that attempted to establish a "pantisocracy" in Northumberland, PA. Priestley moved into the family house with his family and father on the Susquehanna River when his mother died in 1796.

Mary Priestley (1742–1796)—Wife of Joseph Priestley, involved in his work although it is uncertain how much she participated in the lab. She handled everything in Priestley's household, freeing him for his scientific and theological investigations; she was not as forgiving as her husband when they were forced by "Church and King" mobs to flee Birmingham.

Sally Priestley (1764–1803)—The first child of Joseph and Mary Priestley, born while her father taught at Warrington Academy. Sally was always a sickly child, and possibly died of tuberculosis. She was the only one of the Priestleys' four children to stay in England when the rest of the family emigrated to the United States.

Timothy Priestley (1734–1814)—Joseph Priestley's closest brother, a Calvinist minister with whom he endlessly debated doctrine and theology. Joseph's various heresies convinced the younger brother that Joseph, whom he admired for his brilliance and loved for his character, was still going to Hell.

William Priestley (dates of birth and death uncertain)—Priestley's second, troubled son, who stood up to the mobs in Birmingham, but was accused of trying to poison his father and brother's family in Northumberland, PA, in 1800. After this incident, he moved with his family to Louisiana, where he became a sugar planter and slaveowner.

Maximilien-F.-M.-I. de Robespierre (1758–1794)—The "Sea-Green Incorruptible," named for his sallow complexion, Robespierre was the revolutionary who came to personify the Reign of Terror. He was executed two months after Lavoisier.

Guillaume-François Rouelle (1703–1770)—French chemist and "demonstrator" at the Jardin du Roi from 1742 to 1768, he inspired an entire generation of French chemists. Although Rouelle was an adherent of the phlogiston theory, his instruction motivated Lavoisier to choose chemistry as a career.

Jean-Jacques Rousseau (1712–1778)—French *philosophe,* social theorist, and novelist, whose *Contrat Social* (1762) developed the case for civil liberty and prepared the ideological background for the French Revolution by defending the popular will against divine right.

Charles-Henri Sanson (1739–?)—Public executioner of Paris from 1788 to 1795, during the Revolution and height of the Reign of Terror. The post of executioner was a hereditary one: Sanson inherited it from his father, and passed it down to his son Henri, who held the office until his death in 1840. There is no record of the death of Charles-Henri.

Carl Wilhelm Scheele (1742–1786)—Obscure Swedish apothecary who apparently isolated oxygen before anyone else, but was not acknowledged because of his delays in publishing. Scheele, who may have died from the effects of breathing the fumes of chlorine gas, discovered more new compounds than any other chemist of the eighteenth century.

Georg Ernst Stahl (1660–1734)—German chemist whose "phlogiston theory" of matter was the dominant chemical paradigm during the mid-eighteenth century.

abbé Joseph-Marie Terray (1715–1778)—Comptroller General of France during the reign of Louis XV and great-uncle of Marie-Anne Paulze. His pressure to force the thirteen-year-old Marie into an unwanted marriage with the son of a political ally led directly to her marriage to a young and unattached Antoine Lavoisier.

Anne-Robert-Jacques Turgot (1727–1781)—French statesman and economist, one of Priestley's hosts when he visited Paris in 1774. Comptroller General of France under Louis XVI.

James Watt (1736–1819)—British inventor of the steam engine with Matthew Boulton; member of the Lunar Society. During the Great Water Controversy, he talked Priestley out of abandoning the phlogiston theory.

Josiah Wedgwood (1730–1795)—English potter and industrialist, member of the Lunar Society. One of Priestley's greatest supporters, he supplied the chemist with lab equipment in exchange for chemical analyses of clay and other materials used in his china and dinnerware.

John Wilkinson (1728–1808)—Mary Priestley's brother and Joseph's brother-in-law. One of the premier ironmasters in England, Wilkinson along with Abraham Darby designed and built the famous iron bridge at the small town called Ironbridge and the first iron boat in 1787. When he died, he was buried in an iron coffin.

CHRONOLOGY

1723—Georg Stahl theorizes that all substances are composed of water and three kinds of earth, one combustible, which he names *phlogiston.* Phlogiston theory becomes the predominant chemical worldview until the 1780s

March 24, 1733—Joseph Priestley born in Yorkshire—March 13 under "Old Style" calendar

Winter 1739–40—The Great Frost in England kills hundreds. Priestley's mother dies

1740–48—War of the Austrian Succession. Great Britain goes to war with France in 1744

1742—Priestley goes to live with his aunt, Sarah Keighley

August 26, 1743—Antoine-Laurent Lavoisier is born in Paris

1745–46—Second Jacobite Rebellion begins in Scotland and sweeps through the North and Midlands before faltering within reach of London. Priestley nearly dies of tuberculosis; because of his Calvinist faith, he is convinced he will go to Hell

1746—Lavoisier's mother dies. Two years later, Lavoisier's father moves his family into the house of his wealthy mother-in-law, Emilie Punctis, and her daughter Constance, who becomes Lavoisier's doting aunt

1746—Invention of the Leyden jar. The rise of the "electricians"

1751—Priestley enters the Dissenting academy at Daventry

1752—Benjamin Franklin "captures" lightning during a thunderstorm in Philadelphia

1754—Lavoisier inherits a fortune; enters the Collège Mazarin, the best school in Paris

1755—Priestley becomes a minister at Needham Market, Suffolk

1756–63—Seven Years' War. The Treaty of Paris ends the war on February 10, 1763

1758—Priestley moves to Nantwich, Cheshire; begins experimenting with new educational techniques

1759—"The Year of Miracles"—English fleet sweeps the seas; Robert Clive and 350 men conquer Bengal; England controls nearly one-third of the world's surface

1760—George II dies. Ascension of George III

1760—A watershed year in the British Industrial Revolution. Blast furnaces are improved for producing cast iron; the Bridgewater Canal is completed, halving the cost of coal in Manchester; the factory system and use of water-driven machinery is firmly established in the silk industry. England becomes the industrial powerhouse of Europe

1761—Priestley takes a teaching position at Warrington Academy, where he teaches classics and literature

June 1761—Lavoisier graduates from the Collège Mazarin

June 23, 1762—Priestley marries Mary Wilkinson, sister of ironmasters John and William Wilkinson

1763–65—Priestley is introduced to chemistry through Matthew Turner's lectures at Warrington

1763–67—Lavoisier studies chemistry under Rouelle

1764—The Priestleys' first child, Sally, is born

1765—Stamp Act imposes tax in American colonies

December 18, 1765—Priestley travels to London to propose his *History of Electricity* to Benjamin Franklin and other "electricians"

1766—Henry Cavendish discovers hydrogen

January 9, 1766—Priestley's first visit to the Royal Society as a guest of Ben Franklin

March 1766—Priestley's kite trials replicate Franklin's "capture" of lightning fourteen years earlier

June 12, 1766—Priestley elected a fellow of the Royal Society

1767—Priestley moves to Leeds; *History of Electricity* published

June 14, 1767—Lavoisier leaves with Guettard to investigate the geology of the Vosges Mountains.

1768—"Wilkes and Liberty" riots in London; Priestley publishes *First Principles of Government*. Two years later he publishes *Institutes of Natural*

and Revealed Religion. These two publications mark the beginning of the attacks against him by the Church of England, George III, and the Church and King Party.

May 2, 1768—Lavoisier joins the *Ferme générale* (General Farm)

May 18, 1768—Lavoisier is elected to the French Academy of Sciences

Summer 1768—Priestley experiments with gas escaping from fermenting vats of beer; invents soda water

1769—James Watt builds his first steam engine

1771–1772—Swedish apothecary Carl Scheele is the first researcher to isolate oxygen, but does not know what it is

August 17–27, 1771—Priestley's experiments with a sprig of mint and his discovery of the respiration of plants

December 16, 1771—Lavoisier marries Marie Paulze

1772—Lavoisier's experiments with sulfur and phosphorus; Daniel Rutherford discovers nitrogen

March 19, 1772—Priestley presents his experiments with gases to the Royal Society, including his invention of carbonated water. Later these will be published as *Experiments and Observations on different kinds of Airs* and will make him famous throughout Europe.

March 20, 1772—John Hyacinth Magellan informs Trudaine de Montigny, his spymaster in France, of Priestley's discoveries, initiating a scientific rivalry and arms race that will last for nearly two decades

July 14, 1772—Lavoisier is asked to test Priestley's findings, but his English is weak and he must pass on the request. This marks the beginning of Lavoisier's interest in gases

August 14, 1772–October 13, 1772—French Academy conducts 190 experiments to "dissolve" diamonds under a huge burning glass

November 1, 1772—Lavoisier delivers his "sealed note" *(pli cacheté)* on phlogiston and oxidation to the French Academy

1773—A rumor starts in Paris that the astronomer Lalande has calculated that a comet will swing close to the Earth in 1789, virtually destroying life on the planet

June 1773—Priestley arrives to serve as librarian and tutor for William Petty, 2nd Earl of Shelburne

November 30, 1773—Priestley wins the Copley Medal from the Royal Society of London for his invention of carbonated water and his experiments with various kinds of airs

February 1774—Pierre Bayen releases report which suggests he isolated oxygen before either Priestley or Lavoisier

May 10, 1774—Louis XVI ascends the throne

August 1, 1774—Priestley's first conscious isolation of oxygen, though he is not yet aware of its properties as a "purer" kind of air. Later that month, Priestley begins the Grand Tour of the Continent with Lord Shelburne

October 1774—Priestley and Lord Shelburne visit Paris on the Grand Tour. Sometime between October 15 and October 17, Priestley dines with the Lavoisiers and the Paris *philosophes;* during that meal he reveals his first inconclusive experiments with oxygen

November 2, 1774—Priestley returns to London

November 19, 1774—Priestley resumes his experiments with oxygen, but halts the tests in frustration

March 1, 1775—Priestley resumes his oxygen tests

March 8, 1775—Priestley's discovery of oxygen—his first tests with mice, one in "common" air and one in oxygen. The mouse in the jar of oxygen thrives twice as long as the mouse in the jar of common air

March 15, 1775—Priestley writes to inform the Royal Society of his discovery

April 19, 1775—War of American Independence begins

April 26, 1775—Lavoisier's "Easter memoir" on calcination is read to the French Academy

1776—American Declaration of Independence; publication of Adam Smith's *Wealth of Nations* and the first volumes of Edward Gibbon's *Decline and Fall of the Roman Empire*

April 1776—Lavoisier moves to the Arsenal to assume control of the Gunpowder Administration; the French crash course on gunpowder production begins. That same month, he conducts his twelve-day experiment, which links Priestley's "pure air" with combustion and is the beginning of Lavoisier's oxygen theory

1777—Carl Scheele publishes *Air and Fire*, which suggests he was the first to discover oxygen, but the book is released so late that he is out of the running

August 1778—Lavoisier announces his oxygen theory and new theory of combustion to the French Academy

1779—Priestley releases the first volume of his revised *Experiments and Observations on different Kinds of Airs* in three volumes (1779–86)

June 5–10, 1780—Gordon Riots rage in London. They begin when Lord George Gordon marches on Parliament at the head of a mob carrying a petition protesting relief granted to Catholics. By night, the demonstration turns into a full-scale riot that rages for five days

1780–81—Britain fights wars all over globe—against the French in India, West Indies, North America, and Africa; against the Spanish in Gibraltar, the Balearic Islands, the Caribbean, and Florida; against the Dutch in Ceylon and West Indies; against the Americans in the colonies

September 1780—Priestley leaves Shelburne's employment and moves to Birmingham

October 1781—General Cornwallis surrenders at Yorktown

1782—Priestley publishes *History of the Corruptions of Christianity*. Lord Rockingham ministry appointed, which includes Shelburne in the cabinet

1782–83—The Great Water Controversy, during which the composition of water is established by Henry Cavendish. Lavoisier's oxygen theory gains ground among European scientists—the beginning of the end for phlogiston theory

1783—Lavoisier and Laplace test their theory of respiration on a guinea pig. Peace treaty between Great Britain and the United States signed

1785—Priestley's "Gunpowder Sermon"

June–July 1785—Lavoisier reads out his *Reflections on Phlogiston* during two consecutive meetings of the Academy. It is a manifesto for his oxygen theory, and the war against phlogiston officially begins

1786—Watt smuggles from France news of Berthollet's process for the chlorine bleaching of textiles, which is incorporated into the English textile industry

1787—First Constitutional Convention in Philadelphia

May 1787—Lavoisier publishes the *Méthode de Nomenclature*

1788—George III's madness; regency crisis

1789—George Washington elected president of the new United States; William Blake publishes *Songs of Innocence*

January 1789—Louis XVI summons the Estates-General; the Third Estate adopts the title of National Assembly

July 14, 1789—Fall of the Bastille, recognized as the beginning of the French Revolution

August 1789—Publication of Lavoisier's *Traité élémentaire de chimie*, the first clear and systematic synthesis of his Chemical Revolution

November 5, 1789—Richard Price gives his sermon "Tremble, all ye oppressors of the world!" at London's Old Jewry

November 1790—Edmund Burke publishes *Reflections on the Revolution in France*

March 1791—*Ferme générale* dissolved

June 22, 1791—French royal family apprehended at Varennes during an escape attempt

July 14–18, 1791—Birmingham mob burns Priestley's home and lab; "Church and King" riots spread into surrounding countryside

1792—Attack on the Tuileries and the royal family held prisoner; French Republic proclaimed; Thomas Paine publishes *Rights of Man*

March 25, 1792—Guillotine approved as official method of execution

April 20, 1792—France declares war on Austria, the beginning of the French revolutionary wars which will continue, with brief truces, until 1815

August 26, 1792—Priestley made honorary citizen of France

September 2–7, 1792—"September Massacres" in Paris

September–November 1792—Antoine and Marie Lavoisier stay on their estate in Fréschines, far from Paris, and contemplate flight from France. In the end, Lavoisier returns to Paris in an attempt to save the embattled French Academy and head off the "annihilation of French sciences"

January 21, 1793—Louis XVI is executed

February 1793—Food riots ravage Paris

February–March 1793—War declared in Paris against Spain; west of France rises up in counterrevolution; preparations begin in Paris to create a Committee of Public Safety

February 1, 1793—Great Britain declares war on France; the British government begins to arrest anyone publishing criticism of the government

July 13, 1793—Charlotte Corday assassinates Jean-Paul Marat in his bath

August 8, 1793—French Academy of Sciences dissolved

September 17, 1793—The Law of Suspects is passed by the Assembly; Reign of Terror begins

October 17, 1793—Marie Antoinette executed

November 24, 1793—Warrant signed for Lavoisier's arrest; he is imprisoned in Port-Libre prison

April 7, 1794—Priestley emigrates to America

May 8, 1794—Antoine Lavoisier executed

July 1794—Fall and execution of Robespierre; the Reign of Terror ends

December 11, 1795—Harry Priestley dies in Northumberland, Pennsylvania

1796—Priestley founds America's first Unitarian church in Philadelphia

September 19, 1796—Mary Priestley dies in Northumberland, Pennsylvania

1798—Battle of the Nile. Nelson's victory assures supremacy of the British fleet

1799—Napoleon becomes First Consul

April 1800—William Priestley accused of trying to poison his father and brother's family. Joseph and William both deny the accusations, but William flees the state to resurface as a sugar planter and slaveowner in Louisiana

February 17, 1801—Thomas Jefferson is declared third president of the United States

February 6, 1804—Priestley dies in Northumberland, Pennsylvania

1818—Mary Shelley publishes *Frankenstein*

GLOSSARY OF CHEMICAL, HISTORICAL, AND SCIENTIFIC TERMS

Acid air—early term for hydrogen chloride

Alkaline air—ammonia

Aqua fortis (also *Spirit of nitre*)—nitric acid

Calcination—eighteenth-century term for an oxidation reaction

Caloric (French *calorique*)—during the eighteenth century, a theoretical fluid found in all bodies that contained the principle of fire and combustion. First posited by Joseph Black, then incorporated into the original oxygen theory by Antoine Lavoisier

Calx—eighteenth-century term for an oxide of a substance, usually used in reference to the oxide of a metal

Choke-damp—the carbon dioxide found in mines, notorious in the eighteenth century for killing miners

Combustion—a chemical reaction in which a compound, usually containing carbon, combines with the oxygen gas in the air. The reaction's most visible and common manifestation is burning

Dephlogisticated air—oxygen

Dephlogisticated nitrous air—nitrous oxide

Fire-damp—methane

Fixed air—carbon dioxide, first named by Scottish chemist Joseph Black

Fluor acid air—silicon tetrafluoride

Inflammable air—hydrogen, discovered by Henry Cavendish in 1766

Marine acid air—hydrogen chloride

Mercuris calcinatus per se—alchemical term for mercuric oxide, also known as "mercury *calx*" during the eighteenth century

Natural philosopher (see *Savant*)—British equivalent of "scientist" during the seventeenth and eighteenth centuries. The difference between a British natural philosopher and a French *savant* was basically in one's relationship to the question of God. A natural philosopher believed that his purpose was to reveal God's natural laws and use that knowledge for man's improvement and comfort. Although a *savant* was not necessarily an atheist, many professed to be; more accurately, God was an immeasurable force, and as such had no place in the calculations of an increasingly quantifiable science

Nitre—potassium nitrate

Oil of vitriol—strong sulfuric acid

Oxidation—a chemical reaction in which a substance loses electrons in going from reactant to product

Oxygen—the most common element on earth, an odorless, tasteless gas that comprises about 21 percent of the atmosphere, is a constituent of water and most minerals, and is involved in such basic chemical and physiological processes as combustion, oxidation, and human respiration. The eighth element on the Table of Elements, atomic number 16. Discovered in 1774 by Joseph Priestley and named by Antoine Lavoisier; also by Carl Wilhelm Scheele and Pierre Bayen

Philosophe—a writer, scientist, philosopher, or thinker during the French Enlightenment devoted to the use of pure reason for the discovery of truth, establishment of liberty, and dissolution of superstition

Philosopher's stone—an imaginary substance or chemical preparation believed by alchemists to have the power of transmuting baser metals into gold

Phlogisticated air—nitrogen

Phlogisticated nitrous air—nitrous oxide

Phlogiston—during the eighteenth century, the flammable principle common to all fuels and oxidized metals that was released during combustion and oxidation. Fire was the visible sign of phlogiston; fuels that burned with almost no residue were considered pure phlogiston. First expressed by the German chemist Georg Ernst Stahl, phlogiston theory was an accepted worldview holding, like the ancient Greeks, that all matter was composed of some combination of four basic elements; unlike the Greeks, who believed that the four elements were fire, water, air, and earth, phlogiston theory held that matter was made of water and three kinds of earth, one of which was flammable

Pure air—oxygen

Putrid air—common air in which animal or vegetable matter has been allowed to putrefy

Redox reaction (see *Oxidation* and *Reduction*)—a reaction in which there is a simultaneous transfer of electrons from one substance to another. A redox reaction is composed of two "half-reactions": **oxidation** (a loss of electrons) and **reduction** (a gain of electrons)

Reduction—a chemical reaction in which a substance gains electrons in going from reactant to product

Sal ammoniac—ammonium chloride

Saltpeter—a natural source of potassium nitrate, the principal ingredient in gunpowder; often found in damp places like cellars

Savant—the French equivalent of "scientist" during the eighteenth century

Spirit of nitre (also *aqua fortis*)—nitric acid

Vitiated air—air not fit for respiration, as a result of combustion, respiration, or some other "phlogistic" process

Vitriolic acid air—sulfur dioxide

NOTES

Prologue: God in the Air

3 "Scientists are maverick personalities": Jacob Bronowski, *The Origins of Knowledge and Imagination* (New Haven, CT: Yale University Press, 1978), pp. 119, 121–23.

3 "high-piercing . . . rock": This version of the Prometheus myth is from Hesiod, and is quoted extensively in Edith Hamilton, *Mythology* (New York: New American Library, 1942), p. 72.

9 promises to "treat, heal, eradicate and cure": "The House of Oxygen," URL: www.mroxygen.co.za.

9 "pervasive human-centered computing . . . in the air we breathe": "Project Oxygen," URL: http://oxygen.lcs.mit.edu/overview.html.

11 "so delicate that if": Edison's comments to *Scientific American* are quoted in "Edison's Ghost Machine," URL: www.ccspr.njit.edu/Faculty/cnap/EdisonPage2.

11 The essence of life left the body: Edison's belief in a vitalistic "soul" that lingered in the air after death to be eventually recycled, expressed in a letter to his friend John Lewis, is quoted in ibid.

12 "overturn in a moment": Joseph Priestley, in Isaac Kramnick, ed., *The Portable Enlightenment Reader* (New York: Penguin Books, 1995), p. xiii.

12 "that I was born so soon . . . beyond the antediluvian standard": Benjamin Franklin, in ibid., p. xiii.

13 a glance at the era's scientific journals: See Robert Darnton, *Mesmerism and the End of the Enlightenment in France* (Cambridge, MA: Harvard University Press, 1968), p. 18.

13 "I observed that while I thus desired": René Descartes, "I think, therefore I am," in Kramnick, ed., *The Portable Enlightenment Reader*, p. 184.

14 "The time of prophets has passed": Quoted in Ken Alder, *The Measure of All Things: The Seven-Year Odyssey and Hidden Error That Transformed the World* (New York: Free Press, 2002), pp. 77–78.

14 "Drunk for a Penny!": Daniel P. Mannix, *The Hellfire Club* (New York: Simon & Schuster, 1961), p. 18.

14 a "rat's-eye view of life": Ibid., p. 185.

15 "I don't know what effect": Quoted in ibid.

15 "riots about corn . . . riots of smugglers": Franklin quoted in ibid., p. 197.

16 "be able to live in a world out of joint": Thomas S. Kuhn, *The Structure of Scientific Revolutions. International Encyclopedia of Unified Science*, vol. 2, no. 2 (Chicago: University of Chicago Press, 1962, 1970), p. 79.

Chapter 1: The Cloth-Dresser's Son

20 "It was my misfortune": Joseph Priestley, *Autobiography of Joseph Priestley: Memoirs Written by Himself* (1806–1807; Cranbury, NJ: Associated University Presses, 1970; cited hereafter as *Memoirs*), p. 1.

20 Joseph, the eldest, was born: The date is under the Old Style Julian calendar. When the New Style Gregorian calendar was adopted in 1752 at the urging of the Earl of Macclesfield, president of the Royal Society of London, Priestley's birthday changed to March 24. Everyone's birthday changed under the new system that added a leap year to correct for a cumulative error in the old calendar, as well as to mold the year's rhythm to the Christian holy days.

20 "kneel down with him while he prayed": Timothy Priestley, *A Funeral Sermon Occasioned by the Death of the Late Rev. Joseph Priestley, etc.* (London: Alexander Hogg & Co., 1804), p. 36.

20 "the distinction of property": Priestley, *Memoirs*, p. 1.

20 "It is a sin, to steal a pin": Roy Porter, *English Society in the Eighteenth Century* (Harmondsworth, Middlesex: Penguin, 1982), p. 182.

21 "Let me go to that fine place": The death of Priestley's mother is in his *Memoirs*, p. 1.

21 "Last Thursday night, a most Melancholy": The report in the 1771 *Northampton Mercury* is quoted in Porter, *English Society in the Eighteenth Century*, pp. 27–28.

23 As early as 1746, there was a general belief: In 1746, traveler John White wrote: "The main body of Dissenters are mostly found in cities and great towns among the trading part of the people, and their ministers are chiefly of the middle rank of men, having neither poverty nor riches. . . . If I had a son brought up in any trade and had no consideration either for him or myself in another world, I should be ready to say to him at setting up, *My son, get Money and in order to do that, be a Dissenter.*" Quoted in ibid., p. 195.

23 "lost communion with God": W. R. Aykroyd, *Three Philosophers* (Westport, CT: Greenwood Press, 1970), p. 35.

23 "See how a man of God dies": Ibid.

24 she "took me entirely": Priestley, *Memoirs*, p. 70.

24 "who knew no other use" . . . "resort for all": Ibid., p. 70.

25 "with sparkling yet kindly eyes": John F. Fulton and Charlotte H. Peters, "An Introduction to a Bibliography of the Educational and Scientific Works of Joseph Priestley." Unpublished paper read at the December 31, 1936, meeting of the American Bibliographical Society, Providence, R.I. Joseph Priestley Collection, Dickinson College Library, p. 1.

27 "exceedingly good LIMONS and BITTER ORANGES": John Prebble, *Culloden* (New York: Penguin Books, 1977), p. 262.

27 "was nearly carried off by a complaint": Timothy Priestley, *A Funeral Sermon*, p. 36.

27 one good flushing of air: Laurie Garrett, *The Coming Plague: Newly Emerging Diseases in a World Out of Balance* (New York: Farrar, Straus & Giroux, 1994), p. 241.

27 "The first he made was on spiders": Timothy Priestley, *A Funeral Sermon*, p. 42. Priestley later said he did not think he started experimenting before 1765, when he was thirty-two, but Timothy had the better family memory, while Joseph admitted that he was often forgetful of details.

27 "I felt occasionally such distress of mind": Priestley, *Memoirs*, pp. 71–72.

28 "I have crucified [Christ] to myself afresh": John Bunyan, *The Pilgrim's Progress, from this World to That which is to Come*, ed., James Blanton Wharey. Oxford: Clarendon Press, 1960, p. 35.

28 "I swallowed more physick than food": Edward Gibbon quoted in Porter, *English Society in the Eighteenth Century*, p. 27.

28 "never was human creature . . . so early an age": Timothy Priestley, *A Funeral Sermon*, p. 36.

29 "Before I went from home . . . not to be quite orthodox": Priestley, *Memoirs*, p. 73.

31 "Peter Annet, a poor schoolteacher": Annet believed in debate to the extent that in 1762 he was pilloried and sentenced to a year in jail for doubting the divinity of Christ and publishing a deistic journal.

31 "Since he did not like to linger": Priestley's strange habit of pacing is described in his *Memoirs*, pp. 142ff.

31 "Three years, from Sept. 1752 to 1755": Ibid.

32 "good, dull and decent": Porter, *English Society in the Eighteenth Century*, p. 76.

32 "port and privilege": Edward Gibbon and Horace Walpole quoted in Porter, *English Society in the Eighteenth Century*, p. 177.

33 "at the club, the chimney took fire . . . after the service": All journal entries for 1754 are included in Tony Rail and Beryl Thomas, "Joseph Priestley's Journal While at Daventry Academy, 1754," *Enlightenment and Dissent*, 13 (1994), pp. 49–113. The authors deciphered Priestley's journal notes from Annet's shorthand.

34 "My brother, when he began . . . pleased him": Timothy Priestley, *A Funeral Sermon*, p. 42.

Chapter 2: The Sums and Receipts of Parallel Worlds

35 a legacy of 45,600 *livres:* Jean-Pierre Poirier, *Lavoisier: Chemist, Biologist, Economist*, trans. Rebecca Balinski (Philadelphia: University of Pennsylvania Press, 1993), p. 4. Poirier calculated the conversion from eighteenth-century French currency to 1996 U.S. dollars as equaling about $1.8 million; this and all later conversions to present-day or inflation-adjusted dollars are calculated by using an "Inflation Adjustor" developed by Professor Robert Sahr of the Oregon State University Political Science Department and included in the Resources section of the *Columbia Journalism Review*'s Web site, URL: http://www.cjr.org.

35 "was bound to instill a taste": Poirier, *Lavoisier*, p. 5.

37 "the most beautiful and prestigious": Quoted in ibid., p. 5.

38 "increasing the comforts and enjoyments": Humphry Davy, *Elements of Chemical Philosophy*, quoted in Mary Jo Nye, "Philosophies of Chemistry Since the Eighteenth Century," in Seymour H. Mauskopf, ed., *Chemical Sciences in the Modern World* (Philadelphia: University of Pennsylvania Press, 1993), p. 8.

39 "I should pity the man . . . good reception anywhere": Arthur Young's comparisons of the French and English scientific societies are quoted in Aykroyd, *Three Philosophers*, p. 27.

43 "is different in every respect": Casanova's observation of this and the incident on Drury Lane are quoted in Porter, *English Society in the Eighteenth Century*, p. 21.

43 "Anything that looks like a fight": Henri Misson quoted in ibid., p. 23. For example, Drury Lane, the Broadway of London, was wrecked by dissatisfied theater patrons in 1743, 1750, 1755, 1763, 1770, and 1776.

43 "Violence was as English as plum pudding": Porter, *English Society in the Eighteenth Century*, p. 114.

43 "unwanted babies were left out in the street": Hibbert quoted in Colin Wilson, *The Mammoth Book of the History of Murder* (1990; New York: Carroll & Graf, 2000), p. 158.

44 "she was quite stupid and senseless": Porter, *English Society in the Eighteenth Century*, p. 35.

44 "dwelt in some far serene planet": Bessie Rayner Belloc, "Joseph Priestley in Domestic Life," from *In a Walled Garden* (1895), in "The Victorian Women Writers Project" on the Web, p. 36. URL: http://www.indiana.edu/~letrs/wwwp/belloc/walled.html.

44 "He impressed those about him": Ibid., p. 37.

45 "a kind of disjointed bird-like trot": Quoted in Jenny Uglow, *The Lunar Men: Five Friends Whose Curiosity Changed the World* (New York: Farrar, Straus & Giroux, 2002), p. 71.

46 "You are the man who brought me": This scene is recounted in Timothy Priestley, *A Funeral Sermon*, p. 41.

47 "She has spared no expense in my education": Belloc, "Joseph Priestley in Domestic Life," p. 39.

47 "Even my next neighbour": Priestley, *Memoirs*, p. 80.

47 "Without some such check as this": Ibid.

48 "Many more, Sir, than I should like to read": Quoted in J. R. Partington, *A History of Chemistry*, (London: Macmillan & Co., 1962), vol. 3, p. 244.

49 "This is most obvious in situations where society": Alice Weaver Flaherty, "Writing Like Crazy: A Word on the Brain," *Chronicle of Higher Education*, Nov. 21, 2003, p. B9.

50 a goal "which is not taken up" Francis Bacon quoted in F. W. Gibbs, *Joseph Priestley: Adventurer in Science and Champion of Truth* (London: Thomas Nelson & Sons, 1965), p. 13.

50 "fine ear" . . . "be more easily pleased": Priestley, *Memoirs*, p. 86.

51 "Beneath this stone my wife doth lie": Gibbs, *Joseph Priestley*, p. 20.

52 "the great societies of mankind": Quoted in ibid.

53 "upwards of 100 hands": Bernard de Mandeville and the letter in *The Gentleman's Magazine* are quoted in Sarah Jordan, "From Grotesque Bodies to Useful Hands: Idleness, Industry, and the Laboring Class," *Eighteenth Century Life*, vol. 25, no. 3 (Fall 2001), p. 69.

53 "grotesque objects of fear and revulsion": Ibid., p. 62.

53 the free thought of Presbyterians crystallized: Dan Eshet, "Rereading Priestley," *History of Science*, vol. 39, issue 2 (June 2001), p. 135.

54 "we have a fine knot of lassies": Quoted in Uglow, *The Lunar Men*, p. 73.

54 "I do not recollect that he ever shewed": Priestley, *Memoirs*, p. 15.

55 "We both shed tears": Joseph Priestley, Letter to John Seddon, April 9, 1762, Joseph Priestley Collection, Dickinson College Library.

55 "extremely intelligent and original": Belloc, "Joseph Priestley in Domestic Life," p. 41.

55 "like the sun in January": Quoted in Gibbs, *Joseph Priestley*, p. 88.

55 "a fiendish organizer": Uglow, *The Lunar Men*, p. 75.

55 "When he was near being married": Timothy Priestley, *A Funeral Sermon*, p. 40.

55 "I took the advantage" and the account that follows: Ibid.

56 "The hazard of bringing a person": Priestley, *Memoirs*, p. 16.

Chapter 3: The Gas in the Beer

59 "experimental notes and accounts": Boyle's recipe for nitric acid is reproduced in Robert Boyle, *Work-diary XXXVII*, Entry 2, URL: www.bbk.ac.uk/boyle/workdiaries.

59 "was that of a feather": Quoted in Gibbs, *Joseph Priestley*, p. 116.

60 Jean Jacques and Henri Bouasse quoted in Nye, "Philosophies of Chemistry Since the Eighteenth Century," in Mauskopf, ed., *Chemical Sciences in the Modern World*, n.p.

61 "the principal actors in the scene . . . were still living": Priestley, *History and present state*

of electricity. Preface to 1st ed. in 3rd ed. (1775; New York and London: J. Johnson reprint, 1966), p. xi.

62 Matthew Boulton said electricity was the fabled: Uglow, *The Lunar Men*, p. 15.

62 "I would like to tell you": Van Musschenboek quoted in "Capacitors," URL: www.hypertextbook.com/physics/electricity/capacitors.

63 "the kite, with all the twine": Letter from Benjamin Franklin to Peter Collinson, Oct. 19, 1752, quoted in "Franklin and His Electric Kite," URL: www.ushistory.org/franklin/kite.

63 "seized fire from the heavens": Turgot's description of Franklin is quoted in Kramnick, ed., *The Portable Enlightenment Reader*, p. xix.

64 "Do not mistake me": Edgar F. Smith, *Priestley in America, 1794–1804* (Philadelphia: P. Blakiston's Son & Co., 1920), pp. 5–6.

64–65 "fatigued with the incessant charging": Quoted in Gibbs, *Joseph Priestley*, p. 27.

65 "When the metal melted, he called": Timothy Priestley, *A Funeral Sermon*, p. 42.

65 "It was 6 feet, 4 inches wide": Ibid., p. 43.

66 "New worlds may open to our view": Quoted in the Introduction to Priestley's *Memoirs*, p. 19.

66 "We went to Stockport, Liverpool": Quoted in Gibbs, *Joseph Priestley*, p. 34.

67 "Having read, and finding by my own experiments": Quoted in ibid., pp. 30–31.

68 Franklin also sickened: Franklin's trials with bubbling water in New Jersey and England are described in his letter to Joseph Priestley, 1774, quoted in Harlow Shipley, Samuel Rapport, and Helen Wright, eds., *A Treasury of Science* (New York: Harper & Bros., 1943); see also the Web site "Franklin," URL: webserver.lemoyne.edu/faculty/giunta/franklin.html.

69 a growing industrial town: Quoted from a letter by Priestley to Richard Price, October 3, 1771, in Robert E. Schofield, *A Scientific Autobiography of Joseph Priestley (1733–1804): Selected Scientific Correspondence* (Cambridge, MA: MIT, 1966), p. 87.

70 "a liberal, friendly, and harmonious congregation": Quoted in Uglow, *The Lunar Men*, p. 170.

70 charm "away the bitterest prejudices": Lester Kieft, "The Diverse Lives of Joseph Priestley," *Foote Prints*, vol. 46, no. 1 (1983), p. 31.

70 "irradiating hope and enthusiasm": Fulton and Peters, "An Introduction to a Bibliography of the Educational and Scientific Works of Joseph Priestley," Joseph Priestley Collection, Dickinson College Library, p. 17.

71 "My dear woman . . . now quite comfortable": The incident of the woman and her electrical exorcism is described in John T. Rutt, *Collected Theological and Miscellaneous Works of Joseph Priestley*, 2 vols. (London: George Smallfield, 1832), vol. 2, p. 112.

72 "for though he had such natural": Timothy Priestley, *A Funeral Sermon*, p. 42.

72 one of the best-paid Dissenting ministers: Maurice Crosland, "A Practical Perspective on Joseph Priestley as a Pneumatic Chemist," *British Journal for the History of Science*, vol. 16, 3, no. 54 (1983), p. 224.

72 "to my number almost every year": Timothy Priestley, *A Funeral Sermon*, p. 42.

72 "had [Joseph] not called out": Ibid., p. 43.

72 "Though we then differed so wide": Ibid., p. 42.

72 "There is no way of examining air": The *General Dictionary* is quoted in Gibbs, *Joseph Priestley*, p. 67.

74 "treated in the same manner": Quoted in Uglow, *The Lunar Men*, p. 233.

75 "My manner has always been": Priestley, *Memoirs*, p. 54.

75 "My experiments on air": Letter from Joseph Priestley to Richard Price, Jan. 16, 1768, quoted in Gibbs, *Joseph Priestley*, p. 68.

Chapter 4: The Prodigy

76 "the quarries, excavating mines, mineral springs": Guettard's proposal for his *Atlas* to the French Academy is quoted in Poirier, *Lavoisier*, p. 17.

77 "could become one of": Quoted in Henry Guerlac, "Some French Antecedents of the Chemical Revolution," *Chymia*, 5 (1959), p. 73.

78 "All the country girls and women": Arthur Young's observations are quoted in William Doyle, *The Oxford History of the French Revolution*, 2nd ed. (New York: Oxford University Press, 2002), p. 14.

78 The demographics of French crime and poverty during the second half of the eighteenth century are found in ibid., p. 16.

78 by 1789, only one boy in fifty-two would attend college: Ibid., p. 48.

78 "I was accustomed to the rigorous": Poirier, *Lavoisier*, p. 7.

79 a "fighting animal": Poirier, *Lavoisier*, p. 8.

79 "few men have had more quarrels": Condorcet quoted in Douglas McKie, *Antoine Lavoisier: Scientist, Economist, Social Reformer* (New York: Henry Schuman, 1952), p. 60.

79 "What a man this is": Ibid., pp. 59–60.

80 "Your health, my dear mathematician": Letter from Monsieur de Troncq to Lavoisier, quoted in Poirier, *Lavoisier*, p. 7.

80 "most abominably lighted up": William Cole quoted in Aykroyd, *Three Philosophers*, p. 6.

80 "a fine course of experiments to run": Carlton E. Perrin, "Research Traditions, Lavoisier, and the Chemical Revolution," *Osiris*, 2nd series, 4 (1988), p. 61.

81 "When Rouelle spoke, he inspired": Mercier's portrait of Rouelle, taken from his *Tableau de Paris*, is quoted in Rhoda Rappaport, "G.-F. Rouelle: An Eighteenth-Century Chemist and Teacher," *Chymia*, 6 (1960), p. 69.

81 his private course in the rue Jacob: Ibid., p. 73. Determining the date of Lavoisier's attendance required some fine detective work by past scholars. Edouard Grimaux and Henry Guerlac claimed Lavoisier owned a copy of Rouelle's lectures; these were printed in the form of "polycopies" Denis Diderot made when attending the lectures in 1754, and were apparently sold until Rouelle's retirement in 1768. According to Grimaux, Lavoisier annotated Diderot's copies while taking his course, and hints from these handwritten notes suggest the period of young Antoine's attendance.

82 "Gentlemen, all that *monsieur le professeur* has just told you": Rouelle quoted in ibid., p. 74.

82 "So you are going to choose a chemist": Charles Henry, *Correspondence inédite de Condorcet et de Turgot, 1770–1779* (Geneva: Slatkine, 1970), pp. 123–24.

82 "Where are we?": Poirier, *Lavoisier*, p. 9.

82 his "damned nephew": Sidney J. French, *Torch and Crucible: The Life and Death of Antoine Lavoisier* (Princeton, NJ: Princeton University Press, 1941), p. 25.

82 "You can see very well, Gentlemen": Poirier, *Lavoisier*, p. 9.

83 "the blood continued to exude": Jimmy M. Skaggs, *The Great Guano Rush: Entrepreneurs and American Overseas Expansion* (New York: St. Martin's Press, 1994), p. 6.

83 "Nihil est in intellectu": Lavoisier mentions the motto on the wall in his collected letters, *Œuvres de Lavoisier*, 6 vols. (Paris, 1864–93), Vol. 1, p. 246. Some biographers have claimed the motto was displayed in the Jardin's auditorium, but Rappaport says this is unlikely since Rouelle was not the only lecturer to use the room—Rappaport, "G.-F. Rouelle," pp. 75, 22ff.

86 "blurring of a paradigm": Kuhn, *The Structure of Scientific Revolutions*, p. 84.

86 "new fundamentals": Ibid.

87 "picking up the other end of the stick": Herbert Butterfield, *The Origins of Modern Science, 1300–1800* (London, 1949), pp. 1–7.
87 "a change in visual gestalt": Kuhn, *The Structure of Scientific Revolutions*, p. 85.
87 "Scientists do not see something": Ibid.
88 "I will await the post": Aykroyd, *Three Philosophers*, p. 7.
88 "Animals and people are delighted": Quoted in Poirier, *Lavoisier*, p. 17.
88 "nasty and extremely awkward errand": Ibid., p. 18.
89 "One would hardly put a dog outside . . . neither bread nor meat": Guettard's letter is quoted in Poirier, *Lavoisier*, p. 19.
90 "There are two ways to present": Lavoisier quoted in Stephen Jay Gould, "Capturing the Center: Antoine-Laurent Lavoisier's scientific contributions," *Natural History*, 107 (December 1998), pp. 14ff.
91 "a young man with knowledge, brains, and energy": Aykroyd, *Three Philosophers*, p. 11.
91 "various uses made of the different types of coal": Guerlac, "Some French Antecedents of the Chemical Revolution," p. 95.
92 "How splendid that at so early an age": Aykroyd, *Three Philosophers*, p. 12.
93 "60,000–120,000 *livres* a year": Poirier, *Lavoisier*, p. 23.
93 Other estimates placed annual incomes: Donald Greer, *The Incidence of the Terror* (Cambridge, MA: Harvard University Press, 1935).
94 "That's just fine!": Poirier, *Lavoisier*, p. 28.
94 "Those who consider the blood of the people": Aykroyd, *Three Philosophers*, p. 17.
95 "one long rush from the counting house to the laboratory": Ibid., p. 8.
96 A 1931 study: W. Platt and R.A. Baker, "The Relation of the Scientific 'Hunch' to Research," *Journal of Chemical Education*, VIII (1931), pp. 1969–2002.
96 "in typical cases, a hunch appears": Walter Bradford Cannon, "The Role of Hunches in Scientific Thought," in Albert Rothenberg and Carl R. Hausman, eds., *The Creativity Question* (Durham, NC: Duke University Press, 1976), p. 64.
97 "theory and observation intersect": Stephen Jay Gould, "Capturing the Center," p. 1.
97 "elusive inner properties": Carlton E. Perrin, "Research Traditions, Lavoisier, and the Chemical Revolution," *Osiris*, 2nd series (1988), p. 63.
98 there was no appreciable change in weight: There was a difference in weight of a quarter of a grain, but that was within the limits of precision of the balance—McKie, *Antoine Lavoisier*, p. 89.
98 it had lost 17.4 grains: The Paris pound was divided into 16 *onces,* each *once* into 8 *gros,* and each *gros* into 72 *grains.*

Chapter 5: The Goodness of Air

99 "So long as we continue Dissenters": Priestley's sermon is quoted in Porter, *English Society in the Eighteenth Century*, p. 363.
100 "all people live in society": Priestley quoted in Gibbs, *Joseph Priestley*, p. 40.
101 "one seeking to defend the old institutional constellation": Kuhn, *The Structure of Scientific Revolutions*, pp. 92–93.
102 "The Unitarian fiend expel": Quoted in Maurice Crosland, "A Practical Perspective on Joseph Priestley as a Pneumatic Chemist," p. 228. Crosland is quoting *The Monthly Repository*, 1808, 3, p. 103, which in turn quotes the 431st hymn in the Methodist "Large Hymn Book."
103 "I have frequently written until I could hardly": Priestley quoted in Gibbs, *Joseph Priestley*, p. 51.

103 "By one means or another I believe": Letter from Joseph Priestley to Rev. T. Lindsey, July 30, 1770, in John T. Rutt, *Life and Correspondence of Joseph Priestley*, 2 vols. (London: George Smallfield, 1832), Vol. 1, p. 118.
103 "quite done with controversy": Gibbs, *Joseph Priestley*, p. 36.
103 "Though I have made discoveries": Quoted in Smith, *Priestley in America, 1794–1804*, pp. 59–60.
103–4 "If we could content ourselves": Joseph Priestley, *Experiments and Observations relating to Various Branches of Natural Philosophy* (London: J. Johnson, 1779), pp. x–xi.
104 In 1770–71, Priestley was overwhelmed: Crosland, "A Practical Perspective on Joseph Priestley as a Pneumatic Chemist," p. 233.
104 "Phosphorus is too expensive for me": Ibid., pp. 233, 51ff.
104 "exceedingly simple and cheap": Rutt, *Life and Correspondence of Joseph Priestley*, Vol. 1, p. 25.
104 The inventory of Priestley's lab equipment and expenditures is found in Lawrence Badash, "Joseph Priestley's Apparatus for Pneumatic Chemistry," *Journal of the History of Medicine and Allied Sciences*, vol. XIX, no. 2 (1964), pp. 139–55.
105 "Good scientists study the most important problems": Peter B. Medawar, *The Art of the Soluble* (Harmondsworth, Middlesex: Penguin, 1969), p. 11.
106 "Plants, instead of affecting the air in the same manner": Quoted in Frederic L. Holmes, "The Revolution in Chemistry and Physics," *Isis*, vol. 91, issue 4 (December 2000), pp. 735ff.
107 "No man can have a more domestic turn": Quoted in Gibbs, *Joseph Priestley*, p. 55.
107 "Some clergymen in the Board of Longitude": Joseph Banks quoted in ibid., p. 56.
108 "I thought this had been a business": Schofield, *A Scientific Autobiography of Joseph Priestley (1733–1804)*, p. 98.
108 "I own myself": George III's letter to Lord North, dated Feb. 23, 1772, is quoted in B. Dobrée, ed., *The Letters of King George III* (London, 1935), p. 82.
111–12 Air was not "good": Partington, *A History of Chemistry*, Vol. 3, p. 254.
112 "did not quite despair of the philosopher's stone": Ibid., p. 246.
112 "another knot in the 'net of the World' ": R. J. Forbes, "On the Origin of Alchemy," *Chymia*, 4 (1953), p. 4.
112 "by means of his wonderful oil and powder": Letter from M. de Cérisy, Jan. 27, 1707, quoted in Charles MacKay, *Extraordinary Popular Delusions and the Madness of Crowds* 2 vols. (1841, 1852; New York: Harmony Books, 1980), p. 219.
112 "transmuting metals, telling fortunes": Ibid., p. 241.
113 "The world is a chaos of delight": Selections from Darwin's *Beagle Diary* and *Origin of Species* relating to an aesthetic sense are quoted in Howard E. Gruber, "Aspects of Scientific Discovery: Aesthetics and Cognition," in John Brockman, ed., *Creativity* (New York: Touchstone Books, 1993), pp. 48–74.
113 In quick order, Priestley discovered: Priestley called hydrogen chloride gas "acid air"; ammonia gas "alkaline air"; sulfur dioxide "vitriolic acid air"; and silicon fluoride "flour acid air." This comes from a number of sources, but most specifically from Priestley's *Experiments and Observations on different Kinds of Airs* (London: 1774–77) and J. R. Partington's *A History of Chemistry* (London: Macmillan and Co., 1962).
114 "first makes man inspired, and then": Plato's *Ion* is quoted in Teresa M. Amabile and Elizabeth Tighe, "Questions of Creativity," in Brockman, ed., *Creativity*, p 20.
115 There would be eight simultaneous discoveries: Lawrence S. Kubie, "Some Unsolved Problems of the Scientific Career," *American Scientist*, 42 (1954), pp. 111ff. Kubie refers to an article by W. H. Manwaring in *Science* (April 1940), and to B. J. Stern, *Social Factors in Medical Progress* (1968).
116 "I have often taken great pleasure": Scheele quoted in Poirier, *Lavoisier*, p. 77.

120 "*flamula vitalis* which animates the blood": The quote is from John Evelyn, reporting on the Aug. 4, 1681, demonstration of the properties of phosphorus at the Royal Society, quoted in D. M. Knight, "The Vital Flame," *Ambix*, vol. 23, issue 1 (March 1976), p. 5.

120–21 "Without that knowledge" . . . "the whole species": Priestley quoted in J. V. Golinski, "Utility and Audience in Eighteenth-Century Chemistry: Case Studies of William Cullen and Joseph Priestley," *British Journal for the History of Science*, vol. 21, part 1, no. 68 (March 1988), p. 26.

122 "I cannot be said to be settled": Gibbs, *Joseph Priestley*, p. 87.

122 "Lord Shelburne is a statesman": Mary Priestley is quoted from the reminiscence of Mary Anne Galton, the daughter of the Birmingham arms maker Samuel Galton; although she was not present at Bowood, she repeated the tale as an old woman from a conversation overheard between her mother and Mary Priestley—Ibid.

123 "Indeed, my Lord, I find the conduct": Ibid., p. 88.

123 "created a kind of invisibility": Porter, *English Society in the Eighteenth Century*, p. 36.

123 "In marriage husband and wife": Quoted in ibid., p. 38.

124 "I (or rather Mrs. Priestley) am transcribing": Belloc, "Joseph Priestley in Domestic Life," p. 44.

124 "Having taken under our most serious deliberation": *The Articles of Association*, Oct. 20, 1774, in "The Avalon Project at Yale Law School: Journals of the Continental Congress," URL: www.yale.edu/anweb/avalon.Contcong.html.

126 "a remarkably vigorous flame": Joseph Priestley, *The Discovery of Oxygen* (1775), published by the Alembic Club (Edinburgh: E. & S. Livingston, 1961), p. 10.

126 "the observation of *events arising*": Ibid., p. 5 (italics in the original).

127 "The force of prejudice biases not only": Ibid., p. 6.

Chapter 6: The Problem of Burning

129 "Chemistry provides a sure, unequivocal means": Poirier, *Lavoisier*, p. 25.

129 "for sincerity and frankness": Ibid., p. 26.

130 "My daughter has a definite aversion": Ibid., p. 39.

130 "hard, sinister, even frightening": This description of Terray by Baron J. B. Auget de Montyon is quoted in Poirier, *Lavoisier*, p. 38.

131 The Marriage section of the 1735 London broadsheet is found in Porter, *English Society in the Eighteenth Century*, p. 40.

133 "plain of dress, unaffectedly mild": Henry Guerlac, *Lavoisier—The Crucial Year: The Background and Origin of His First Experiments on Combustion in 1772* (Ithaca, NY: Cornell University Press, 1961), p. 37.

133 "an extensive correspondence and may circulate": Ibid., p. 40.

135 "May I ask you to repeat": Poirier, *Lavoisier*, p. 53.

139 "Since there are no vessels": Lavoisier quoted in Bernard Jaffe, *Crucibles: The Story of Chemistry, from Ancient Alchemy to Nuclear Fission*, 4th ed. (New York: Dover Publications, 1976), p. 69.

139 "This increase in weight comes": Lavoisier's note has been printed in many publications, with slight variations in translation. This version is found in French, *Torch and Crucible*, p. 67.

140 "he might inadvertently disclose": Henry Guerlac, "A Curious Lavoisier Episode," *Chymia*, 7 (1961), p 104. Guerlac prints the original, unsanitized version in French: *"et comme il est difficile de ne pas laisser entrevoir à ses amis dans la conversation quelque chose qui puisse les mettre sur la voye de la verité."* The 1792 revision spreads the blame *"entre les savants de France et ceux d'Angleterre."*

Chapter 7: The Sentimental Journey

143 The phrase itself first surfaced: Robert Shackleton, "The Grand Tour in the Eighteenth Century," in Louis T. Milic, ed., *Studies in Eighteenth-Century Culture*, Vol. 1: *The Modernity of the Eighteenth Century* (Cleveland: Case Western University, 1971), p. 127.

143 "I left London with so much precipitation": Laurence Sterne, *A Sentimental Journey* (1768; Harmondsworth, Middlesex: Penguin, 1975), p. 92.

143–44 "Once you are known to be a foreigner": Quoted in Shackleton, "The Grand Tour in the Eighteenth Century," p. 130.

144 "it is by means of such traveling philosophers": Thomas Nugent, *The Grand Tour* (London, 1756), I, p. iii, quoted in Shackleton, "The Grand Tour in the Eighteenth Century," p. 133.

144 "who was the most brilliant": French, *Torch and Crucible*, p. 71.

144–45 "No piece of recent work has made me appreciate": Aykroyd, *Three Philosophers*, p. 63.

146 "in slippers, without any thing": Priestley, *Memoirs*, p. 100.

146 "The pain that I felt": Ibid., p. 101.

146 "I can hardly express how very low": Ibid., p. 105.

147 "Our guide, who, no doubt, was in league": Ibid., p. 103.

147 "We had the Rhine to the left": Ibid., p. 107.

147 "which threw me, who was the occasion of it": Ibid.

147 "In the art of living": Arthur Young, *Travels During the Years 1787, 1788, 1789: Undertaken More Particularly with a View of Ascertaining the Cultivation, Wealth, Resources, and National Prosperity of the Kingdom of France* (1794; New York: Doubleday, 1969), pp. 323–25.

149 "The French are clean": Ibid., p. 324.

149 "seeing and conversing with every person": Priestley, *Memoirs*, p. 111.

150 "As far as I can judge": Priestley, *Memoirs*, p. 110.

150 "They could not possibly . . . show more respect": Priestley quoted in French, *Torch and Crucible*, p. 87.

151 "Reason is to the *philosophe*": César Chesneau Dumarsaid, "Definition of a Philosophe," in Kramnick, ed., *The Portable Enlightenment Reader*, p. 21.

151 "Who are the two men across the table from us?": The exchange between Priestley and Turgot is described in French, *Torch and Crucible*, pp. 87–88.

153 "Sir, I have no need of that hypothesis": Laplace's famous line is quoted in Roger Hahn, "Laplace and the Vanishing Role of God in the Physical Universe," in Harry Woolf, ed., *The Analytic Spirit: Essays in the History of Science* (Ithaca, NY: Cornell University Press, 1981), p. 85.

154 "God is like an excess wheel": Jacques-André Naigea, *Adresse à l'Assemblée Nationale sur la liberté des opinions* (Paris, 1790), p. 31, quoted in Hahn, "Laplace and the Vanishing Role of God in the Physical Universe," p. 92.

156 "The man who went into battle": Lawrence S. Kubie, "Some Unsolved Problems of the Scientific Career, Part II," *American Scientist*, 42 (1954), p. 109.

156 "This is the ultimate gamble": Ibid., p. 111.

157 "the rapid production of, I believe": Priestley quoted in French, *Torch and Crucible*, p. 88.

157 "The world has great expectations": Ibid.

157 "I shall not even think": Ibid.

157 "they lose one eighth of their weight": Bayen quoted in French, *Torch and Crucible*, p. 73.

158 it arrived fifteen to seventeen days later: Anthony R. Butler, "Lavoisier: A Letter from Sweden," *Chemistry in Britain*, 20 (1984), p. 619.

158 "decline and conjugate for my own pleasure": The 1777 letter from Marie Lavoisier to her brother Balthazar Paulze is quoted in Denis I. Duveen, "Madame Lavoisier, 1758–1836," *Chymia*, 4 (1953), pp. 16, 7ff.

159 a supper given for Marie Antoinette: Raymond Oliver, *Gastronomy of France*, trans. Claude Durrell (Cleveland: Wine and Food Society, 1967), pp. 300–1.

160 the "active, feral nature": Dena Goodman, *The Republic of Letters: A Cultural History of the French Enlightenment* (Ithaca, NY: Cornell University Press, 1994), p. 55.

160 "That the girls and women showed less discomfort": Deborah Tannen, *Gender and Discourse* (New York: Oxford University Press, 1994), p. 128.

160 "I heated *plumb rouge*": The conversation of Priestley and the *philosophes* is recounted in Aykroyd, *Three Philosophers*, p. 65.

161 "At this, all the company": Ibid.

161 "Perhaps the proof": Lennard J. Davis, "A Scholarly Appreciation of Irrational Inspiration," *Chronicle of Higher Education*, Nov. 28, 2003, p. B10.

Chapter 8: The Mouse in the Jar

163 "I leave the Doctor to give you a long list": Letter from Theophilus Lindsey to Mr. Turner, quoted in Gibbs, *Joseph Priestley*, p. 91.

163–64 "a candle still burned in it": Priestley, *The Discovery of Oxygen*, p. 13.

164 "The mind delights in a static environment": Wilfred Trotter quoted in Bernard Barber, "Resistance by Scientists to Scientific Discovery," *Science*, CXXXIV (1961), p. 597.

164–65 "There is in all of us a psychological tendency": W. I. B. Beveridge quoted in ibid.

165 "Probably we do not adequately appreciate": Sigmund Freud, *Moses and Monotheism* (1939; New York: Vintage Books, 1967), pp. 83–84.

165 "It cannot be removed, it cannot be suppressed": Edmund Burke quoted in Barbara W. Tuchman, *The March of Folly: From Troy to Vietnam* (New York: Ballantine Books, 1969), p. 206.

166 "To me every thing looks": Priestley quoted in Clarke Garrett, "Joseph Priestley, the Millennium, and the French Revolution," *Journal of the History of Ideas*, vol. 34, no. 1 (January–March 1973), p. 56.

166 he took "so strongly for granted": Priestley, *The Discovery of Oxygen*, p. 6.

167 "I wish my reader be not quite tired": Ibid., p. 15.

168 "Suppose we were able": William James, *Principles of Psychology*, quoted in Oliver Sacks, "Speed," *The New Yorker*, Aug. 23, 2004, p. 62.

169 "Oh! Hear a pensive captive's prayer": Anna Laetitia Barbauld's poem is in the Joseph Priestley Collection, Dickinson College Library.

171 "though one mouse would live only a quarter of an hour": Priestley, *The Discovery of Oxygen*, p. 16.

171 "not having taken care": Ibid., p. 54.

172 "a large jar full of it to the standard": Ibid., p. 54.

172 "a laboratory might be constructed": Ibid., p. 51.

173 "so we might . . . *live out too fast*": Ibid., p. 54.

173 "as loud as that of a small pistol": Ibid., p. 52.

173 "I should not . . . think it difficult": Ibid., p. 53.

173 "supply a pair of bellows": Ibid., pp. 52–53.

173–74 "You will blame me for not having waited on you": Letter from Joseph Priestley to Jeremy Bentham, March 19, 1775, in the Joseph Priestley Collection, Dickinson College Library.

Chapter 9: The Twelve Days

175 "threw out such a brilliant light": Aykroyd, *Three Philosophers*, pp. 65ff.
175–76 "Atmospheric air is not a simple body": Poirier, *Lavoisier*, p. 78.
176 "more respirable, more combustible": Lavoisier's "Easter memoir" is quoted in Aaron Ihde, "Priestley and Lavoisier," in Erwin N. Hiebert, Aaron J. Ihde, and Robert E. Schofield, *Joseph Priestley: Scientist, Theologian, and Metaphysician* (Cranbury, NJ: Associated University Presses, 1980), p 72.
176–77 "I have no intention of taking over": Lavoisier quoted in Poirier, *Lavoisier*, p. 79.
177 "something that until now has never been understood": Ibid., p. 82.
177 "a vague assertion, thrown out by chance": Ibid.
178 "After I left Paris": Priestley quoted in James Bryant Conant, ed., *The Overthrow of the Phlogiston Theory: The Chemical Revolution of 1775–1789* (Cambridge, MA: Harvard University Press, 1966), pp. 32–33. (Priestley's italics.)
178–79 "rendering the phenomena, which all practical chemists": Bryan Higgins quoted in Gibbs, *Joseph Priestley*, p. 95.
179 "It may be my fate to be a kind of comet": Ibid., p. 96.
179 "It is pleasant when we can be": Conant, *The Overthrow of the Phlogiston Theory*, p. 34.
181 "You know the tender friendship": French, *Torch and Crucible*, p. 108.
181 Nobles formed a separate social order or estate: Details of France's noble class are found in Doyle, *Oxford History of the French Revolution*, p. 26.
183 "For him it was a blissful day": Poirier, *Lavoisier*, p. 95.
183–84 "M. Lavoisier is a pleasing looking young man": Anders-Johann Lexell quoted in Poirier, *Lavoisier*, p. 96.
184 *"l'air déphlogistique de M. Prisley":* Lavoisier's log is quoted in Ihde, "Priestley and Lavoisier," p. 72.
184 "nothing noteworthy happened": Lavoisier's lab notes for the twelve-day experiment are included in Aykroyd, *Three Philosophers*, pp. 65–69.
185 "I carefully collected the 45 grains": Ibid., p. 68.
186 "I have established in the foregoing memoirs": Ibid., p. 69.
186 "I feel obliged . . . to distinguish": Ibid.
187 "It can be taken as an axiom": French, *Torch and Crucible*, p. 177.

Chapter 10: The Language of War

192 "If man could learn from history": Samuel Taylor Coleridge quoted in Tuchman, *The March of Folly*, p. 383.
192 an "arch-magician" who "imposed upon our credulity": Jaffe, *Crucibles*, p. 73.
192 "was neither friendly nor agreeable to the rules": Letter from Joseph Priestley to Benjamin Smith Barton, Sept. 14, 1797, Priestley Collection, Dickinson College.
192 "I thank you for your ingenious": Joseph Priestley, "A Letter to Dr. Mitchell, in reply to the preceding, by JOSEPH PRIESTLEY, LL. D. &c," *Medical Repository* (1798), p. 511.
192 "We hope you had rather gain us": Priestley's "Open Letter to the French Chemists" of June 15, 1796, is quoted in the Introduction to his *Memoirs*, p. 44.
192–93 "Lavoisier had to use Priestley": French, *Torch and Crucible*, p. 125.
193 "I also had the advantage": Edmond Genet quoted in Poirier, *Lavoisier*, p. 81.

193 "You triumph over my doubts, Madame": Carlton E. Perrin, "The Triumph of the Antiphlogistians," in Woolf, ed., *The Analytic Spirit*, p. 55.

194 "Imagine what can be the temper of the soul of a man": Rousseau's *Politics and the Arts*, p. 103, is quoted in Goodman, *The Republic of Letters*, p. 55.

194 "treated me like a ninny": Denis Diderot's complaint is found in his *Correspondence*, Vol. VII, p. 132, quoted in Arthur M. Wilson, " 'Treated Like Imbecile Children' (Diderot): The Enlightenment and the Status of Women," in Paul Fritz and Richard Morton, eds., *Women of the 18th Century and Other Essays* (Toronto: Samuel Stevens Hakkert & Co., 1976), p. 95.

194 "absently out of a window": French, *Torch and Crucible*, p. 115.

195 "I am personally acquainted with Mr. Lavoisier": Perrin, "The Triumph of the Antiphlogistians," p. 51.

195 "Madame Lavoisier, a lively, sensible": Arthur Young quoted in French, *Torch and Crucible*, p. 116.

195 "I do not forget Paris": Benjamin Franklin's letter to Marie Lavoisier is quoted in ibid.

196 "still, safe world of science": Uglow, *The Lunar Men*, p. 431.

197 "a man void of affectation": Young quoted in French, *Torch and Crucible*, p. 168.

198 "Ah, Priestley. An evil man, Sir": Samuel Johnson quoted in Isaac Kramnick, "Eighteenth-Century Science and Radical Social Theory: The Case of Joseph Priestley's Scientific Liberalism," *Journal of British Studies*, 25 (January 1986), p. 4.

198 "If Dr. Priestley applies to my librarian": Dobrée, ed., *The Letters of King George III*, p. 139.

199 "Resort to prophecy is a universal response": Samuel Eddy quoted in John Dominic Crossan, *The Historical Jesus: The Life of a Mediterranean Peasant* (New York: HarperCollins, 1991), p. 104.

199 "the society of persons eminent": Belloc, "Joseph Priestley in Domestic Life," p. 49.

200 "one of the best philosophical clubs in the Kingdom": William Withering's praise of his club is quoted in Simon Schaffer, "Learned Insane," *London Review of Books*, April 17, 2003, p. 34 (review of Uglow's *The Lunar Men*).

200 "What inventions, what wit": Erasmus Darwin quoted on the Web at "Joseph Priestley," URL: www.woodrow.org/teachers/chemistry/institutes/1992/Priestley.html.

200–1 "I have raised a confidence in Dr. Priestley": Matthew Boulton's response to James Watt is quoted in J. V. Golinski, "Utility and Audience in Eighteenth-Century Chemistry: Case Studies of William Cullen and Joseph Priestley," *British Journal for the History of Science*, vol. 21, part 1, no. 68 (March 1988), p. 23.

201 "It has often struck us": The anonymous reviewer is quoted in Schaffer, "Learned Insane," p. 35.

201 "licking some great man's arse": Ibid.

202 "Unitarian principles are gaining ground": Joseph Priestley, "The Importance and Extent of Free Enquiry" (1785), quoted in Gibbs, *Joseph Priestley*, p. 173.

203 "the good town of Birmingham": Samuel Horsley's comment is quoted in Uglow, *The Lunar Men*, p. 408.

204 "consisted *solely* of a multitude of objects": Cecil J. Schneer, *Mind and Matter: Man's Changing Concepts of the Material World* (New York: Grove Press, 1969), p. 97.

204–5 "He did not love; he did not hate,": Quoted in French, *Torch and Crucible*, p. 155.

206 "I was for a long time of the opinion": Priestley quoted in Ihde, "Priestley and Lavoisier," p. 83.

207 "You do not surely expect that chemistry should be able": Quoted in the Introduction to Priestley's *Memoirs*, p. 56.

208 "I know that I am doing my duty": George III's comment to Lord North quoted in Tuchman, *The March of Folly*, p. 217.
208 "Universal dejection": Ibid.
209 "the author of the plan should be hanged": From the lampoon *Réclamation d'un citoyen contre la nouvelle enceinte de Paris élevée par les fermiers généraux*, quoted in Poirier, *Lavoisier*, pp. 172–73.
210 "three sets of spies": Uglow, *The Lunar Men*, p. 397.
211 "a manly, brave hero is killed": Quoted in Hans Delbrück, *The Dawn of Modern Warfare*. Vol. IV of *The History of the Art of War*, trans. Walter J. Renfroe Jr. (Lincoln, NB: University of Nebraska Press, 1990, originally published in German in 1920), p. 31.
214 "far from having brought light to bear": Lavoisier's revisions are quoted in Perrin, "The Triumph of the Antiphlogistians," p. 44.
214 a "fatal error in chemistry": Lavoisier's *Reflections on Phlogiston* is quoted in ibid., p. 46.
215 "Then violent objections were made": Ibid.
215 "I do not expect that my ideas will be accepted": Ibid.
215 "Surely anyone who has not made a specialized study": Ibid., p. 48.
215 "This work . . . is a tissue of non sequiturs and absurdities": Lavoisier, *Œuvres*, Vol. IV, p. 368, quoted in Perrin, "The Triumph of the Antiphlogistians," pp. 49, 18ff.
216 zinc oxide was "philosophic wool": Poirier, *Lavoisier*, pp. 182–83.
216–17 "Let us ask ourselves": Ibid., p. 183.
217 "We think only through the medium of words": Lavoisier's collection of Condillac's maxims and his own reasoning is quoted in Schneer, *Mind and Matter*, p. 100.
217 "The impossibility of separating the nomenclature": Lavoisier quoted in ibid., p. 100.
219 "confined to discussions of a metaphysical nature": Ibid., p. 101.
219 "The establishment of a new nomenclature in any science": Jaffe, *Crucibles*, p. 74.
219 "So Lavoisier has substituted the word 'oxide' for the calx of metal": Ibid., p. 75.
220 "no man ought to surrender his own judgment": Priestley quoted in Maurice Crosland, "Lavoisier, the Two French Revolutions, and 'The Imperial Despotism of Oxygen,'" *Ambix*, vol. 42, part 2 (1995), p. 110.
221 "Linguistic chaos": Karin Littan, "The Primal Scattering of Languages: Philosophies, Myths, and Genders," on the Web at URL: www.bu.edu/wcp/Papers/Lite/LiteLitt.htm
221 "so strong, indeed, that no philologer": Jones quoted in Roy Harris and Talbot J. Taylor, *Landmarks in Linguistic Thought I: The Western Tradition from Socrates to Saussure* (London and New York: Routledge, 1989), p. xx.
221 "they almost always exert over time": Bill Bryson, *The Mother Tongue: English and How It Got that Way* (New York: William Morrow, 1990), p. 139.
222 "unsuitable to the genius of a free nation": Priestley quoted in Albert C. Baugh and Thomas Cable, *A History of the English Language* (London: Routledge & Kegan Paul, 1978), p. 269.
222 "It is *my* theory": French, *Torch and Crucible*, p. 178.
223 "air-tight vessels to ascertain whether the weight": Jaffe, *Crucibles*, p. 72.
223 "as simple as algebra": French, *Torch and Crucible*, p. 178.
224 "In early infancy, our ideas spring": Antoine-Laurent Lavoisier, *Elements of Chemistry* (1789; Chicago: Encyclopaedia Britannica, 1952), p. 1.

Chapter 11: "King Mob"

228 "temper and high spirits": Gibbs, *Joseph Priestley*, p. 194.
228 "with a fine meadow on one side": Gerald M. Moser, *Seven Essays on Joseph Priestley* (State College, PA: Unitarian-Universalist Foundation of Centre County, 1994), p. 15.

229 "There is not only most virtue": Priestley, *Memoirs*, p. 109.

229 The city directory of 1770 listed: The demographics and vital statistics of Birmingham in the 1770s and 1780s are quoted in Porter, *English Society in the Eighteenth Century*, p. 97. The full list of tradesmen from the 1770 Birmingham city directory includes:

248 innkeepers	46 brassfounders
129 buttonmakers	39 shopkeepers
99 shoemakers	39 bucklemakers
77 merchants	36 gunmakers
74 tailors	35 jewelers
64 bakers	26 maltsters
56 toymakers	24 drapers
52 platers	23 gardeners
49 butchers	21 plumbers/glaziers
48 carpenters	21 ironmongers
46 barbers	

229–30 "I was surprised at the place but more": William Hutton quoted in Porter, *English Society in the Eighteenth Century*, p. 214.

230 "From a handsome entrance the ladies are now led": William Hutton, *An History of Birmingham*. (London, 1783; E. P. Publishing, 1976), p. 32.

230 Mary was content: The observations of the caretaker Isaac Whitehouse are quoted in Moser, *Seven Essays on Joseph Priestley*, p. 15.

231 "writing on any subject by the parlor fire": Priestley, *Memoirs*, pp. 125–26.

231 "In the pulpit he is mild": Catherine Hutton quoted in Uglow, *The Lunar Men*, pp. 406–07.

231 "It has been a singular happiness to me": Priestley, *Memoirs*, pp. 122–23.

231–32 the "cardinal esteem": Ibid., p. 127.

232 "I esteem it a singular happiness": Ibid.

232 "as the happiest event": Ibid., p. 120.

232 "reduces the information it gets": Frederick Turner and Ernst Poppel, "The Neural Lyre," reprinted from *Poetry* (1983), URL: www.pacifer.com/~starling/lyre.html.

233 "fountain of erroneous opinions": Aykroyd, *Three Philosophers*, p. 131.

234 "The King is *insane*": Burke quoted in Christopher Reid, "Burke, the Regency Crisis, and the 'Antagonist World of Madness,' " *Eighteenth Century Life*, vol. 16, no. 2 (May 1992), p. 70.

235 "the Bretons are provisionally carrying out": Quoted in Elias Canetti, *Crowds and Power* (New York and London: Penguin Group, 1992), p. 67.

238 "plump as ease can render him": Young, *Travels During the Years 1787, 1788, 1789*, p. 309.

238 "a mob followed her talking very loud": Ibid.

239 "pretty good-natured-looking boy": Ibid.

239 "Liberty, Reason, Brotherly Love forever!" Mary Anne Galton's recollection of Harry Priestley's enthusiasm is quoted in Uglow, *The Lunar Men*, p. 434.

240 "Tremble, all ye oppressors of the world!": Richard Price, *A Discourse on the Love of our Country*, quoted in "The Language of Politics," URL: www.historyguide.org

241 "When an old hound misleads the pack": Gibbs, *Joseph Priestley*, p. 183.

241 "How would you like it yourself": John Nott quoted in ibid.

241 "huddled about you with their smelling-bottles": Ibid., p. 185.

242 "Damn Priestley, damn him, damn him forever": British and Foreign Unitarian

Association, *The Priestley Memorial at Birmingham, 1874* (London: Longman, Green, Reader, & Dyer, 1875), pp. 16ff.

243 approximately two hundred titles pouring from the English press: "The Language of Politics," URL: www.historyguide.org.

243 another survey estimated 1,086 titles: Gayle Trusdel Pendleton, "The English Pamphlet Literature in the Age of the French Revolution Anatomized," *Eighteenth Century Life*, vol. 5, no. 1 (Fall 1978), p. 29.

243 Burke's work was a hit: Doyle, *Oxford History of the French Revolution*, p. 169.

243 "You appear to me not to be sufficiently cool": Priestley quoted in Anne Holt, *The Life of Joseph Priestley* (Westport, CT: Greenwood Press, 1931), p. 150.

243 "a vehicle for the same poison": Quoted in Kramnick, "Eighteenth-Century Science and Radical Social Theory: The Case of Joseph Priestley's Scientific Liberalism," p. 5.

244 "the wild gas, the fixed air is plainly broken loose": Ibid.

244 "The age of chivalry is gone": Burke quoted in Maurice Crosland, "The Image of Science as a Threat: Burke versus Priestley and the 'Philosophic Revolution,' " *British Journal for the History of Science*, vol. 20, part 3, no. 66 (July 1987), p. 279.

244 "alchymist or empiric": Ibid., p. 284.

244 "considered man in their experiments no more": Ibid., p. 294.

245 The demographics of Birmingham in 1791 are from R. B. Rose, "The Priestley Riots of 1791," *Past and Present*, 18 (November 1960), pp. 65–73.

245–46 "Ask your parents for a description of the country": Josiah Wedgwood quoted in Porter, *English Society in the Eighteenth Century*, p. 355.

246 "The poor are crowded in offensive": Ibid.

246 "Bread or blood!": Marvin Harris, *Cannibals and Kings: The Origins of Modern Cultures* (New York: Vintage Books, 1977), p. 277.

247 "beggarly, brass-making, brazen-faced": From a 1789 letter by Rev. J. Bartham to Samuel Parr, first printed in J. Johnstone, *Works of Samuel Parr* (London, 1828), I, p. 336, and quoted in Rose, "The Priestley Riots of 1791," p. 70.

248 "Such was the furious loyalty and entire devotion": John Binns, *Recollections of the Life of John Binns* (Philadelphia: Parry & M'Millan, 1854), p. 68.

249 "Remember that on the 14th of July": The suspect handbill is reprinted in the Introduction to Priestley, *Memoirs*, p. 28.

249 "a riot was expected on Thursday": Belloc, "Joseph Priestley in Domestic Life," pp. 50–51.

250 "he would not have a hair of his head injured": Holt, *Life of Priestley*, p. 157.

250 For most of the day he entertained: R. E. W. Maddison and Francis R. Maddison, "Joseph Priestley and the Birmingham Riots," *Notes and Records of the Royal Society of London*, 12 (1957), pp. 100ff.

250 "Nonsense, my dear": E. Robinson, "New Light on the Priestley Riots," *The Historical Journal*, vol. 3, no. 1 (1960), pp. 73–74.

251 "never trouble themselves with anything": Rose, "The Priestley Riots of 1791," p. 77.

252 "our supreme governors": Quoted in Porter, *English Society in the Eighteenth Century*, p. 102.

253 "an unexpected obstacle will be destroyed": Gustave Le Bon, *The Crowd, A Study of the Popular Mind* (1896; London: Ernest Benn, 1938), pp. 42–43.

253 "Fire is the same wherever it breaks out": Canetti, *Crowds and Power*, p. 89.

254 "That they should think of molesting me": Joseph Priestley, *An Appeal to the Public on the Riots at Birmingham* (Birmingham and London, 1791).

254 "she would not go, abandoning her pleasant, orderly rooms": Belloc, "Joseph Priestley in Domestic Life," p. 53.

254 The account of the Birmingham riot and Priestley's flight is contained in many

sources, but many of the quotes used in these pages come from: "To the INHABITANTS of the TOWN of BIRMINGHAM," in broadsheet, with the account of the riots and Captain James Keir's letter "To the PRINTER of the BIRMINGHAM & STAFFORD CHRONICLE," West Bromwich, July 20, 1791. Joseph Priestley Collection, Dickinson College Library; Holt, *The Life of Joseph Priestley*, chapter X, pp. 145–78; Priestley, *An Appeal to the Public on the Riots at Birmingham*; Aykroyd, *Three Philosophers*; James Belcher, *An Authentic Account of The Riots in Birmingham, On the 14th, 15th, 16th, and 17th days of July 1791; also, The Judge's Charge. The Pleadings of the Counsel, and the Substance of the Evidence given on the Trials of the Rioters. And an Impartial Collection of Letters, &c. Written by the Supporters of the Establishment and the Dissenters, in Consequence of the Tumult* (London, 1791); Robinson, "New Light on the Priestley Riots," pp. 73–75; Rose, "The Priestley Riots of 1791," pp. 68–88; Maddison and Maddison, "Joseph Priestley and the Birmingham Riots," pp. 98–113; and *Views of Buildings Destroyed during Riots at Birmingham in 1791*, illus. P. H. Witton, engraved by Wm. Ellis (Birmingham, 1792).

258 "We now passed several houses": Martha Russell's recollection of her flight is quoted in Holt, *The Life of Joseph Priestley*, pp. 163–69.

259 "I cannot but feel better pleased that Priestley": Quoted in Porter, *English Society in the Eighteenth Century*, p. 369.

259 "He is very well": W. P. Griffith, "Priestley in London," *Notes and Records of the Royal Society of London*, 38 (1984), p. 9.

260 "God can require it of us as a duty": Quoted in Aykroyd, *Three Philosophers*, p. 137.

260 "they will scarcely find so many respectable characters": British and Foreign Unitarian Assoc., *The Priestley Memorial at Birmingham, 1874*, pp. 10ff.

Chapter 12: The World Out of Joint

261 "Escape may be checked by water and land": Edith Hamilton, *Mythology* (New York: New American Library, 1940), quoting Apollodorus, p. 139.

262 "of those whose function it is to obey": Lavoisier's letter to Franklin is quoted in Aykroyd, *Three Philosophers*, p. 119.

263 "At a time when everything, good and evil alike": Ibid., p. 121.

264 "They are tame, healthy creatures, easy to feed": Ibid., p. 108.

265 "As long as we considered respiration": Ibid., pp. 110–11.

267 "The J-curve is this": James C. Davies, "The J-Curve of Rising and Declining Satisfactions as a Cause of Some Great Revolutions and a Contained Rebellion," in Hugh Davis Graham and Ted Robert Gurr, eds., *Violence in America: Historical and Comparative Perspectives (A Report Submitted to the National Commission on the Causes and Prevention of Violence)* (New York: Bantam Books, 1969), p. 690.

267 "I have ever sought the truth": Marat's autobiography is quoted in Stanley Loomis, *Paris in the Terror: June 1793–July 1794*. (New York: J. P. Lippincott Co., 1964), p. 85.

269 "Physically, Marat had the burning" . . . "His countenance": Ibid., p. 92.

269 "Rise up, you unfortunates of the city": Ibid., p. 90.

269 "A man who is starving": Ibid.

269 "From my earliest years": Loomis, *Paris in the Terror*, p. 86.

272 "old absurdities long since proved false": French, *Torch and Crucible*, p. 138.

273 "M. Franklin, having exposed his bald head": Poirier, *Lavoisier*, p. 111.

273 "The revolution that M. Marat has just produced": I. Bernard Cohen, "Scientific Revolution and Creativity in the Enlightenment," *Eighteenth Century Life*, vol. 7, no. 2 (January 1982), p. 47.

275 "generalized hostile beliefs": Neil J. Smelser, *Theory of Collective Behavior* (New York: Free Press of Glencoe, 1963), pp. 67–130.

275 The idea of a group mind is hard to reject: Such an idea is almost instantaneous when observing an emotion-filled crowd. I was amazed in winter 1991 when covering the revolt in Vilnius, Lithuania, against Soviet Russia: a huge crowd had gathered before the Parliament building when word spread that tanks were coming from the military base outside the city. As one, the mass seemed to swivel to face the path of the tanks; there was a boulevard that turned from up the hill onto the plaza, and the tanks would either have to stop before the crowd, swivel on their axis for a right-angled turn, or plow straight into those demanding their freedom. As the caravan of fast-moving light tanks crested the rise, the crowd pressed forward at this meeting point; the movement seemed almost amoebic from above, as if a group mind were at work to present a human barrier to the drivers. As the tanks sped forward, the barrier of bodies added to itself. Luckily, the tanks were agile enough to make that turn at the last minute, or the numbers of dead and injured would have been horrific.

275 "there is no doubt that something exists": Sigmund Freud, *Group Psychology and the Analysis of the Ego* (1921; New York: W.W. Norton & Co., 1975), p. 25.

275 "I'll tear the heart out of that infernal Lafayette": Loomis, *Paris in the Terror*, p. 91.

277–78 "But today, with France given over": Letter dated Jan. 6, 1793, from Lavoisier to Kerr, quoted in Aykroyd, *Three Philosophers*, p. 142.

279 "The Commune of Paris hastens to inform": Loomis, *Paris in the Terror*, p. 96.

280 "Louis cannot be judged": Doyle, *Oxford History of the French Revolution*, p. 195.

281 "seal the decree which declares France": Daniel Arasse, *The Guillotine and the Terror*, trans. Christopher Miller (London: Penguin Books, 1989), p. 61.

281 "The blood of Capet": Ibid., quoting the revolutionary newspaper *Révolutions de Paris*, Jan. 26, 1793.

281–82 "the head blocke wherein the ax is fastened": From "Hallifax and its Gibbet-Law Placed in a True Light" (1708), quoted in Alfred Marks, *Tyburn Tree, Its History and Annals* (London: Brown, Langham & Co., 1908), p. 23.

282 "stain the hand of a man with the slaughter": G. Lenotre, *The Guillotine and Its Servants* (London: Hutchinson & Co., 1929), p. 160.

282–83 "As he got out of the carriage": Sanson's official report is included in Arasse, *The Guillotine and the Terror*, pp. 58–59.

284 French chemistry had "given law to all nations": Aykroyd, *Three Philosophers*, p. 151.

284–85 "There is not an academician who": Lavoisier's appeal to the Commune is quoted in Sir Edward Thorpe, *Essays in Historical Chemistry* (London: Macmillan & Co., 1923), pp. 144–45.

285 "protected it against despotic interference": Aykroyd, *Three Philosophers*, p. 151.

285 "the annihilation of the sciences": Thorpe, *Essays in Historical Chemistry*, p. 145.

285–86 "A collection of vain men": Marat quoted in Poirier, *Lavoisier*, p. 329.

286 "these dismal academies . . . are incompatible": Quoted in Aykroyd, *Three Philosophers*, p. 153.

286 "I killed him in cold blood": Marat's assassination is contained in eyewitness statements and police and the coroner's report, quoted in Loomis, *Paris in the Terror*, pp. 128–31.

287 "It seems like a long trip, doesn't it?": Corday's execution is described in ibid., pp. 146–47.

287 "by their conduct, their contacts, their words": Doyle, *Oxford History of the French Revolution*, p. 251.

287 "Terror is nothing else than justice": Maximilien Robespierre, "Terror is nothing else than justice," in Brian MacArthur, ed., *The Penguin Book of Historic Speeches* (New York: Penguin Books, 1995), pp. 183–84.

288 "Martin Alleaume, seventeen, a hairdresser's apprentice": The list of victims is contained in Loomis, *Paris in the Terror*, p. 329.

288 "I am not well": Ibid.

288 But there was a window of confusion, which gave him some time: The confusion during Lavoisier's arrest is detailed in Denis I. Duveen and Herbert S. Klickstein, "Some New Facts Relating to the Arrest of Antoine-Laurent Lavoisier," *Isis*, 49 (1958), pp. 347–48, and in Duveen, "Antoine-Laurent Lavoisier 1743–1794: A Note Regarding His Domicile During the French Revolution," *Isis*, 42 (1951), pp. 233–34.

289 Such mixups could be deadly: These mistakes and others are detailed in Loomis, *Paris in the Terror*, pp. 333–34.

290 "I foresaw that the commissioners would accuse": Poirier, *Lavoisier*, p. 353.

290 The Farmers were accused of stealing 107,819,033 *livres*: Aykroyd, *Three Philosophers*, p. 169.

290 the number of executions steadily increased: Ibid., p. 171.

290–91 "My dear one, you are giving yourself": Poirier, *Lavoisier*, pp. 357–58.

291 Lavoisier reconciled with the Church and took communion: See Lucien Scheler and William A. Smeaton, "An Account of Lavoisier's Reconciliation with the Church a Short Time Before His Death," *Annals of Science*, 14 (1958), pp. 148–53.

291 "I have not come to humble myself": Marie Lavoisier's interview with Antoine Dupin is described in Poirier, *Lavoisier*, pp. 370–71, and Aykroyd, *Three Philosophers*, p. 173.

292 "I made efforts for the unfortunate Lavoisier": Fourcroy's account and that of his student, André Augier, are quoted in Poirier, *Lavoisier*, p. 384.

293 "The crimes of these vampires cry aloud for justice": Aykroyd, *Three Philosophers*, p. 186.

293–94 *"La République n'a pas besoin de savants":* Ibid., p. 187.

295 "brought back to life": Arasse, *The Guillotine and the Terror*, pp. 152, 11ff. The experiment, quoted from Auberive, *Anecdotes sur les décapités*, Paris, Year V, pp. 7–8, is worth quoting in its entirety simply because it seems to have assumed the force of urban myth, and thus, a popular truth. According to the text:

> It is easier to cut a rope neatly than to replace a head. The latter experiment has nevertheless been attempted . . . to find out if it was possible to detain the departing soul and prolong life for a few moments after the fatal blow. The subject of the experiment was a young man, sentenced to be decapitated for a crime. No sooner had he been executed, than surgeons stopped the flow of blood from the trunk with styptics, while others who had held up the head placed it back on the neck with all possible precision and dexterity, vertebra on vertebra, muscle on muscle, artery on artery. The incision was wrapped around with compresses, which were mechanically held in place. Finally strong spirits were placed beneath his nose. The head then seemed to come back to life. A perceptible movement of the face muscles occurred, and the eyelids twitched. A cry of surprise and wonder was heard. The young man was gently lifted up and taken to a nearby house where, after having given some very slight signs of life, he expired. This much seems to me to be established. But it appears that the experiment was ill-conducted and the arrangements most unsatisfactory.

295 "I am convinced . . . the heads would speak": Sue's assertion is quoted in Arasse, *The Guillotine and the Terror*, p. 39.

296 "Death is not instantaneous." From "Henri Landru: Bluebeard," URL: www.crimelibrary.com/serial_killers/history/landru/guillotine_7.html?sect=6. The controversy remains contemporary, if somewhat underground. In 1983, Harold Hillman, a

reader of physiology at the University of Surrey, wrote in the *New Scientist* that "Consciousness is probably lost within 2–3 seconds due to a rapid fall of intracranial perfusion of blood"—Oct. 27, 1983, p. 276. Another doctor, Ron Wright, estimated that consciousness could last thirteen seconds—"Does the Head of a Guillotined Individual Remain Briefly Alive?" *The History Net*, URL: http://europeanhistory.about.com/library/bldyk10.htm.

297 needed only "a minute per person": Canetti, *Crowds and Power*, p. 58.

297 "The law has been passed": Sanson quoted in Arasse, *The Guillotine and the Terror*, p. 125.

298 "What grieves me most is to have": Poirier, *Lavoisier*, p. 381.

298 "filled with people running with all their might": Arasse, *The Guillotine and the Terror*, p. 112.

298 to see "what it was like": The only complete account of an execution, from the loading of the victims in the tumbrils to the final loading of their separated remains, was written by the abbé Carrichon after attending the execution of two friends. It is included in Loomis, *Paris in the Terror*, p. 349.

299 There is a story that persists today: The various but similar versions come from "Ain't No Way to Go: Does Beheading Hurt?," URL: www.aarrgghh.com/no-way/beheading.htm; "The Straight Dope: Does the head remain briefly conscious after decapitation?," URL: www.straightdope.com/classico/a5_262.html; conversation with Dr. Clay Drees, Virginia Wesleyan College, Virginia Beach, VA.

Chapter 13: The New World

300 "a theatre of calamity": William Godwin, *Caleb Williams* (1794; New York: Penguin Books, 1988), p. 5.

300 "the chased deer is avoided by all the herd": Gibbs, *Joseph Priestley*, p. 212.

301 "In this business, we are the sheep": Quoted in Priestley, *Memoirs*, p. 29.

301 "almost all great minds in all ages of the world": Uglow, *The Lunar Men*, p. 447.

301 "that should direct you not to risk a life": Ibid.

301 "Long have you been the Danger of this Country": J. Edmund White, "The Priestley Memorial Volume," *Notes and Records of the Royal Society of London*, vol. 48, no. 1 (1994), p. 89.

301 "the King's Head upon a Charger": This and the other political satires of Priestley are included in Arthur Sheps, "Public Perception of Joseph Priestley, the Birmingham Dissenters, and the Church-and-King Riots of 1791," *Eighteenth-Century Life*, vol. 13, no. 2 (May 1989), p. 57, and M. Fitzpatrick, "Priestley in Caricature," *Third BOC Priestley Conference*, pp. 347–69.

302 The fact that he tried to reclaim £4,083 10*s* 3*d*: Douglas McKie, "Priestley's Laboratory and Library and Other of His Effects," *Notes and Records of the Royal Society of London*, 12 (1957), p. 129.

302 "I shall be glad to take refuge in your country": Schofield, *A Scientific Autobiography of Joseph Priestley (1733–1804)*, pp. 263–64. Schofield, who must be considered one of the leading scholars on Priestley and his correspondence, says that "it is hard to believe that they [Priestley and Lavoisier] did not correspond, but no letters other than this one printed here have been located" (p. 364).

302 "Remember you are to be a *man of business*": Gibbs, *Joseph Priestley*, pp. 214–15.

303 "Fly from the country you hate": White, "The Priestley Memorial Volume," p. 89.

303 "And the bright flame ascends the kindling skies": Ibid., p. 90.

304 The poet William Blake: By 1793 and 1794, Blake so feared government persecution that he published *The French Revolution* anonymously and only distributed it to friends

and political sympathizers. Ironically, when Blake was charged with treason and tried in Chichester, it was not over his work but because of an argument in 1803 with a drunken soldier whom he had removed from his garden. The angered man said Blake "damned the King and said that soldiers were all slaves," but the testimony of drunken soldiers was suspect even in Georgian England and Blake was acquitted.

305 "Capt. Smith for New York": Gibbs, *Joseph Priestley*, p. 222.

305 "I do not pretend to leave this country": Ibid.

306 "all the arts and sciences are put under a state of requisition": Smith, *Priestley in America, 1794–1804*, pp. 14–19.

307 "the only great nursery of free men": Jonathon Shipley, *Works* (London, 1792), Vol. II, pp. 191–92, quoted in Arthur Sheps, "Ideological Immigrants in Revolutionary America," in Paul Fritz and David Williams, eds., *City and Society in the 18th Century* (Toronto: Hakkert, 1973), pp. 231.

307 "have contracted for a large quantity of land": *The Gentleman's Magazine*, 64, II, pp. 1170–72, quoted in Mary Catherine Park, "Joseph Priestley and the Problem of Pantisocracy," Ph.D. Dissertation, University of Pennsylvania, 1947, p. 16.

307–8 "men of wisdom and information should organize the plan": Brissot's report is quoted in ibid., p. 14.

308 there were aborted plans to set up an asylum: Interview with Andrea Bashore, administrator and historian, Joseph Priestley House, Northumberland, PA. Even today, one can find on detailed state maps tiny Asylum Township, Pennsylvania, located 60–70 miles north of Northumberland in Bradford County, bordering the state line with New York. The 2000 Census lists its population as 1,097. Web sites for the township state that it received its name because it was conceived as an asylum for French refugees of the Revolution.

308 "To live in a beautiful country": Coleridge quoted in Park, "Joseph Priestley and the Problem of Pantisocracy," p. 37.

308 "wielding the axe, now to cut down the tree": Ibid., p. 40.

308 "Let us be free ourselves": Ibid., p. 28.

308 "God Save the King!, and may he be the last": Ibid., p. 36.

309 "I would recommend every one to have" . . . "shipwreck and famine": Mary Priestley to Rev. Thomas Belsham, New York, June 15, 1794, in Moser, *Seven Essays on Joseph Priestley*, pp. 20–21.

310 "I found myself more vexed than frightened": Ibid., p. 21.

310 "We had many things to amuse us in the passage": Priestley, *Memoirs*, p. 38.

310 "thick fog and rain": Moser, *Seven Essays on Joseph Priestley*, p. 20.

310 a "safe and honourable retreat": Derek Davenport, "Joseph Priestley in America: 1794–1804." Unpublished manuscript in Joseph Priestley Collection, Dickinson College Library, p. 4.

311 "wonderfully pleased with everything": Moser, *Seven Essays on Joseph Priestley*, p. 21.

311 "Probably in no other place on the Continent": Smith, *Priestley in America*, p. 55.

311 "almost out of the world": Gibbs, *Joseph Priestley*, p. 227.

312 "Dr. Priestley . . . has left off his periwig": Smith, *Priestley in America*, p. 53.

312 "Here every housekeeper has a garden": Ibid., pp. 60–61.

313 although of those whose stories were recorded, many did well: See George A. Frick, "Names of Persons Constituting the Colony Brought Over from England by Dr. Jos. Priestley in the year 17—." Unpublished manuscript in the Joseph Priestley Collection, Dickinson College Library.

313 "He works as hard as any farmer in the country": Smith, *Priestley in America*, p. 61.

313 "I am happy and thankful": Mary Priestley quoted in Holt, *The Life of Joseph Priestley*, p. 188.

314 "and there I saw the good old father": Smith, *Priestley in America*, pp. 70–71.

315 "died apparently without any pain": Joseph Priestley Jr., Letter to Joseph Rayner Priestley, May 18, 1816, Joseph Priestley Collection, Dickinson College Library.

315 "I was only a lodger in her house": Priestley's letter to John Wilkinson, Sept. 19, 1796, is quoted in Gibbs, *Joseph Priestley*, p. 231.

315 "I feel quite unhinged": Ibid., p. 233.

315 "I never stood in more need of friendship": Aykroyd, *Three Philosophers*, p. 210.

316 "the agreeable Miss Peggy Foulke": William Priestley's wedding announcement was published in *The American Advertiser*, February 13, 1796, quoted in Smith, *Priestley in America*, p. 79.

316 "The life they lead, quite solitary in the woods": Letter from Joseph Priestley to John Wilkinson (date not given), quoted in Gibbs, *Joseph Priestley*, p. 230.

316 "in his sickness spoke of his second son": Benjamin Rush quoted in Smith, *Priestley in America*, p. 80.

318 "Is it a crime to doubt the capacity": Dumas Malone, *The Public Life of Thomas Cooper: 1783–1839* (Columbia, SC: University of South Carolina Press, 1961), p. 92.

318 "He is as weak as water": Derek Davenport, "Joseph Priestley in America: 1794–1804," p. 12.

319 "On Monday last, Doctor Priestley": *Reading Advertiser*, April 26, 1800, Files of the Historical Society of Pennsylvania, in the Archives of the Joseph Priestley House, Northumberland, PA.

320 the township of Middle Paxton: Now Halifax, in Dauphin Co., PA.

320 resettling eventually in St. James Parish, Louisiana: Marianna Griswold van Rensselaer, *Henry Hobson Richardson and His Works* (1888; New York: Dover Publications, 1969), pp. 2–5.

320–21 "I feel more compassion than resentment": Letter from Joseph Priestley to John Wilkinson, June 1800, quoted in Gibbs, *Joseph Priestley*, p. 241.

321 "I hope, however, I feel solid consolation": Letter from Joseph Priestley to Theophilus Lindsey (Jan. 1, 1798), quoted in Moser, *Seven Essays on Joseph Priestley*, p. 74.

321 "Something favorable is promised in Egypt": Aykroyd, *Three Philosophers*, p. 211.

322 "As the storm is now subsiding": Davenport, "Joseph Priestley in America: 1794–1804," p. 13.

322 "Yours is one of the few lives precious to mankind": Jefferson quoted in Gibbs, *Joseph Priestley*, p. 242.

322 "I have lived a little beyond the usual term of human life": Quoted in "Obituary of Joseph Priestley," *Medical Repository* (1804), p. 433; Joseph Priestley Collection, Dickinson College Library.

323 "I am going to sleep as well as you": Aykroyd, *Three Philosophers*, p. 216.

323 "the people of America have a saying": Quoted in Kramnick, ed., *The Portable Enlightenment Reader*, p. xix.

324 "The American people have a wonderful talent": Binns, *Recollections of the Life of John Binns*, p. 173.

325 "He died in about ten minutes after we had moved him": Joseph Priestley Jr. is quoted in an addendum to Priestley's *Memoirs*, p. 139. The story of Priestley's death is drawn primarily from this source.

Epilogue: The Burning World

327 "I fancy there is a certain way of making": Michel de Montaigne's "Use Makes Perfect" is quoted in Sherwin B. Nuland, *How We Die: Reflections on Life's Final Chapter* (New York: Vintage Books, 1993), pp. 129–30.

328 "The mad *philosophes*": Joseph de Maistre quoted in Kramnick, ed., *The Portable Enlightenment Reader*, p. xx.

328 "Man is but a foundling in the cosmos": Carl Becker quoted in the Web site www.en.thinkexist.com.

328 "The fully enlightened earth": Theodor W. Adorno and Max Horkheimer, *Dialectic of Enlightenment: Philosophical Fragments* (1944; Stanford, CA: Stanford University Press, 2002), p. 3.

329 "The ancient teachers of this science": Mary Shelley, *Frankenstein, or, The Modern Prometheus* (1818; New York: New American Library, 1965), p. 47.

329 "so much has been done": Ibid.

330 "The time will therefore come": Marquis de Condorcet, "The Future Progress of the Human Mind," in Kramnick, ed., *The Portable Enlightenment Reader*, p. 30.

331 "both sin and neurosis represent the individual": Ernest Becker, *The Denial of Death*, quoted in Lacey Baldwin Smith, *Fools, Martyrs, Traitors: The Story of Martyrdom in the Western World* (Evanston, IL: Northwestern University Press, 1997), p. 19.

331 "an act of supreme egotism": Smith, *Fools, Martyrs, Traitors*, p. 17.

331 "pale martyr in his shirt of fire": Alexander Smith, *A Life Drama* (sc. 2, line 225), quoted in "GIGA Quote Topic Page for Martyrdom," www.giga-usa.com.

333 "This great, excellent, and extraordinary": John Adams quoted in Clarke Garrett, "Joseph Priestley, the Millennium, and the French Revolution," *Journal of the History of Ideas*, vol. 34, no. 1 (Jan.–March, 1973), p. 51.

333–34 "Carry yourself back to that time": Fourcroy's memorial is quoted in Schneer, *Mind and Matter*, p. 94.

335 "peace no longer dwells in my habitation": Count Rumford's letters to his daughter and the description of the incident with the flowers are in Aykroyd, *Three Philosophers*, pp. 203–04.

336 "It's true that by blundering about we stumbled": Francis Crick quoted in Nicholas Wade, "Crick, DNA Structure Co-Discoverer, Dies at 88," *New York Times*, June 30, 2004, p. A1.

336 "She often gave elegant balls": Adrien Delahante quoted in Poirier, *Lavoisier*, pp. 410–11.

336 "You have to have lived under the vacuum pump": Marie Lavoisier quoted in François Guizot, "Madame de Rumford (1758–1836)," in Guizot, *Mélanges biographiques et littéraires* (Paris: Michel Lévy, 1868), p. 83.

BIBLIOGRAPHY

Print

Abrahams, Harold J. "Priestley Answers the Proponents of Abiogenesis," *Ambix*, vol. 12, part 1 (1964), pp. 44–71.

Adorno, Theodor W., and Max Horkheimer, *Dialectic of Enlightenment: Philosophical Fragments*. 1944; Stanford, CA: Stanford University Press, 2002.

Alder, Ken. *The Measure of All Things: The Seven-Year Odyssey and Hidden Error That Transformed the World*. New York: Free Press, 2002.

Amabile, Teresa M., and Elizabeth Tighe, "Questions of Creativity," in John Brockman, ed., *Creativity*. New York: Touchstone Books, 1993, pp. 7–27.

Andrews, Jean Marie. "Joseph Priestley, Icon of Enlightenment," *Early American Homes*, vol. XXXI, no. 2 (April 2000), pp. 36–43.

Arasse, Daniel. *The Guillotine and the Terror*, trans. Christopher Miller. London: Penguin Books, 1989.

Ariès, Philippe. *Centuries of Childhood: A Social History of Family Life*, trans. Robert Baldick. New York: Vintage Books, 1962.

Aristophanes. *The Clouds*, trans. and ed. William Arrowsmith. Ann Arbor, MI: University of Michigan Press, 1970.

Asimov, Isaac. *A Short History of Chemistry: An Introduction to the Ideas and Concepts of Chemistry*. Garden City, NY: Anchor Books, 1965.

Aykroyd, W. R. *Three Philosophers*. Westport, CT: Greenwood Press, 1970.

Bacon, Francis. "The New Science," in Isaac Kramnick, ed., *The Portable Enlightenment Reader*. New York: Penguin Books, 1995, pp. 39–42.

Badash, Lawrence. "Joseph Priestley's Apparatus for Pneumatic Chemistry," *Journal of the History of Medicine and Allied Sciences*, vol. XIX, no. 2 (1964), pp. 139–55.

Barber, Bernard. "Resistance by Scientists to Scientific Discovery," *Science*, CXXXIV (1961), pp. 596–602.

Barron, Frank. "The Psychology of Imagination," *Scientific American*, vol. 199, no. 3 (September 1958), pp. 151–66.

Baugh, Albert C., and Thomas Cable. *A History of the English Language*. London: Routledge & Kegan Paul, 1978.

Belcher, James. *An Authentic Account of the Riots in Birmingham, On the 14th, 15th, 16th, and 17th days of July 1791; also, The Judge's Charge. The Pleadings of the Counsel, and the Substance of the Evidence given on the Trials of the Rioters. And an Impartial Collection of Letters, &c. Written by the Supporters of the Establishment and the Dissenters, in Consequence of the Tumult*. London, 1791.

Bell, David A. "Enlightenment's Errand Boy," *London Review of Books*, May 22, 2003, pp. 28–29.

Belloc, Bessie Rayner. "Joseph Priestley in Domestic Life," from *In a Walled Garden* (1895), in "The Victorian Women Writers Project." URL: http://www.indiana.edu/~letrs/wwwp/belloc/walled.html.

Binns, John. *Recollections of the Life of John Binns*. Philadelphia: Parry & M'Millan, 1854.

Bradley, John. "On the Operational Interaction of Classical Chemistry," *British Journal of the Philosophy of Science*, 6 (1955–56), pp. 32–42.

British and Foreign Unitarian Association, *The Priestley Memorial at Birmingham, 1874*. London: Longman, Green, Reader, & Dyer, 1875.

Brockman, John, ed. *Creativity*. New York: Touchstone Books, 1993.

Bronowski, Jacob. "The Creative Process," *Scientific American*, vol. 199, no. 3 (September 1958), pp. 59–65.

———. *The Origins of Knowledge and Imagination*. New Haven, CT: Yale University Press, 1978.

Bronowski, Jacob, and Bruce Mazlish. *The Western Intellectual Tradition: From Leonardo to Hegel*. 1960; New York: Dorset Press, 1986.

Bruner, J. S., and Leo Postman, "On the Perception of Incongruity: A Paradigm," *Journal of Personality*, XVIII (1949), pp. 206–23.

Bryson, Bill. *The Mother Tongue: English and How It Got That Way*. New York: William Morrow, 1990.

Buehrens, John A., and F. Forrester Church. *Our Chosen Faith: An Introduction to Unitarian Universalism*. Boston: Beacon Press, 1989.

Bunyan, John. *The Pilgrim's Progress, from this World to That Which Is to Come*. ed., James Blanton Wharey. Oxford: Clarendon Press, 1960.

Burr, Sandra J. "Inspiring Lunatics: Biographical Portraits of the Lunar Society's Erasmus Darwin, Thomas Day, and Joseph Priestley," *Eighteenth Century Life*, vol. 24, no. 2 (2000), pp. 111–27.

Butler, Anthony R. "Lavoisier: A Letter from Sweden," *Chemistry in Britain*, 20 (1984), pp. 617–19.

Butterfield, Herbert. *The Origins of Modern Science, 1300–1800*. New York: Macmillan, 1957.

Canetti, Elias. *Crowds and Power*. 1960; New York and London: Penguin Group, 1992.

Cannon, Walter Bradford. "The Role of Hunches in Scientific Thought," in Albert Rothenberg and Carl R. Hausman, eds., *The Creativity Question*. Durham, NC: Duke University Press, 1976, pp. 63–69.

Chandler, Charles Lyon. Letter to Rev. Earl Morse Wilbur, President, Pacific Unitarian School for the Ministry, Berkeley, CA, April 1, 1927. Joseph Priestley Collection, Dickinson College Library.

Christie, William. *A Speech delivered at the grave of the Revd. Joseph Priestley, LL.D., F.R.S.* Northumberland, PA: Andrew Kennedy, 1804.

Clarke, John, and Jasper Ridley. *The Houses of Hanover and Saxe-Coburg-Gotha*. Berkeley: University of California Press, 2000.

Clifford, James. "Some Aspects of London Life in the Mid-18th Century," in Paul Fritz and David Williams, eds., *City and Society in the 18th Century*. Toronto: Samuel Stevens Hakkert & Co., 1973, pp. 19–38.

Cohen, I. Bernard. "Scientific Revolution and Creativity in the Enlightenment," *Eighteenth Century Life*, vol. 7, no. 2 (January 1982), pp. 41–54.

Colley, Linda. "I am the Watchman," *London Review of Books*, November 20, 2003, pp. 16–17.

Conant, James Bryant, ed. *The Overthrow of the Phlogiston Theory: The Chemical Revolution of 1775–1789*. Cambridge, MA: Harvard University Press, 1966.

de Condorcet, marquis. "The Future Progress of the Human Mind," in Kramnick, ed., *The Portable Enlightenment Reader*, pp. 26–38.

Conlin, Michael F. "Joseph Priestley's American Defense of Phlogiston Reconsidered," *Ambix*, vol. 43, part 3 (September 1996), pp. 129–45.

Crosland, Maurice. *Historical Studies in the Language of Chemistry*. Cambridge, MA: Harvard University Press, 1962.

———. "The History of Chemistry Seen in a Broader Context," *Impact of Science on Society*, 159 (1989), pp. 227–36.

———. "The Image of Science as a Threat: Burke versus Priestley and the 'Philosophic Revolution,' " *British Journal for the History of Science*, vol. 20, part 3, no. 66 (July 1987), pp. 277–307.

———. "Lavoisier, the Two French Revolutions, and 'The Imperial Despotism of Oxygen,' " *Ambix*, vol. 42, part 2 (1995), pp. 101–18.

———. "A Practical Perspective on Joseph Priestley as a Pneumatic Chemist," *British Journal for the History of Science*, vol. 16, part 3, no. 54 (1983), pp. 223–38.

Crossan, John Dominic. *The Historical Jesus: The Life of a Mediterranean Peasant*. New York: HarperCollins, 1991.

Darnton, Robert. *Mesmerism and the End of the Enlightenment in France*. Cambridge, MA: Harvard University Press, 1968.

Daumas, Maurice. "Precision of Measurement and Physical and Chemical Research in the Eighteenth Century," in A. C. Crombie, ed., *Scientific Change: Historical studies in the intellectual, social and technical conditions for scientific discovery and technical invention, from antiquity to the modern era*, a collection of papers presented at the Symposium on the History of Science, University of Oxford, July 9–15, 1961. New York: Basic Books, 1963, pp. 418–30.

———. *Scientific Instruments of the Seventeenth and Eighteenth Centuries*, trans. Mary Holbrook. New York: Praeger Publishers, 1972; originally published in French, 1953.

Davenport, Derek. "Joseph Priestley in America: 1794–1804." Unpublished manuscript in Joseph Priestley Collection, Dickinson College Library, n.d.

Davies, James C. "The J-Curve of Rising and Declining Satisfactions as a Cause of Some Great Revolutions and a Contained Rebellion," in Hugh Davis Graham and Ted Robert Gurr, eds., *Violence in America: Historical and Comparative Perspectives (A Report Submitted to the National Commission on the Causes and Prevention of Violence)*. New York: Bantam Books, 1969, pp. 690–730.

Davis, Lennard J. "A Scholarly Appreciation of Irrational Inspiration," *Chronicle of Higher Education*, Nov. 28, 2003, pp. B10–B11.

Delbrück, Hans. *The Dawn of Modern Warfare*. Vol. IV of *The History of the Art of War*, trans. Walter J. Renfroe, Jr. Lincoln, NB: University of Nebraska Press, 1990, originally published in German, 1920.

Descartes, René. "I Think, Therefore I Am," in Kramnick, ed., *The Portable Enlightenment Reader*, pp. 181–85.

Diderot, Denis. "Encyclopédie," in Kramnick, ed., *The Portable Enlightenment Reader*, pp. 17–21.

Dobrée, B., ed., *The Letters of King George III*. New York: Funk & Wagnalls, 1968; London, 1935.

Dormandy, Thomas. *The White Death: A History of Tuberculosis*. New York: New York University Press, 2000.

Doyle, William. *The Oxford History of the French Revolution*. 2nd ed. New York: Oxford University Press, 2002.

Dull, Jonathan R. "France and the American Revolution Seen as a Tragedy," in Ronald T.

Hoffman and Peter J. Albert, eds., *Diplomacy and Revolution: The Franco-American Alliance of 1778*. Charlottesville, VA: University of Virginia Press, 1981), pp. 73–106.
Dumarsaid, Cesar Chesneau. "Definition of a Philosophe," in Kramnick, ed., *The Portable Enlightenment Reader*, pp. 21–22.
Duveen, Denis I. "Antoine-Laurent Lavoisier 1743–1794: A Note Regarding His Domicile During the French Revolution," *Isis*, 42 (1951), pp. 233–34.
———. "Lavoisier," *Scientific American*, vol. 194, no. 5 (May 1956), pp. 84–94.
———. "Madame Lavoisier, 1758–1836," *Chymia*, 4 (1953), pp. 13–29.
Duveen, Denis I., and Herbert S. Klickstein. "Some New Facts Relating to the Arrest of Antoine-Laurent Lavoisier," *Isis*, 49 (1958), pp. 347–48.
Eisenstadt, Shmuel N."Heterodoxies and Dynamics of Civilizations," *Proceedings of the American Philosophical Society*, vol. 128, no. 2 (1984), pp. 104–13.
Eshet, Dan. "Rereading Priestley," *History of Science*, vol. 39, issue 2 (June 2001), pp. 127–59.
Faure, Edgar. *La disgrâce de Turgot*. Paris: Gallimard, 1961.
Fitzgerald, Debra. "Translating Ideas: What Scientists Can Teach Fiction Writers About Metaphor," *The Writer's Chronicle*, vol. 36, no. 5 (March–April 2004), pp. 12–20.
Fitzpatrick, M. "Priestley in Caricature." *Oxygen and the Conversion of Future Feedstock, the Proceedings of the Third BOC Priestley Conference*. London: Royal Society of Chemistry, 1985, pp. 347–69.
Flaherty, Alice Weaver. "Writing Like Crazy: A Word on the Brain," *Chronicle of Higher Education*, Nov. 21, 2003, pp. B6–B9.
Forbes, R. J. "On the Origin of Alchemy," *Chymia*, 4 (1953), pp. 1–11.
Franklin, Benjamin. A Letter from Benjamin Franklin to Joseph Priestley, 1774, in Harlow Shapley, Samuel Rapport, and Helen Wright, eds., *A Treasury of Science*. New York: Harper & Brothers, 1943.
French, Sidney J. "The Du Ponts and the Lavoisiers: A Bit of Untold History with an Accent on America." *Journal of Chemical Education*, vol. 56, no. 12 (December 1979), pp. 791–93.
———. *Torch and Crucible: The Life and Death of Antoine Lavoisier*. Princeton, NJ: Princeton University Press, 1941.
Freud, Sigmund. *Group Psychology and the Analysis of the Ego*. 1921; New York: W. W. Norton & Co., 1975.
———. *Moses and Monotheism*. 1939; New York: Vintage Books, 1967.
Frick, George A. "Names of Persons Constituting the Colony Brought Over from England by Dr. Jos. Priestley in the year 17—." Unpublished manuscript in the Joseph Priestley Collection, Dickinson College Library, n.d.
Fritz, Paul, and Richard Morton, eds. *Women of the 18th Century and Other Essays*. Toronto: Samuel Stevens Hakkert & Co., 1976.
Fruchtman Jr., Jack. "The Revolutionary Millennialism of Thomas Paine," in O. M. Brack Jr., ed., *Studies in the Eighteenth Century*, vol. 13. Madison, WI: University of Wisconsin Press, 1984, pp. 65–77.
Fulton, John F., and Charlotte H. Peters."An Introduction to a Bibliography of the Educational and Scientific Works of Joseph Priestley." Unpublished paper read at the December 31, 1936, meeting of the American Bibliographical Society, Providence, R.I. Joseph Priestley Collection, Dickinson College Library.
———. "The Warrington Academy and Its Influence upon Medicine and Science." *Bulletin of the Institute of the History of Medicine*, vol. 52, no. 2 (1933), pp. 50–80.
Gardner, Howard. "Seven Creators of the Modern Era," in Brockman, ed., *Creativity*, pp. 28–47.

Garrett, Clarke. "Joseph Priestley, the Millennium, and the French Revolution," *Journal of the History of Ideas*, vol. 34, no. 1 (January–March, 1973), pp. 51–66.

Garrett, Laurie. *The Coming Plague: Newly Emerging Diseases in a World Out of Balance*. New York: Farrar, Straus & Giroux, 1994.

Gibbs, F. W. *Joseph Priestley: Adventurer in Science and Champion of Truth*. London: Thomas Nelson & Sons, 1965.

Gillispie, Charles Coulston. *Science and Polity in France at the End of the Old Regime*. Princeton, NJ: Princeton University Press, 1980.

———. "Science in the French Revolution," *Behavioral Sciences*, 4 (1959), pp. 67–73.

Godwin, William. *Caleb Williams*. 1794; New York: Penguin Books, 1988.

Golinski, J. V. "Utility and Audience in Eighteenth-Century Chemistry: Case Studies of William Cullen and Joseph Priestley," *British Journal for the History of Science*, vol. 21, part 1, no. 68 (March 1988), pp. 1–31.

Goodman, Dena. *The Republic of Letters: A Cultural History of the French Enlightenment*. Ithaca, New York: Cornell University Press, 1994.

Gopnik, Adam. "American Electric," *The New Yorker*, June 30, 2003, pp. 96–100.

Gould, Stephen Jay. "Capturing the Center: Antoine-Laurent Lavoisier's Scientific Contributions," *Natural History*, vol. 107 (December 1998), pp. 14ff.

Greenbaum, Louis S. O. "The Humanitarianism of Antoine-Laurent Lavoisier," *Studies on Voltaire and the Eighteenth Century*, 88 (1972), pp. 651–75.

Greer, Donald. *The Incidence of the Terror*. Cambridge, MA: Harvard University Press, 1935.

Griffin, Rev. Frederick R. "Joseph Priestley, 1733–1804." Address to the First Unitarian Church of Philadelphia, 1933. Joseph Priestley Collection, Dickinson College Library.

Griffith, W. P. "Priestley in London," *Notes and Records of the Royal Society of London*, 38 (1984), pp. 1–16.

Gruber, Howard E. "Aspects of Scientific Discovery: Aesthetics and Cognition," in Brockman, ed., *Creativity*, pp. 48–74.

Guerlac, Henry. "A Curious Lavoisier Episode," *Chymia*, 7 (1961), pp. 103–08.

———. *Lavoisier—The Crucial Year: The Background and Origin of His First Experiments on Combustion in 1772*. Ithaca, NY: Cornell University Press, 1961.

———. "Some French Antecedents of the Chemical Revolution," *Chymia*, 5 (1959), pp. 73–112.

Guizot, François. "Madame de Rumford (1758–1836)," in Guizot, *Mélanges biographiques et littéraires*. Paris: Michel Lévy, 1868, pp. 49–88.

Guy, Christian. *An Illustrated History of French Cuisine: From Charlemagne to Charles de Gaulle*, trans. Elisabeth Abbott. New York: Bramhall House, 1962.

Hahn, Roger. "Laplace and the Vanishing Role of God in the Physical Universe," in Harry Woolf, ed., *The Analytic Spirit: Essays in the History of Science*. Ithaca, NY: Cornell University Press, 1981, pp. 85–95.

Hall, A. Rupert. *The Scientific Revolution, 1500–1800: The Formation of the Modern Scientific Attitude*. London: Longmans, Green, 1954, 1983.

Hamilton, Edith. *Mythology*. New York: New American Library, 1940.

Hammond, J., and B. L. Hammond. *The Town Laborer, 1760–1832: The New Civilization*. London: Longman, Green, 1966.

Harris, Marvin. *Cannibals and Kings: The Origins of Modern Cultures*. New York: Vintage Books, 1977.

Harris, Roy, and Talbot J. Taylor. *Landmarks in Linguistic Thought I: The Western Tradition from Socrates to Saussure*. London and New York: Routledge, 1989.

Hellholm, David, ed. *Apocalypticism in the Mediterranean World and the Near East*.

Proceedings of the International Colloquium on Apocalypticism, Uppsala, Sweden, August 12–17, 1979. Tübingen: Mohr, Siebeck, 1980.
Henry, Charles, ed. *Correspondence inédite de Condorcet et de Turgot, 1770–1779*. Geneva: Slatkine, 1970.
Henry, William. "An Estimate of the Philosophical Character of Dr. Priestley," *American Journal of Science and Arts* (1833), pp. 28–39.
Heseltine, George C. *Great Yorkshiremen*. Freeport, NY: Books for Libraries Press, 1932.
Hibbert, Christopher. *King Mob: The Story of Lord George Gordon and the London Riots of 1780*. Cleveland: World Publishing Co., 1958.
Hiebert, Erwin N., Aaron J. Ihde, and Robert E. Schofield. *Joseph Priestley: Scientist, Theologian, and Metaphysician*. Cranbury, NJ: Associated University Presses, 1980.
Holmes, Frederic L. *Lavoisier and the Chemistry of Life: An Exploration of Scientific Creativity*. Madison, WI: University of Wisconsin Press, 1985.
———. "The Revolution in Chemistry and Physics," *Isis*, vol. 91, issue 4 (December 2000), pp. 735ff.
Holt, Anne. *The Life of Joseph Priestley*. Westport, CT: Greenwood Press, 1931.
Horne, R. A. "Aristotelian Chemistry," *Chymia*, 11 (1966), pp. 21–27.
Hughes, Mary Joe. "Child-Rearing and Social Expectations in Eighteenth-Century England: The Case of the Colliers of Hastings," in O. M. Brack Jr., ed., *Studies in the Eighteenth Century*, vol. 13, pp. 79–100.
Hunter, Jean E. "The 18th-Century Englishwoman: According to *Gentleman's Magazine*," in Fritz and Morton, eds., *Women of the 18th Century and Other Essays*, pp. 73–88.
Hutton, William. *An History of Birmingham*. 1783; London: E. P. Publishing Ltd., 1976.
Ihde, Aaron J. *The Development of Modern Chemistry*. New York: Harper & Row, 1964.
———. "Priestley and Lavoisier," in Hiebert, Ihde, and Schofield, *Joseph Priestley: Scientist, Theologian, and Metaphysician*, pp. 62–91.
Ingle, Dwight J. "Psychological Barriers in Research," *American Scientist*, 42 (1954), pp. 283–93.
Jaffe, Bernard. *Crucibles: The Story of Chemistry, from Ancient Alchemy to Nuclear Fission*. New York: Dover Publications, 1976.
Jefferson, Thomas. *The Jefferson Bible: The Life and Morals of Jesus of Nazareth*. Boston: Beacon Press, 1989.
Jordan, Sarah. "From Grotesque Bodies to Useful Hands: Idleness, Industry, and the Laboring Class," *Eighteenth Century Life*, vol. 25, no. 3 (Fall 2001), pp. 62–79.
Kieft, Lester. "The Diverse Lives of Joseph Priestley," *Foote Prints*, vol. 46, no. 1 (1983), pp. 24–36.
Knight, D. M. "The Vital Flame," *Ambix*, vol. 23 part 1 (1976), pp. 5–15.
Kramnick, Isaac. "Eighteenth-Century Science and Radical Social Theory: The Case of Joseph Priestley's Scientific Liberalism," *Journal of British Studies*, 25 (January 1986), pp. 1–30.
Kramnick, Isaac, ed. *The Portable Enlightenment Reader*. New York: Penguin Books, 1995.
Kubie, Lawrence S. "Some Unsolved Problems of the Scientific Career," *American Scientist*, 41 (1953), pp. 596–613, and 42 (1954), pp. 104–12.
Kuhn, Thomas S. *The Essential Tension: Selected Studies in Scientific Tradition and Change*. Chicago: University of Chicago Press, 1977.
———. "Historical Structure of Scientific Discovery," *Science*, 136, June 1, 1962, pp. 760–64.
———. *The Structure of Scientific Revolutions. International Encyclopedia of Unified Science*, vol. 2. Chicago: University of Chicago Press, 1962, 1970.
Lavoisier, Antoine-Laurent. *Elements of Chemistry*. 1789; Chicago: Encyclopaedia Britannica, 1952.

Le Bon, Gustave. *The Crowd, A Study of the Popular Mind.* 1896; London: Ernest Benn, 1938.

Lenotre, G. *The Guillotine and Its Servants.* trans. Mrs. Rudolph Stawell. London: Hutchinson & Co., 1929.

Levin, A. "Venel, Lavoisier, Fourcroy, Cabanis and the Idea of Scientific Revolution: The French Political Context and the General Patterns of the Conceptualization of Scientific Change," *History of Science*, 22 (1984), pp. 303–20.

Linn, John Blair. *A Letter to Joseph Priestley, L.L.D., F.R.S., in Answer to His Letter in Defence of his Pamphlet entitled 'Socrates and Jesus Compared.'* Philadelphia: John Conrad & Co., 1803.

Littan, Karin. "The Primal Scattering of Languages: Philosophies, Myths, and Genders." URL: www.bu.edu/wcp/Papers/Lite/LiteLitt.htm.

Loomis, Stanley. *Paris in the Terror: June 1793–July 1794.* New York: J. P. Lippincott Co., 1964.

Lopez, Claude A. "Saltpeter, Tin, and Gunpowder: Addenda to the Correspondence of Lavoisier and Franklin," *Annals of Science*, 16 (1960), pp. 83–94.

Lukehart, Peter M., ed. *Joseph Priestley in America, 1794–1804.* Catalogue compiled for an exhibition on Priestley's life in America, The Trout Gallery, Emil R. Weiss Center for the Arts, Dickinson College, Carlisle, PA, September 14–November 12, 1994.

MacArthur, Brian, ed. *The Penguin Book of Historic Speeches.* New York: Penguin Books, 1995.

Mackay, Charles. *Extraordinary Popular Delusions and the Madness of Crowds.* 1841; New York: Harmony Books, 1980.

Maddison, R. E. W., and Francis R. Maddison, "Joseph Priestley and the Birmingham Riots," *Notes and Records of the Royal Society of London*, 12 (1957), pp. 98–113.

Malone, Dumas. *The Public Life of Thomas Cooper: 1783–1839.* Columbia, SC: University of South Carolina Press, 1961.

Mannix, Daniel P. *The Hellfire Club.* New York: Simon & Schuster, 1961.

Mantel, Hilary. "Is it still yesterday?" *London Review of Books*, April 17, 2003, pp. 12–16.

Marks, Alfred. *Tyburn Tree, Its History and Annals.* London: Brown, Langham & Co., 1908.

Mason, Stephen F. "Jean Hyacinthe de Magellan, F.R.S., and the Chemical Revolution of the Eighteenth Century," *Notes and Records of the Royal Society of London*, 45 (2) (1991), pp. 155–64.

Mauskopf, Seymour H. *Chemical Sciences in the Modern World.* Philadelphia: University of Pennsylvania Press, 1993.

——— "Gunpowder and the Chemical Revolution," *Osiris*, 4 (1988), pp. 93–118.

——— "Lavoisier and the Improvement of Gunpowder Production," *Revue d'Histoire des Sciences*, 48 (1995), pp. 95–124.

McEvoy, John G. "Joseph Priestley, Scientist, Philosopher, and Divine," *Proceedings of the American Philosophical Society*, vol. 128, no. 3 (1984), pp. 193–99.

——— "Priestley Responds to Lavoisier's Nomenclature: Language, Liberty, and Chemistry in the English Enlightenment," in Bernadette Bensaude-Vincent and Ferdinando Abbri, eds., *Lavoisier in European Context: Negotiating a New Language for Chemistry.* Canton, MA: Science History Publications, 1995, pp. 123–45.

McKie, Douglas. *Antoine Lavoisier: Scientist, Economist, Social Reformer.* New York: Henry Schuman, 1952.

——— "Joseph Priestley and the Copley Medal," *Ambix*, vol. 9, part 1 (1961), pp. 1–22.

——— "Priestley's Laboratory and Library and Other of His Effects," *Notes and Records of the Royal Society of London*, 12 (1957), pp. 114–36.

Medawar, Peter B. *The Art of the Soluble.* Harmondsworth, Middlesex: Penguin, 1969.

Mee, Jon. "Apocalypse and Ambivalence: The Politics of Millennarianism in the 1790s," *South Atlantic Quarterly*, vol. 95, no. 3 (Summer 1996), pp. 671–97.

Miles, Wyndham. "Early American Chemical Societies," *Chymia: Annual in the History of Chemistry*, Vol. 3. Philadelphia: University of Pennsylvania Press, 1950, pp. 95–113.

Milic, Louis T., ed. *Studies in Eighteenth Century Culture, Vol. I: The Modernity of the Eighteenth Century*. Cleveland: Case Western University, 1971.

Morris, Richard. "Inventing the Universe," in Brockman, ed., *Creativity*, pp. 130–48.

Moser, Gerald M. *Seven Essays on Joseph Priestley*. State College, PA: Unitarian-Universalist Foundation of Centre County, 1994.

Mutthauf, Robert. "The French Crash Program in Saltpeter Production, 1776–1794," *Technology and Culture*, 12 (1971), pp. 163–81.

Newell, Lyman C. "Peter Porcupine's Persecution of Priestley," *Journal of Chemical Education*, vol. 10, no. 3 (March 1933), pp. 151–59.

Nuland, Sherwin B. *How We Die: Reflections on Life's Final Chapter*. New York: Vintage Books, 1993.

Nye, Mary Jo. "Philosophies of Chemistry Since the Eighteenth Century,"in Seymour H. Mauskopf, ed., *Chemical Sciences in the Modern World*. Philadelphia: University of Pennsylvania Press, 1993.

"Obituary of Joseph Priestley," *Medical Repository* (1804), pp. 429–34.

"Old Northumberland's Romance," *The Philadelphia Press*, Sun., Sept. 18, 1892, p. 22.

Oliver, Raymond. *Gastronomy of France*, trans. Claude Durrell. Cleveland: Wine and Food Society, 1967.

Orange, A. D. "Oxygen and One God," *History Today*, vol. 24, no. 11 (November 1974), pp. 773–81.

Park, Mary Catherine. *Joseph Priestley and the Problem of Pantisocracy*. Ph.D. Dissertation, University of Pennsylvania, 1947.

Partington, J. R. *A History of Chemistry*. Vol. 3. London: Macmillan & Co., 1962.

———. *A History of Greek Fire and Gunpowder*. Cambridge: W. Heffer & Sons, 1960.

———. *A Short History of Chemistry*. 3rd ed. London: Macmillan & Co., 1960.

Pendleton, Gayle Trusdel. "The English Pamphlet Literature in the Age of the French Revolution Anatomized," *Eighteenth Century Life*, vol. 5, no. 1 (Fall 1978), pp. 29–37.

Perrin, Carlton E. "Research Traditions, Lavoisier, and the Chemical Revolution," *Osiris*, 2nd series, 4 (1988), pp. 53–81.

———. "The Triumph of the Antiphlogistians," in Woolf, ed., *The Analytic Spirit: Essays in the History of Science*, pp. 40–63.

Petroski, Henry. *The Evolution of Useful Things: How Everyday Artifacts—from Forks and Pins to Paper Clips and Zippers—Came to Be as They Are*. New York: Vintage Books, 1994.

Piper, Herbert. "The Pantheistic Sources of Coleridge's Early Poetry," *Journal of the History of Ideas*, 20 (1959), pp. 47–59.

Platt, W., and R. A. Baker, "The Relation of the Scientific 'Hunch' to Research," *Journal of Chemical Education*, VIII (1931), pp. 1969–2002.

Poirier, Jean-Pierre. *Lavoisier: Chemist, Biologist, Economist*, trans. Rebecca Balinski. Philadelphia: University of Pennsylvania Press, 1993.

Porter, Roy. *English Society in the Eighteenth Century*. Harmondsworth, Middlesex: Penguin, 1982.

Prebble, John. *Culloden*. New York: Penguin Books, 1977.

Price, Richard. "Tremble all ye oppressors of the world!" in Brian MacArthur, ed., *The Penguin Book of Historic Speeches*. New York: Penguin Books, 1995, pp. 156–59.

Priestley, Hannah Taggert. "Family Traditions of Dr. Jos. Priestley." Speech, n.d., Joseph Priestley Collection, Dickinson College Library.

Priestley, Joseph. *An Appeal to the Public on the Riots at Birmingham*. Birmingham and London, 1791.

———. *Considerations on the Doctrine of Phlogiston, and The Decomposition of Water* (1796). "Joseph Priestley on Phlogiston." URL: http://webserver.lemoyne.edu/faculty/giunta/phlogiston.html.

———. *Directions for Impregnating Water with Fixed Air; In order to communicate to it the peculiar Spirit and Virtues of Pyrmont Water, and other Mineral Waters of a similar Nature*, 1771.

———. *The Discovery of Oxygen* (1775). The Alembic Club, Edinburgh: E. & S. Livingstone, 1961.

———. "Education for Civil and Active Life," in Kramnick, ed., *The Portable Enlightenment Reader*, pp. 235–42.

———. *An Essay on the First Principles of Government, and on the Nature of Civil, Political, and Religious Liberty*. London, 1768; second edition, 1771.

———. *Experiments and Observations on different Kinds of Airs*. 3 vols. London, 1774–77.

———. *Experiments and Observations relating to Various Branches of Natural Philosophy*. London: J. Johnson, 1779.

———. *History and present state of electricity*. 3rd ed. London: J. Johnson, 1775.

———. *An History of the Corruptions of Christianity*. Birmingham, 1782.

———. *History of the Early Opinions Concerning Jesus Christ* (1786).

———. *Institutes of Natural and Revealed Religion*. London, 1772–74; second edition, 1782.

———. "A Letter to Dr. Mitchell, in reply to the preceding, by Joseph Priestley, LL. D. &c," *Medical Repository* (1798), pp. 511–12.

———. "Letter to Citizen Péregaux in Paris, April 12, 1801," Joseph Priestley Collection, Dickinson College Library.

———. *The Memoirs of Dr. Joseph Priestley, LL. D., F.R.S., Written by Himself; With a Continuation to the Time of His Decease by His Son Joseph Priestley*. London, 1806–07. Modern Edition: *Autobiography of Joseph Priestley*. Cranbury, NJ: Associated University Presses, 1970.

———. "The Organization of Scientific Research," a selection from Priestley's *History and present state of electricity*, in Kramnick, ed., *The Portable Enlightenment Reader*, pp. 69–73.

———. *A Sermon on the Subject of the Slave Trade; delivered to a Society of Protestant Dissenters at the New Meeting, in Birmingham, and Published at Their Request*. London: Pearson & Rollason, 1788.

———. *Socrates and Jesus Compared*. Philadelphia: P. Bryne, 1803.

———. "To the INHABITANTS of the TOWN of BIRMINGHAM," broadsheet, with the account of the riots and Capt. James Keir's letter "To the PRINTER of the BIRMINGHAM & STAFFORD CHRONICLE," West Bromwich, July 20, 1791. Joseph Priestley Collection, Dickinson College Library.

———. "Unitarianism," a selection from Priestley's *Letters to Dr. Horsley* (1783), in Kramnick, ed., *The Portable Enlightenment Reader*, pp. 155–60.

Priestley, Timothy. *A Funeral Sermon Occasioned by the Death of the Late Rev. Joseph Priestley, etc*. London: Alexander Hogg & Co., 1804.

Pyne, Stephen J. *Fire: A Brief History*. Seattle: University of Washington Press, 2001.

Rail, Tony, and Beryl Thomas. "Joseph Priestley's Journal While at Daventry Academy, 1754," *Enlightenment and Dissent*, 13 (1994), pp. 49–113.

Rappaport, Rhoda. "G.-F. Rouelle: An Eighteenth-Century Chemist and Teacher," *Chymia*, 6 (1960), pp. 68–101.

Read, John. *Through Alchemy to Chemistry: A Procession of Ideas and Personalities*. London: G. Bell & Sons, 1961.

Reading Advertiser, April 26, 1800. Files of the Historical Society of Pennsylvania, Archives of the Joseph Priestley House, Northumberland, PA.

Reid, Christopher. "Burke, the Regency Crisis, and the 'Antagonist World of Madness,' " *Eighteenth Century Life*, vol. 16, no. 2 (May 1992), pp. 59–75.

Riley, Mark T. "Antoine-Laurent Lavoisier," in Ian P. McGreal, ed., *Great Thinkers of the Western World*. New York: HarperCollins, 1992, pp. 297–301.

Ritcheson, Charles R. "The Fragile Memory: Thomas Jefferson at the Court of George III," *Eighteenth Century Life*, vol. 6, nos. 2–3 (January and May 1961), pp. 1–16.

Robespierre, Maximilien. "Terror is nothing else than justice," in MacArthur, ed., *The Penguin Book of Historic Speeches*, pp. 182–84.

Robinson, E. "New Light on the Priestley Riots," *The Historical Journal*, vol. 3, no. 1 (1960), pp. 73–75.

Rose, R. B. "The Priestley Riots of 1791," *Past and Present*, 18 (November 1960), pp. 68–88.

Rossotti, Hazel. *Fire: Servant, Scourge, and Enigma*. Mineola, NY: Dover Publications, 1993.

Rothenberg, Albert, and Carl R. Hausman, eds. *The Creativity Question*. Durham, NC: Duke University Press, 1976.

Rudé, George. *The Crowd in History: A Study of Popular Disturbances in France and England, 1730–1848*. New York: John Wiley & Sons, 1964.

———. *The Crowd in the French Revolution*. London: Oxford University Press, 1967.

———. "The London 'Mob' of the Eighteenth Century," *Historical Journal*, 2 (1959), pp. 1–18.

———. "Popular Protest in 18th Century Europe," in Fritz and Williams, eds., *The Triumph of Culture: 18th Century Perspectives*, pp. 277–97.

Ruston, Alan. "Priestley and The Gentleman's Magazine," *Transactions of the Unitarian Historical Society*, vol. XVIII, no. 1 (April 1983), pp. 9–13.

Rutt, John T., ed. *Collected Theological and Miscellaneous Works of Joseph Priestley*. Vol. 2. London: George Smallfield, 1832.

———. *Life and Correspondence of Joseph Priestley*. 2 vols. London: George Smallfield, 1831.

Sacks, Oliver. "Speed," *The New Yorker*, August 23, 2004, pp. 60–69.

Sayers, Sean. "Images of the French Revolution," *Radical Philosophy*, 53 (Autumn 1989), pp. 50–51.

Schaffer, Simon. "Learned Insane," *London Review of Books*, April 17, 2003, pp. 34–35 (review of Uglow's *The Lunar Men*).

Schama, Simon. "The Unloved American: Two Centuries of Alienating Europe," *The New Yorker*, March 10, 2003, pp. 34–39.

Scheler, Lucien, and William A. Smeaton. "An Account of Lavoisier's Reconciliation with the Church a Short Time Before His Death," *Annals of Science*, 14 (1958), pp. 148–53.

Schneer, Cecil J. *Mind and Matter: Man's Changing Concepts of the Material World*. New York: Grove Press, 1969.

Schneider, Hans-Georg. "The 'Fatherland of Chemistry': Early Nationalistic Currents in Late Eighteenth Century German Chemistry," *Ambix*, vol. 36 (1989), pp. 14–21.

Schofield, Robert E. *The Enlightenment of Joseph Priestley: A Study of His Life and Works from 1733 to 1773*. University Park, PA: Pennsylvania State University Press, 1997.

———. "Joseph Priestley, Natural Philosopher," *Ambix*, vol. 14, part 1 (1967), pp 1–15.

Schofield, Robert E., ed. *A Scientific Autobiography of Joseph Priestley (1733–1804): Selected Scientific Correspondence*. Cambridge, MA: MIT, 1966.

Shackleton, Robert. "The Grand Tour in the Eighteenth Century," in *Studies in Eighteenth-Century Culture*. Vol. 1: *The Modernity of the Eighteenth Century*, ed. Louis T. Milic. Cleveland: Case Western University, 1971, pp. 127–42.

Sharpe, Emily. *A Sketch of the Life of Joseph Priestley, LL.D.* London: British and Foreign Unitarian Association, 1892. Joseph Priestley Collection, Dickinson College Library.

Shelley, Mary. *Frankenstein, or, The Modern Prometheus.* 1818; New York: New American Library, 1965.

Sheps, Arthur. "Ideological Immigrants in Revolutionary America," in Fritz and Williams, eds., *City and Society in the 18th Century*, pp. 231–46.

———. "Public Perception of Joseph Priestley, the Birmingham Dissenters, and the Church-and-King Riots of 1791," *Eighteenth-Century Life*, vol. 13, no. 2 (May 1989), pp. 46–64.

Siegfried, Robert. "Lavoisier's Table of Simple Substances: Its Origin and Interpretation," *Ambix*, vol. 29, part 1 (March 1982), pp. 29–47.

Skaggs, Jimmy M. *The Great Guano Rush: Entrepreneurs and American Overseas Expansion.* New York: St. Martin's Press, 1994.

Smeaton, William A. "Some Large Burning Lenses and Their Use by Eighteenth Century French and British Chemists," *Annals of Science*, 44 (1987), pp. 265–76.

Smelser, Neil J. *Theory of Collective Behavior.* New York: Free Press of Glencoe, 1963.

Smith, Edgar F. *Priestley in America, 1794–1804.* Philadelphia: P. Blakiston's Son & Co., 1920.

Smith, Lacey Baldwin. *Fools, Martyrs, Traitors: The Story of Martyrdom in the Western World.* Evanston, IL: Northwestern University Press, 1997.

Stark, Rodney. *The Rise of Christianity: A Sociologist Considers History.* Princeton, NJ: Princeton University Press, 1996). Cf. chapter 8, "The Martyrs," a discussion of the strategic rewards of martyrdom for a new religion as seen through the lens of rational choice theory, pp. 167–89.

Sterne, Laurence. *A Sentimental Journey.* 1768; Harmondsworth, Middlesex: Penguin, 1975.

Stevenson, John. *Popular Disturbances in England: 1700–1870.* London: Longman, Green, 1979.

Tannen, Deborah. *Gender and Discourse.* New York: Oxford University Press, 1994.

Teich, Mikuláš. "Circulation, Transformation, Conservation of Matter and the Balancing of the Biological World in the Eighteenth Century," *Ambix*, vol. 29, part 1 (March 1982), pp. 17–28.

Thompson, E. P. "The Moral Economy of the English Crowd in the Eighteenth Century," *Past and Present*, 50 (February 1971), pp. 76–136.

Thorpe, Sir Edward. *Essays in Historical Chemistry.* London: Macmillan & Co., 1923.

Toulmin, Priestley. "The Descendants of Joseph Priestley, LL.D., F.R.S. (A Progress Report)," in *The Northumberland County Historical Society Proceedings*, vol. XXXII, part IIB (Sunbury, PA, 1994), pp. 1–126.

Toulmin, Stephen E. "Crucial Experiments: Priestley and Lavoisier," *Journal of the History of Ideas*, 18 (1957), pp. 205–20.

Tuchman, Barbara W. *The March of Folly: From Troy to Vietnam.* New York: Ballantine Books, 1969.

Turgot, Anne-Robert-Jacques. "On Progress," in Kramnick, ed., *The Portable Enlightenment Reader*, pp. 361–63.

Turner, Frederick, and Ernst Poppel, "The Neural Lyre," reprinted from *Poetry* (1983). URL: www.pacifer.com/~starling/lyre.html.

Twining, Thomas. *Notes and Reminiscences by Thomas Twining.* New York: Harper & Bros., 1894.

Uglow, Jenny. *The Lunar Men: Five Friends Whose Curiosity Changed the World.* New York: Farrar, Straus & Giroux, 2002.

Van Rensselaer, Marianna Griswold. *Henry Hobson Richardson and His Works*. 1888; New York: Dover Publications, 1969.
Views of Buildings Destroyed during Riots at Birmingham in 1791, illus. P. H. Witton, engraved Wm. Ellis. Birmingham, 1792.
Wade, Nicholas. "Crick, DNA Structure Co-Discoverer, Dies at 88," *New York Times*, June 30, 2004, p. A1.
Walker, W. Cameron. "The Beginnings of the Scientific Career of Joseph Priestley," *Isis*, XXI (1934), pp. 81–97.
Weatherill, Lorna. "A Possession of One's Own: Women and Consumer Behavior in England, 1660–1740," *Journal of British Studies*, vol. 25, no. 2 (April 1986), pp. 131–56.
Westfall, Richard S. *The Construction of Modern Science: Mechanisms and Mechanics*. New York: John Wiley & Sons, 1971.
White, J. Edmund. "The Priestley Memorial Volume," *Notes and Records of the Royal Society of London*, vol. 48, no. 1 (1994), pp. 85–96.
Willey, Basil. *The Eighteenth Century Background: Studies on the Idea of Nature in the Thought of the Period*. London: Chatto & Windus, 1940; chapter X is on Priestley.
Wilson, Arthur M. " 'Treated Like Imbecile Children' (Diderot): The Enlightenment and the Status of Women," in Fritz and Morton, eds., *Women of the 18th Century and Other Essays*, pp. 89–105.
Wilson, Colin. *The Mammoth Book of the History of Murder*. 1990; New York: Carroll & Graf, 2000.
Woolf, Harry, ed. *The Analytic Spirit: Essays in the History of Science*. Ithaca, NY: Cornell University Press, 1981.
Wordsworth, William. "The Prelude, or, Growth of a Poet's Mind, " in Russell Noyes, ed., *English Romantic Poetry and Prose*. New York: Oxford University Press, 1956, pp. 269–302.
Wright, Sam. *Crowds and Riots: A Study in Social Organization*. Beverly Hills, CA: Sage Publications, 1978.
Young, Arthur. *Travels During the Years 1787, 1788, 1789: Undertaken More Particularly with a View of Ascertaining the Cultivation, Wealth, Resources, and National Prosperity of the Kingdom of France*. 1794; New York: E. P. Dutton & Co., xxxx.

Archives and Collections

Joseph Priestley Collection, Dickinson College, Carlisle, PA
Joseph Priestley House, Northumberland, PA

Interviews

Andrea Bashore, administrator and historian, Joseph Priestley House, Northumberland, PA
Connie Donadieu, tour guide and regional historian, St. James Parish, LA

Web Sites

"The AFU and Urban Legend Archive: Medical: decapitated head blinking more." www.urbanlegends.com/medical/decapitated_head_blinking_more
"Age of the Enlightenment." http://mars.wnec.edu/~grempel/courses/wc2/lectures/enlightenment.html
"Ain't No Way to Go: Does Beheading Hurt?" www.aarrgghh.com/no_way/beheading
"Antoine Laurent Lavoisier." www.english.upenn.edu/~jlynch/FrankenDemo/People/lavois.html
"Capacitors." www.hypertextbook.com/physics/electricity/capacitors

"Carl Becker Quotes." www.en.thinkexist.com
"Chemistry: Periodic Table: Oxygen." www.webelements/elements/text/O/key.html
"Columbia Journalism Review: Resources: Inflation Adjustor." http://www.cjr.org
"Court TV's Crime Library: Henri Landru: Bluebeard." www.crimelibrary.com/serial_killers/history/landru/guillotine_7
"Davis Black Powder." www.internet.cybermesa.com/~sam1/powder
"Does the Head of a Guillotined Individual Remain Briefly Alive?" *The History Net*, http://europeanhistory.about.com/library/bldyk10.htm
"The Enlightenment." www.wsu.edu:8080/~brians/hum_303/enlightenment.html
"Franklin." webserver.lemoyne.edu/faculty/giunta/franklin.html
"Franklin and His Electric Kite." www.ushistory.org/franklin/kite
"Free Speech Philosophers—Joseph Priestley." www.uark.edu/depts/comminfo/freespeech/priestley.html
"GIGA Quote Topic Page for Martyrdom." www.giga-usa.com
"Henri Landru: Bluebeard." www.crimelibrary.com/serial_killers/history/landru/ guillotine_7.html?sect=6
"Inventors, with Mary Bellis." www.inventors.about.com/library/inventors/blguillotine
"Joseph Priestley." www.chemheritage.org
"Joseph Priestley." www.spartacus.schoolnet.co.uk/PRpriestley.htm
"Joseph Priestley." www.woodrow.org/teachers/chemistry/institutes/1992/Priestley.html
"Joseph Priestley, 1733–1804." www.historyguide.org/intellect/priestley.html
"Joseph Priestley, by Don Carter." http://home.ptd.net/~sjrubin2/uucsv/carter.htm
"Joseph Priestley: Discoverer of Oxygen." http://center.acs.org/landmarks/landmarks/priestley/index.html
"The Language of Politics—England and the French Revolution." www.historyguide.org/intellect/lecture14a.html
"Lavoisier on the Atmosphere." This is a translation of Lavoisier's *Memoir on the Nature of the Principle which Combines with Metals During Calcination and Increases Their Weight*, also known as the "Easter Memoir," read to the Académie des Sciences, Easter, 1775. http://dbhs.wvusd.k12.ca.us/Chem-History/Lavoisier-1777.html
"Lavoisier on Water." Translation by Carmen Giunta of "Report of a memoir read by M. Lavoisier at the public session of the Royal Academy of Sciences on November 12, on the nature of water and on experiments which appear to prove that this substance is not strictly speaking an element but that it is susceptible of decomposition and recomposition," *Observations sur la Physique*, 23 (1783): 452–55. http://webserver.lemoyne.edu/faculty/giunta/laveau.html
"Max Weber—The Protestant Ethic." Transcription of Weber's entire *The Spirit of Capitalism and the Protestant Ethic*; chapter 2 is most relevant to Priestley and his times. www.hewett.norfolk.sch.uk/curric/soc/weber
"NOVA Online/Everest/How the Body Uses O_2." www.pbs.org/wgbh/nova/everest/exposure/body.html
"Robert Boyle Work-diaries." www.bbk.ac.uk/boyle/workdiaries
"The Straight Dope: Does the head remain briefly conscious after decapitation?" www.straightdope.com/classics
"Thomas Cooper—Biography" and "Cooper's Account of the Trial of Thomas Cooper." http://deila.dickinson.edu/theirownwords
"William Blake." www.spartacus.schoolnet.co.uk/PRblake.htm

ACKNOWLEDGMENTS

PROJECTS LIKE THIS ARE NEVER POSSIBLE WITHOUT THE AID OF experts and well-informed friends and acquaintances, and over the years, I've been fortunate to develop an informal brain trust of professors at Virginia Wesleyan College in Norfolk, Virginia, close to where I live. They are patient scholars, accustomed by now to the strange questions of writers, and have been a source of information and inspiration for each of my nonfiction narratives. *A World on Fire* was no exception. I would like to thank Dr. Clayton J. Drees, Professor of History, for his knowledge of the "urban legends" springing up around Antoine Lavoisier; Dr. Steven M. Emmanuel, Professor of Philosophy, for his guidance through the pitfalls of Enlightenment era philosophy; Dr. Deborah E. Otis, Professor of Chemistry, for her explanation of the mysteries of phlogiston; Dr. Lynn Sawlivich, Assistant Professor of Classics, for help in various Latin translations, as well as a primer on Greek science; and finally, Dr. Craig S. Wansink, Professor of Religious Studies, for his thoughts on how to be an effective martyr.

I would also like to thank the community of Priestley scholars and archivists hidden away in central Pennsylvania. The archivists at the Joseph Priestley Collection, Dickinson College, Carlisle, Pennsylvania, seemed to know everything ever written about Joseph Priestley, Antoine Lavoisier, and the discovery of oxygen, and this book would not have been possible without their continued assistance. In addition, Andrea Bashore, administrator and historian at the Joseph Priestley House, Northumberland, Pennsylvania, gave valuable insight into the minds of Priestley, his wife, and his children, a perspective that wasn't available elsewhere.

Some subjects are so esoteric that a writer despairs of finding information, and yet, as a newspaper editor once told me, every subject has its expert. The challenge is in finding them. I'd like to thank Connie Donadieu, tour guide and regional historian in St. James Parish, Louisiana, for her knowledge and generosity regarding the fate of William Priestley after he left Pennsylvania. And I'd like to thank L. Anderson Orr, gourmet cook, whose private library on French gastronomy was a real godsend.

As always, I'd like to thank my literary agent, Noah Lukeman, who helps me make a living. And I'd like to thank my editor at Viking Penguin, Wendy Wolf, who believed in this little tale of science, revolution, and tragedy.

Finally, I always dedicate my books to my wife, Kathy Merlock Jackson, and my son, Nick, but for this book Kathy went beyond the call of duty. Due to tight deadlines, I was unable to go on a school trip to the Costa Rican rainforest, so Kathy and Nick went alone. Kathy is an acknowledged city girl from Pittsburgh, unaccustomed to the joys of typhoons, poison frogs, and stinging insects that Floridians like me take as a matter of course, so her attendance was courageous enough; when she broke her shoulder during a hike through a torrential rainstorm, the experience was transformed from the uncomfortable into the heroic and legendary. Here's to you, Sheena of Pittsburgh—I couldn't have done it without you.

INDEX